Adhesion Measurement of Films & Coatings, Volume 2

T0173957

ADHESION MEASUREMENT OF FILMS & COATINGS

VOLUME 2

Editor:
K.L. Mittal

CRC Press
Taylor & Francis Group
Boca Raton London New York

CRC Press is an imprint of the
Taylor & Francis Group, an **informa** business

Contents

Adhesion Measurement of Films and Coatings, Vol. 2, pp. vii–viii
Ed. K.L. Mittal
© VSP 2001

Preface

This book documents the proceedings of the Second International Symposium on Adhesion Measurement of Films and Coatings held under the auspices of MST Conferences in Newark, New Jersey, October 25–27, 1999. We labelled it the Second Symposium as the first event on this topic was held in Boston, December 5–7, 1992, under the aegis of Skill Dynamics, an IBM company, the proceedings of which have been properly chronicled [1]. However, for historical reasons it should be recorded that the premier symposium on this topic (with a slight change in title) was held in Philadelphia under the auspices of the American Society for Testing and Materials (ASTM), the proceedings of which were published as STP-640 by the ASTM in 1978. Because of the long hiatus between the ASTM Symposium and the one held in Boston, we deemed it more appropriate to label the Boston event as the First Symposium.

Films and coatings are used for a variety of purposes – functional, decorative, protective, etc. – in a host of applications. Irrespective of the purpose or application of a film or a coating, their adequate adhesion to the underlying substrates is of paramount importance. Concomitantly, the need to develop techniques for quantitative assessment of adhesion of films and coatings is all too obvious.

Since the Boston Symposium in 1992 there has been considerable activity in devising new, more reliable and more efficient ways to measure adhesion of films and coatings. In the opening article in the proceedings volume of the Boston Symposium, yours truly had listed 355 techniques for measuring adhesion of films and coatings. Why are there so many techniques? Apparently, there is no single technique which will apply in every situation or which everyone will be happy with. A more important question is: What do these techniques measure? And the answer is: Practical Adhesion. Quite often the question asked is: What is the best method for adhesion measurement? And the answer is: The best method is the one that simulates the actual usage stress conditions as closely as possible. Concomitantly, the best method will be different, depending on the conditions to which a film–substrate system will be exposed.

As a result of the brisk activity and tremendous interest in the topic of adhesion measurement of films and coatings, we decided to hold this second event. The technical program for this symposium contained a total of 29 papers and many different techniques were discussed. There were very lively and illuminating (not exothermic) discussions, both formally and informally.

Apropos, the third symposium on this topic is planned to be held in Newark, New Jersey, November 5–7, 2001.

Now coming to this volume, it contains a total of 20 papers. It must be recorded here that all manuscripts were rigorously peer reviewed and suitably modified (some twice or thrice) before inclusion in this book. So this book is not merely a collection of unreviewed papers but represents the highest standard of a publication. The topics covered include: measurement and analysis of interface adhesion; relative adhesion measurement for thin film structures; adhesion testing of hard coatings by a variety of techniques; challenges and new directions in scratch adhesion testing of coated substrates; application of scratch test to different films and coatings; evaluation of coating- and substrate adhesion by indentation experiments; measurement of interfacial fracture energy in multifilm applications; laser induced decohesion spectroscopy (LIDS) for measuring adhesion; pulsed laser technique for assessment of adhesion; blade adhesion test; JKR adhesion test; coefficient of thermal expansion measurement; and residual stresses in diamond films.

Yours truly sincerely hopes that this book, along with its predecessor [1], will provide a commentary on the current state of the art anent adhesion measurement of films and coatings and will further provide a fountainhead for new ideas.

Acknowledgements

First, I am thankful to Dr Robert H. Lacombe, my colleague and friend, in helping to organize this symposium by taking care of a myriad of details entailed in such an endeavor. My sincere thanks go to the reviewers for their time and efforts in providing valuable comments which are *sine qua non* to maintain the highest standard of a publication. Without the contribution, interest and enthusiasm of the authors, this book could not be embodied and my thanks to all the contributors. Last, but not least, my appreciation is extended to the staff of VSP for the job well done in producing this book.

K.L. Mittal

[1] K.L. Mittal (ed.), *Adhesion Measurement of Films and Coatings*, VSP, Utrecht, The Netherlands (1995)

Adhesion Measurement of Films and Coatings, Vol. 2, pp. 1–18
Ed. K.L. Mittal
© VSP 2001

Interface adhesion: Measurement and analysis

A.G. EVANS*

Princeton Materials Institute, Princeton University, Princeton, NJ 08540, U.S.A.

Abstract—A protocol for quantitative adhesion measurements would allow implementation of design codes and durability models for multi-layer devices. The mechanics underlying the adhesion energy is largely complete. Test methods capable of providing the necessary information are at an advanced stage of development, but future innovation and analysis are still needed before a fully-integrated methodology can be prescribed. The status of the test methods is described and related to the overall goal. Models that allow adhesion to be related to the fundamentals of bond rupture and plasticity are examined, inclusive of concepts that inter-relate quantum mechanics results to adhesion measurements made at the continuum level. The additional effort needed to coalesce these models into a predictive tool is discussed.

Keywords: Adhesion measurement; interface toughness; oxide/metal interfaces; adhesion models; adhesion test methods.

1. INTRODUCTION

Two types of measurements are relevant to adhesion [1-4]: (i) the stress at which the interface separates, σ_c, and (ii) the energy dissipated per unit area upon extending a crack along the interface, Γ_i (in J/m^2). The latter has the same role as the fracture toughness in homogeneous materials [5-7]. The former includes effects of defects and of stress concentrations (especially at free edges) [3] and is thus test specific and inherently stochastic. While both are important, here, the energy density is emphasized, since it is amenable to quantitative comparison with mechanisms and models [2,8-10] and moreover, in principle, the measurements can be used explicitly in design codes and durability models for multi-layer systems. That is, a methodology similar to fracture mechanics-based design of structural components [5-7] could be used subject to the construct of the appropriate numerical code [11]. This prospect can only be realized if the test methods yield quantitative measures of Γ_i. Accordingly, the emphasis of this brief overview is on a pathway toward a quantitative design strategy applicable to multi-layer systems.

Since the likelihood of establishing a successful strategy would be enhanced if a mechanistic basis is established, a complementary theme is the development of

* Phone: 609 258 4762, E-mail: anevans@princeton.edu

mechanism-based models. Much of the quantitative information about mechanisms has been gained from metal/oxide interfaces [2]. A summary of this information (Fig.1) provides a perspective. Clean interfaces devoid of reaction products are inherently tough and ductile. Such high adhesion is realized even though the metals are polycrystalline and non-epitaxial. When failure occurs, it does so either by brittle cracking in the oxide or by ductile fracture in the metal. Only measurements for Al/Al_2O_3 indicate *consistently* tough, ductile interfaces. Broad ranges have been cited for most other interfaces, because of embrittling effects of contaminants and segregants. Stress corrosion due to the presence of moisture in the test environment exacerbates weakening in some cases.

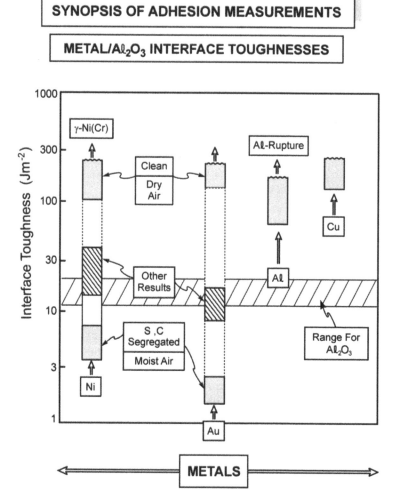

Figure 1. The range of toughness found between various metals and Al_2O_3. Note that many results reside in discrete domains that depend on "cleanliness".

Two fundamentally important factors cause cracks at interfaces to differ from those in homogeneous materials [4]. (i) The elastic property mismatch causes the energy release rate, G, and the mode mixity angle, ψ, to differ at the same loading. These differences are fundamentally related to the first Dundurs' parameter:

$$\alpha_D = \frac{\overline{E}_1 - \overline{E}_2}{\overline{E}_1 + \overline{E}_2} \tag{1}$$

where $\overline{E}$ is the plane strain Young's modulus. The subscripts 1 and 2 refer to the two adjoining materials. (ii) Mixed mode cracks ($\psi \neq 0$) may extend along interfaces. Accordingly, the fracture toughness must be specified as a function of ψ. A useful phenomenological relation is [4]:

$$\Gamma_i / \Gamma_i^o = 1 + \tan^2 (1 - \lambda) \psi \tag{2}$$

where Γ_i^o is the mode I toughness and λ is a mixity parameter that reflects the role of the interface non-planarity, as well as the plasticity in the adjoining materials. There is no mixed mode effect if $\lambda = 1$, but a strong dependence when λ is small. Interface adhesion is only meaningful when addressed with specific reference to λ.

With this background the overview is laid out as follows. The basic philosophy of quantitative adhesion measurement is addressed. The merits of specific test methods for thin films are discussed. Some adhesion results are reviewed and related to mechanisms.

2. MEASUREMENT PHILOSOPHY

Measuring interface adhesion is more challenging than the corresponding measurements on their homogeneous counterparts. There are two main issues. The geometrical configurations encompassing interfaces of practical interest often constrain specimen design. Large-scale inelastic/plastic deformations limit options, because of the vastly different thermo-mechanical properties of the adjoining materials. For interfaces made by diffusion bonding or brazing [2,12-17], many different configurations are available (Fig. 2) [1,2, 18-24]. The main restriction is that residual stresses often exist and these must be taken into account in determining the energy release rate, G, and mode mixity angle, ψ. Among these configurations, those that exhibit stable crack growth are preferred [21,22], wherein G *decreases* with increase in interface crack length, a, at specified load, P (or remains invariant with crack length). Such configurations greatly facilitate the introduction of well-defined pre-cracks before conducting the adhesion measurement. For mode I, the Double Cantilever Drilled Compression (DCDC) specimen [21] has this feature. It has been used to test metal/oxide and polymer/inorganic interfaces (Fig.2b). For mixed-mode loading, bending configu-

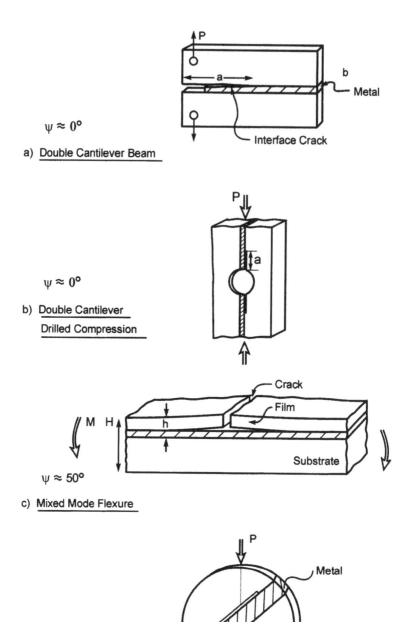

Figure 2. Test configurations used to measure the interface toughness on bonded bi-material systems.

rations are applicable (Fig.2c) [22]. In both cases, when one of the constituents is transparent, optical imaging may be used to monitor crack growth and study the mechanisms [12,15].

When one of the constituents is a thin film or coating, few quantitative methods exist. There are two basic approaches.

(i) Loads are applied to the film and the displacements are measured as the interface de-adheres. Such methods are exemplified by the peel test (Fig. 3) [25,26].

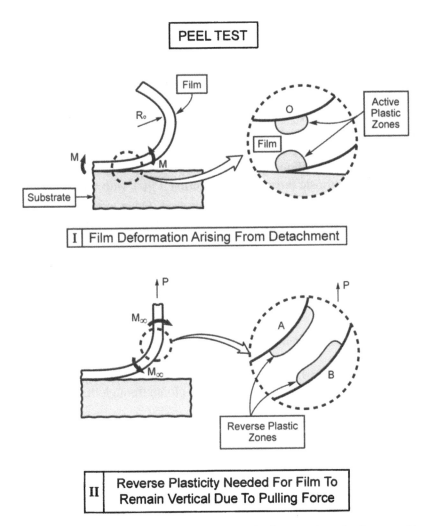

Figure 3. The peel test with a schematic indicating the plastic zones that arise because of bending (near-tip and reverse), which complicate interpretation of the measurements. In step I, film curvature arises because of the moment M near the tip that causes delamination, with consequent plasticity. In step II, the straightening of the film required by the applied loading is demonstrated, along with the zones of reverse yielding.

This test (and others like it) has the advantage of testing simplicity, but the interpretation is complex. The problem is that the work done is not solely governed by the energy expended around the crack (Fig.3). Deconvoluting the measurements in a manner that isolates the adhesion is challenging[26].

(ii) Residual strains are introduced by various means and the extent of the interface delamination caused by these strains is measured. There are at least three variants. (a) When the *substrate is ductile*, strain can be introduced into the film by deforming the substrate: multi-strain [19] and impression tests [23,24,27] exemplify this approach (Fig.4). In such tests, the strains can be imposed precisely and measured accurately. The precision is limited by the ability to detect and measure the dimensions of the interface delamination. (b) For films on *brittle substrate*, a strain energy density sufficient to de-adhere the interface can be induced by depositing an adherent overlayer. This layer must have sufficient intrinsic strain and thickness to achieve the critical strain energy density. This approach has been referred to as the superlayer test[20] (Fig.5). (c) Alternatively, localized strains can be generated by indenting or scratching the film [18]. Quantification requires calibration of the strains induced in the film.

There have been several reviews of thin film adhesion measurement [28-32]. The differing purpose of the following section is to provide perspective on the subset of tests that have the intrinsic prospect of measuring $\Gamma_i(\psi)$ with the precision needed to use the results for design and durability purposes. Other tests will continue to be used for quality control since they are straightforward to implement.

3. THIN FILM ADHESION MEASUREMENT METHODS

3.1. Peel tests

These tests have been applied primarily to flexible films. The test has the attribute that the peeling force is measured in a steady-state condition, wherein the shape of the strip remains invariant with displacement. This force is used as a measure of interfacial quality. The responses may be explored through the parameter [25],

$$\eta = 6EP / \sigma_o^2 h \tag{3}$$

where P denotes the applied force per unit width of the strip, E its Young's modulus, h its thickness and σ_o is the yield strength of the film. When $\eta \ll 1$, the film deforms elastically and the peel force P becomes a direct measure of the interfacial fracture resistance: $\Gamma_i \equiv P$. However, the minimum film thickness h^* needed to assure elastic peeling (obtained from (3) by equating η to unity), is typically quite large, and given by:

$$h^* = 6E\Gamma_i / \sigma_o^2 \tag{4}$$

For example, to be in the elastic range, Cu films having $\sigma_o \approx 100MPa$ and $\Gamma_i \approx 100Jm^{-2}$, require that $h^* \geq 1cm$. Thinner films deform plastically, whereupon a large scale yielding analysis is needed to interpret the measurements.

Then, the problem is that the work done by the peel force is not solely governed by the energy expended around the crack, for the following reason [26] (Fig.3). As the strip peels, it experiences a bending moment that induces yielding not only around the crack front but also on the upper surface. Such deformations cause the film to curl. Moreover, the reaction force induced by interface bonding requires that the film be bent back as the crack extends. This is realized through an opposite bending moment that causes reverse yielding. The yield zones O (near-tip bending), A and B (reverse bending) comprise redundant plastic work which convolutes with the interfacial work of rupture. De-convoluting the measurements in a manner that isolates the adhesion energy is required and, often, is not possible.

3.2. Microscratch and impression tests (Fig.4b)

These tests are variants on the same basic idea. That is, forces imposed through an indentation introduce a residual stress field that, in turn, causes the interface to delaminate. Measurements of the force and the size of the induced delamination permit estimates to be made of the interface toughness[23,24,27].The difference between the two methods relates to the scale of the indentation and the nature of the substrate. *Impression tests* are used with ductile substrates having well-established stress/strain characteristics. The indenter is pushed through the film into the substrate to an extent sufficient to assure that the indentation force is dominated by the substrate. In such cases, the strain distribution induced in the film can be determined with precision by using a finite element analysis: whereupon, the energy release rate and the mode mixity can be determined as a function of the delamination size. Consequently, provided that the latter can be measured, $\Gamma_i(\psi)$ can be determined, explicitly.

The Microscratch method confines the impression to the film [18]. It may thus be used with both brittle and ductile substrates. However, the constitutive properties of the film are rarely known with any accuracy and, moreover, the deformation fields are quite complex. Consequently, the strains in the film and the energy release rate can only be approximated. As analyses of the test improve, the approximations become less restrictive, but to the authors' knowledge, significant uncertainties still remain. Nevertheless, given the relative simplicity of the test, further efforts at improving the analysis are clearly warranted.

3.3. The blister test

This test is commonly used for thin polymeric films spun onto a substrate having a circular or square perforation [33-36]. The blister is created by fluid pressure applied through the perforation. The pressure needed to cause debonding can be explicitly related to Γ_i through the mechanics of a pressurized elastic, circular blister.

The effects of residual stress can be readily included. When applicable, this test provides measurements amenable to usage in design. The limitations include the inability to measure Γ_i for adherent interfaces that cause the film to yield.

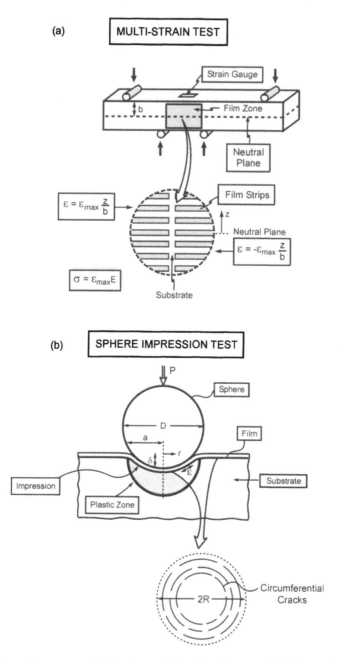

Figure 4. Tests used to measure the toughness of interfaces between thin films and a ductile substrate: (a) multi-strain test, (b) sphere impression test.

3.4. The superlayer test (Fig. 5)

This test allows a steady-state loading of a thin film system at the mode mixities relevant to system failure [20]. The approach involves the use of a residual stress that duplicates the problem of interest. Since, for typical thin films and representative residual stresses, the induced energy release rate is below the fracture toughness of most interfaces having practical interest, a procedure that substantially increases G is required. Deposition of a superlayer increases the effective film thickness and also elevates the residual stress. Implementation of this test requires micro-machining by photolithography, wherein the film of interest is deposited in the usual manner onto the substrate. Thereafter, the superlayer is electron beam evaporated onto the film. Subsequent lift-off defines the geometry. The film is patterned to form narrow strips. A through-cut is made in the bi-layer by either etching or milling. When the strips delaminate after severing, the energy release rate exceeds the debond energy, $G_{ss} > \Gamma_i$. However, the debond arrests before reaching the end of the strip. The length of the remnant ligament is a measure of the adhesion energy (Fig. 5) [2].

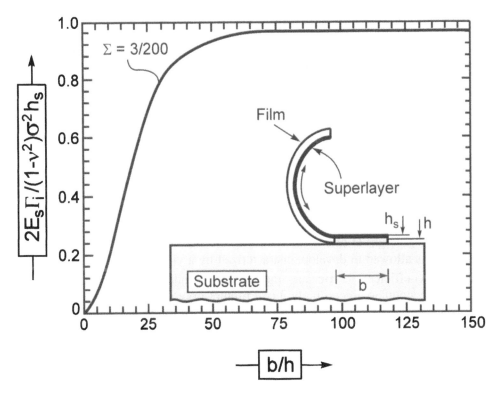

Figure 5. Superlayer method for measuring the interface adhesion between thin films and brittle substrates. Also shown is the energy release rate as a function of the length of the remnant ligament.

3.5. The multistrain test

This test requires a ductile template in the form of a beam that can be deformed after the films have been deposited [19,37] (Fig. 4a). Stainless steel has been used for this purpose. One surface is polished to an optical finish and a thin layer of polyimide spun onto this surface. The film is then deposited onto the poly-imide and patterned into strips, parallel to the long axis of the beam, with a gap at the center. The beam is subjected to bending, with the coated surface on the side. As bending occurs, each strip experiences a different strain: zero at the neutral axis and a maximum adjacent to the tensile surface. There is a corresponding variation in the strain energy. Upon testing, those strips located near the tensile surface are most susceptible to de-adhesion, while those closer to the neutral plane remain attached. Accordingly, in tests that exhibit delamination, a critical strain at which it occurs can be identified. Then, if the film behaves elastically and its Young's modulus is known, the steady-state energy release rate and the interface toughness can be obtained with acceptable accuracy. However, adherent films do not de-adhere in this test, limiting its range of utility.

4. INTERFACE RUPTURE MODELS

A complementary goal of interface adhesion investigations is the development of mechanism-based models that provide understanding of the issues that control Γ_i [2]. The inherent challenge is to provide a connection between quantum level re-sults for the salient bond rupture parameters [2,38-40] (the work-of-adhesion, W_{ad}, and the bond strength, $\hat{\sigma}$) and the practical, or engineering, adhesion en-ergy, Γ_i, through models of plastic deformation that incorporate length scale ef-fects (Fig. 6) [2,41-43]. This section examines the current status. Approaches have been attempted that address this challenge at two limits. These limits and a transition between them are reviewed.

In one limit, designated as the Suo, Shih and Varias (SSV) model[9], the inter-face cracks are assumed to be atomically-sharp and surrounded by an elastic en-clave, having height D above the interface. Outside the elastic zone, plastic de-formation is allowed to develop, characterized by a yield strength, σ_o and strain hardening coefficient N. Because the tip is surrounded by an elastic region the stresses are singular. Accordingly, bond rupture only requires that the energy re-lease rate attain the work-of-adhesion. This criterion fully prescribes the model, with the understanding that D is a fitting parameter. The primary dependencies of Γ_i/W_{ad} are on D/R_o and N, where R_o is the plastic zone size, defined as [2,9,40]:

$$R_o = (1/3\pi)EW_{ad}/\sigma_o^2 \qquad (5)$$

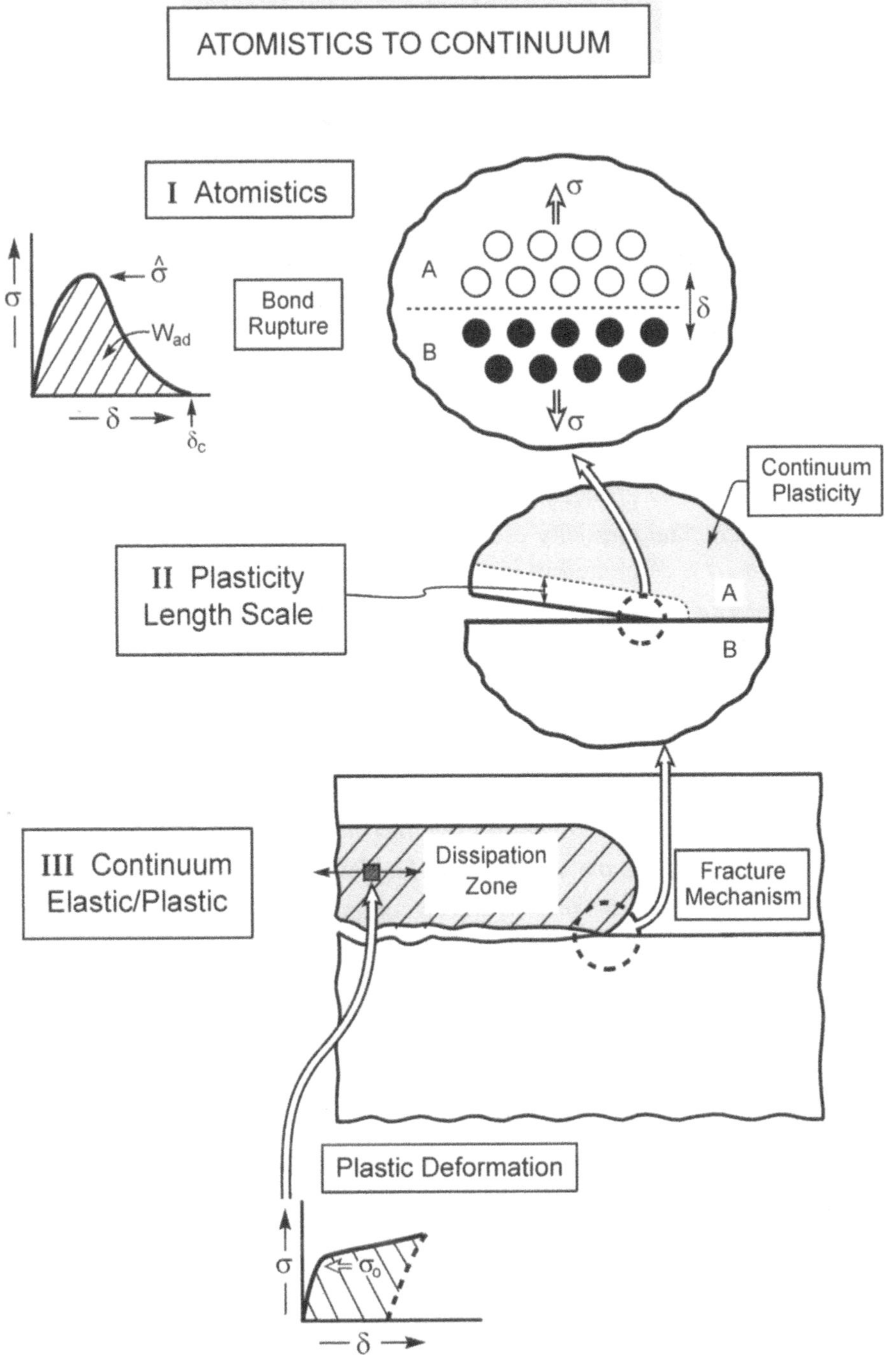

Figure 6. A schematic illustrating the zones of inelastic deformation that occur around interface cracks, with associated length scales.

If plasticity is imagined to extend to the crack tip, the energy release rate is always zero, whereupon the SSV model becomes unbounded as D approaches zero. To obviate this limitation, the embedded process zone (EPZ) model adopts a traction-separation relation as the description of the interface [8,10,40]. The relation is prescribed using a potential with W_{ad} as the work of separation and $\hat{\sigma}$ as the peak separation stress. The metal above the interface is described by continuum plasticity with parameters σ_o and N, supplemented by the material length scale ℓ if the strain gradient theory is used [40-43]. The model is fully specified such that the dependencies of the crack growth resistance $\Gamma_R(\Delta a)$ can be computed in terms of the interface parameters, W_{ad} and $\hat{\sigma}$ and the continuum properties of the metal and oxide. The non-dimensional terms involved are [40]:

$$\Gamma_{ss}/W_{ad}, N, \hat{\sigma}/\sigma_o, \ell/R_o$$

An approximate way to account for the dependence of the toughness on ℓ/R_o is to use the conventional plasticity predictions with a scale-adjusted yield strength, $\sigma_o^*/\sigma_o = g(\ell/R_o, N)$, as plotted in Fig. 7 [2,40]. The approximation uses the conventional plasticity result with σ_o^* replacing σ_o. The EPZ model predicts essentially unbounded toughness if the interface strength exceeds a critical value ($\hat{\sigma}/\sigma_o^* \cong 5$ for $N = 0.2$), indicative of a brittle-to-ductile transition [2,44,45].

A unified model provides the transition between the SSV and EPZ models [2,40]. It employs both the elastic zone *and* the traction-separation relation at the interface. Accordingly, the model employs all the parameters from the other two models. That is, $W_{ad}, \hat{\sigma}$ and D specify the interface and the elastic zone, whereas σ_o, N and ℓ specify the metal plasticity. The length of the zone over which interface separation occurs, denoted by d, is a computed quantity. If d<<D, the process zone lies deep within the elastic zone, such that attainment of a critical stress, $\hat{\sigma}$, ceases to be controlling and the SSV limit pertains. At the other limit, d>>D, the elastic zone has essentially no influence and the EPZ model holds. These two limits and the transition between them are depicted in Fig. 8, where curves of Γ_{ss}/W_{ad} are plotted as a function of $\hat{\sigma}/\sigma_o$ for fixed values of D/R_o using conventional plasticity $(\ell = 0)$, with $N = 0.2$.

The dominant features are revealed in a schematic form (Fig. 9), visualized as a "fracture toughness surface" with distinct demarcations of the domains wherein the EPZ and SSV models predominate. Interfaces with high R_o/D and low $\hat{\sigma}/\sigma_o$ are described by the EPZ limit, while the SSV limit applies to interfaces with low R_o/D and high $\hat{\sigma}/\sigma_o$. The transition between the two comprises a large portion of parameter space. The plasticity length scale would modify these predictions when ℓ is comparable to R_o.

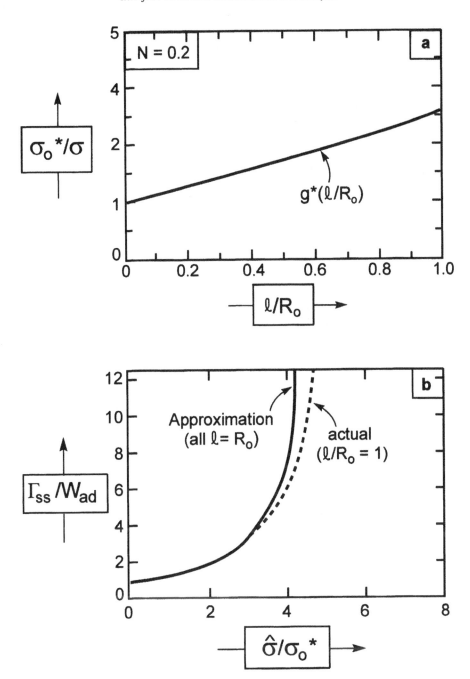

Figure 7. (a) Rescaling factors for the yield strength in the embedded process zone model. (b)The effect of the plasticity length scale on the steady-state interface toughness for a process zone mechanism.

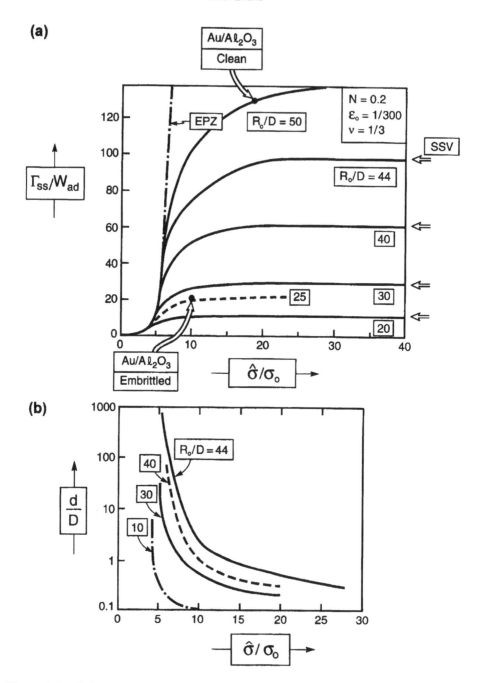

Figure 8. Predictions of the unified zone model (a) Effects of strength and scale indices on the steady-state toughness. (b) The relative sizes of the process zone, d and the dislocation free zone, D.

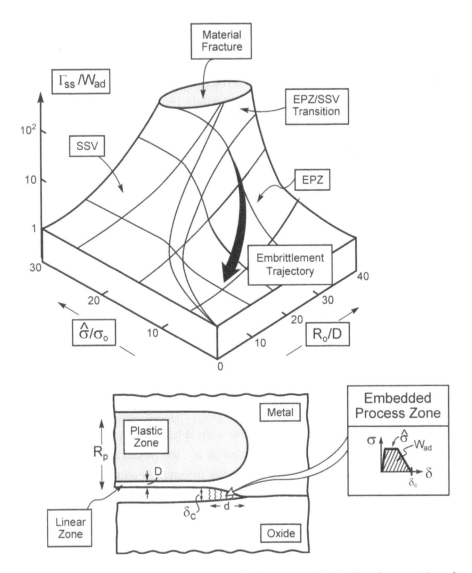

Figure 9. Fracture landscape predicted by the unified zone model including the truncation when material failure intervenes and the embrittlement trajectory is caused by segregants.

5. ADHESION ENERGIES AND DEBONDING MECHANISMS

5.1. Embedded interfaces

Relatively few measurements performed on metal/oxide or metal/inorganic interfaces satisfy the rigorous dictates of valid fracture toughness determination (Fig.1). A significant fraction of the tests exhibits extensive crack blunting with tip-opening displacements exceeding $1\mu m$ [2,44-46]. Each of these systems, when ultra-clean, is able to sustain energy release rates larger than $200 Jm^{-2}$ without rupturing the interfaces. When these systems fail, the rupture occurs in one of the adjoining materials rather than at the interface. For example, at the interface with $\gamma - Ni$, cracks diverge into the sapphire substrate [45]. In bonds with Al and Au, failure occurs by plastic flow in the metal layer [2,12,44,46].

Some of these interfaces can be dramatically embrittled by segregation. A quantified illustration involves the infusion of carbon into Au/Al_2O_3 interfaces, which reduces the toughness from ~ 250 to $2 Jm^{-2}$ [46]. This happens without changing the yield strength of the Au. The dramatic reduction in toughness is attributed to the effects of carbon segregation on the work-of-adhesion, W_{ad}, and the bond strength, $\hat{\sigma}$. Similarly, in $\gamma - Ni/Al_2O_3$, while the "clean" interface has toughness $> 300 Jm^{-2}$ reductions to $< 5 Jm^{-2}$ are caused by S release [47]. This degradation must be attributed to effects of carbon at the interface on some combination of W_{ad} and $\hat{\sigma}$. Making distinctions based solely on W_{ad} does not suffice. For the Au/Al_2O_3 system, this quantity changes by less than a factor 2 during the segregant infusion process. Such reductions would at most diminish Γ_{ss} by a factor of 5, whereas the measurements indicate changes in excess of a factor of 100 [2,46]. Consequently, the bond strength, $\hat{\sigma}$, must have a major role. The first principles calculations suggest that the reduction in $\hat{\sigma}$ upon segregation should be about the same as the reduction in W_{ad} [38,39](Fig.8), because both are similarly affected by the increase in the equilibrium atom spacing across the interface. Accordingly, large changes in Γ_{ss} can occur with relatively small increases in both W_{ad} and $\hat{\sigma}$. Since W_{ad} is specified, as well as σ_o, the model allows limited scope in the choices of D and $\hat{\sigma}$ that can be made in order to compare with measurement. It requires that $D \approx 200 nm$ and, absent segregation, $\hat{\sigma} \approx 1 GPa$, with $W_{ad} = 0.6 Jm^{-2}$. Upon carbon infusion, $\hat{\sigma} \rightarrow 500 MPa$, while $W_{ad} \rightarrow 0.3 Jm^{-2}$. Based on these reductions, the zone model predicts a toughness that changes from 70 to $5 Jm^{-2}$, compared with a measured change from 200 to $2 Jm^{-2}$.

5.2. Thin films

Most interfaces between thin films and thick substrates have adhesion energies in the range $2 - 20 Jm^{-2}$ [2,20,48-50]. This toughness range seems to apply when the film is ductile and the substrate brittle, and vice versa. There appear to be three principal reasons. (i) The thinness of the metal restricts the plastic zone size and reduces the plastic dissipation [1,51] (Fig. 10). (ii) The interfaces are typically made at low homologous temperatures so that thermal equilibration does not occur and the bonds may not be in their equilibrium (lowest energy) state. (iii) In a related sense, contaminants remain confined to the interface and inhibit the development of the metal/oxide bonds. It remains to quantify these effects.

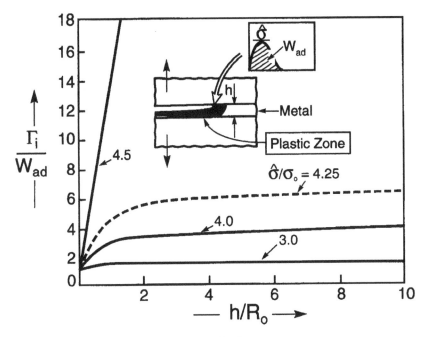

Figure 10. Effect of metal layer thickness on the adhesion energy.

6. CONCLUSION

The quantification of adhesion measurements toward a level that allows them to be used as input to design codes and durability models for multi-layer systems has progressed steadily during the last decade: however, the effort has been sporadic. Nevertheless, the issues are clear and the potential exists for creating a methodology within the framework of a focused research initiative. The mechanics underlying the adhesion energy is largely complete. Test methods capable of providing the necessary information are at an advanced stage of development, but innovation and analysis must continue. Models that allow adhesion to be related to the fundamentals of bond rupture and plasticity are being pursued, inclusive of concepts that interelate quantum mechanics results to the continuum response. Considerable additional effort is needed to combine these models into a predictive tool.

REFERENCES

1. A. Bagchi and A.G. Evans, *Interface Sci*, **3**, 169-193 (1996).
2. A.G. Evans, J.W. Hutchinson and Y.G Wei, *Acta Mater.*, **47**, 4093-4113 (1999).
3. M.Y. He and A.G. Evans, *Acta Metall Mater.*, **38**, 1587-1593 (1991).
4. J. W. Hutchinson and Z. Suo, *Adv. Appl. Mech*, **29**, 63-191 (1992).
5. J. R. Rice, *J. Appl.Mech.*, **55**, 98-103 (1988).
6. J. R. Rice, W. J. Drugan and T.-L. Sham, *ASTM STP* **708**, 189-221 (1980).
7. M.F. Kanninen and C.H. Popelar, *Advanced Fracture Mechanics*, Oxford University Press (1985).

8. Y. Wei and J. W. Hutchinson, *J. Mech. Phys. Solids*, **45**, 1253-1273 (1997).
9. Z. Suo, F. Shih and A. Varias, *Acta Metall Mater.*, **41**, 1551 (1993).
10. V. Tvergaard and J. W. Hutchinson, *J. Mech. Phys. Solids*, **41**, 1119-1135 (1993).
11. A. G. Evans and B. J. Dalgliesh, *Acta Metall Mater.*, **40**, S295 (1992).
12. M. Turner and A. G. Evans, *Acta Metall Mater.*, **44**, 863-871 (1996).
13. D. Korn, G. Elsnner, H. F. Fischmeister and M. Rühle, *Acta Metall. Mater.*, **40**, S335 (1992).
14. G. Elsnner, D. Korn, and M. Rühle, *Scripta Metall. Mater.*, **31**, 1037 (1994).
15. I. Reimanis, B. J. Dalgleish, and A. G. Evans, *Acta Metall Mater.*, **39**, 3133(1991).
16. T.S. Oh, R.M. Cannon, and R.O. Ritchie, *J. Am. Ceram. Soc*, **70**, 253 (1987).
17. N. P. O'Dowd, M. G. Stout and C. G. Shih, *Phil. Mag. A*, **66**, 1037 (1992).
18. N. R. Moody, R. Q. Hwang, S. Venkataraman, J. E. Angelo, D. P. Norwood and W. W. Gerberich, *Acta Mater,* **46**, 1170 (1998).
19. D. K. Leung, M. Y. He and A. G. Evans, *J. Mater. Res.*, **10**, 1693-1699 (1995).
20. A. Bagchi, G. E. Lucas, Z. Suo and A. G. Evans, *J. Mater. Res.*, **9**, 1734-1741 (1994).
21. M. Y. He, M. R. Turner and A. G. Evans, *Acta Metall. Mater.*, **43**, 3453-3458 (1995).
22. P. G. Charalambides, J. Lund, A. G. Evans and R. M. McMeeking, *J. Appl. Mech.*, **111**, 77-82 (1989).
23. M. D. Drory and J.W. Hutchinson, *Proc. Roy. Soc.*, **452**, 2319 (1996).
24. J. J. Vlassak, M. D. Drory and W. D. Nix. *J. Mater. Res.*, **12**, 1900 (1997).
25. K-S. Kim and N. Aravas, *Int. J. Solids Struct.*, **24**, 417-435 (1988).
26. Y. Wei and J.W. Hutchinson, *Int. J. Fracture*, **93**, 315-333 (1998).
27. M.R. Begley, A.G. Evans and J.W. Hutchinson, *Int. J. Solids Struct.*, **36**, 2773 (1999).
28. R.W. Hoffman, *Phys. Thin Films*, **3**, 211 (1966).
29. D.S. Campbell, in *Handbook of Thin Film Technology*, L.I. Maissel and R. Glang (eds) Chapter 12, McGraw Hill (1970), pp. 12.3-12.50.
30. (a) K.L. Mittal, *Electrocomponent Sci. Technol.* **3**, 21-42 (1976).
    (b) K.L. Mittal, in: *Adhesion Measurement of Films and Coatings,* K.L. Mittal (ed.) pp.1-13, VSP, Utrecht, The Netherlands (1995).
31. P.A. Steinmann and H.E. Hintermann, *J. Vac. Sci. Technol.* **A7**, 2267-2272 (1989).
32. A.G. Evans and J.W. Hutchinson, *Acta Metall. Mater.*, **43**, 2507-2530 (1995).
33. M.G. Allen and S.D. Senturia, *J. Adhesion*, **29**, 219-231 (1989).
34. H.M. Jensen, *Engr. Fracture Mech.* **40**, 475-86 (1991).
35. H.M. Jensen and M.D. Thouless, *Int. J. Solids Struct.* **30**, 779-95 (1993).
36. H.S. Jeong and R.C. White, *J. Vac. Sci. Technol.*, **A11**, 1373-76 (1993).
37. J.S. Wang, Y. Sugimura, and A.G. Evans, *Thin Solid Films*, **325**, 163 (1998).
38. J.E. Reynolds, J.E. Smith, G.L. Zhao, and D.J. Srolovitz, *Phys. Rev. B*, **53**,13883 (1996).
39. M.W. Finnis, *J. Phys.: Condens. Matter*, **8**, 5811 (1996).
40. Y. Wei and J.W. Hutchinson, *Int. J. Fracture*, **95**, 1-17 (1999).
41. N. A. Fleck and J W. Hutchinson, *Adv. Appl. Mech.* **33**, 295-361 (1997).
42. J. S. Stölken and A. G. Evans, *Acta Mater.*, **46**, 5109 (1998).
43. N. A. Fleck, G. M. Muller, M. F. Ashby and J. W. Hutchinson, *Acta Metall. Mater.*, **42**, 475-487 (1994).
44. J. M. McNaney, R. M. Cannon and R. O. Ritchie, *Acta Mater.*, **44**, 4713-4728 (1996).
45. F. G. Gaudette, S. Suresh, A. G. Evans, G. Dehm and M. Rühle, *Acta Mater.*, **45,** 3503-3514 (1997).
46. D. M. Lipkin, D. R. Clarke and A. G. Evans, *Acta Mater.*, **46**, 4835 (1998).
47. F. G. Gaudette, S. Suresh and A. G. Evans, *Metall. Mater. Trans.* **31**A, 1977-1983 (2000).
48. A. V. Zhuk, A. G. Evans, J. W. Hutchinson and G. M. Whitesides, *J. Mater. Res.*, **13**, 3555 (1998).
49. Q. Ma, H. Fujimoto, P. Flinn, V. Jain, F. Adibi-Rizi and R. H. Dauskardt, *Mater. Res. Soc. Symp. Proc.*, **391**, 91-96 (1995).
50. M. Lane, R. H. Dauskardt, R. Ware, Q. Ma and H. Fujimoto, *Mater. Res. Soc. Symp. Proc.*, **473**, 3-14 (1997).
51. V. Tvergaard and J.W. Hutchinson, *Phil. Mag.* **A70**, 641-56 (1994).

Adhesion Measurement of Films and Coatings, Vol. 2, pp. 19–47
Ed. K.L. Mittal

Relative adhesion measurement for thin film microelectronic structures. Part II

L.P. BUCHWALTER*

IBM T.J. Watson Research Center, Yorktown Heights, NY 10598, U.S.A.

Abstract—The present discussion on the relative adhesion measurement for thin film microelectronic structures will emphasize experimental results on the two adhesion test methods compared: the peel test and the modified edge lift-off test (MELT). The data presented show that the fracture behaviour in the modified edge lift-off test can be significantly more complex than what is detected in the peel test, and further emphasizes the importance of careful locus of failure analyses. A comparison of present results with the published literature underlines the fact that it is impossibile to determine the fundamental adhesion using a practical adhesion test. This is due to the inability to account for all energy dissipating processes during the testing. The peel test, however, can result in reliable practical adhesion measurement. MELT needs more characterization and possibly modification in sample preparation before the same can be said of this new test method.

Keywords: Adhesion; fundamental adhesion; practical adhesion, adhesion measurement.

1. INTRODUCTION

The reader may feel that this paper offers a pessimistic view on thin film adhesion measurement. It is true that fundamental adhesion cannot presently be measured or calculated. However, practical adhesion can be measured reliably, if certain constraints are observed in sample preparation and analysis [1]. The author solicits and will be greatful for any solutions to overcome the inability to determine the fundamental adhesion for thin film microelectronic structures.

The measurement of adhesion in microelectronics applications is of critical importance, as it is in many other technologies. When two dissimilar materials are brought together, the issue of adhesion arises. There are primarily two different directions from which the problem of adhesion can be approached: one is that of chemistry and the other that of mechanics [1-3]. It is interesting to read the literature from these two points of view, which often discuss the same matter, though it may not be immediately obvious. Many different terms, for example, are used for the fundamental adhesion depending on the disciplines of the authors:

---

* Phone: 914-945-3454, Fax: 914-945-2141, E-mail: paivikki@us.ibm.com

1) Interfacial fracture resistance, interfacial fracture toughness, or interfacial fracture energy [4-6]
2) Fracture energy [7, 8]
3) Interface decohesion [9, 10]
4) Debond energy [10-12]
5) Critical adhesion energy [13]
6) Work of adhesion or thermodynamic work of adhesion [14, 15]
7) Intrinsic adhesion [16]

For clarity it would be beneficial that the nomenclature be standardized. However, that will not be a topic of the present discussion of adhesion.

It has been shown that chemical bonding is important in adhesion [17-20]. A particularly clear example can be found in Ahagon and Gent paper from 1975 [20]. They discuss the adhesion of polybutadiene to glass surface, which had been treated with a mixture of vinyl- and ethylsilane with a steadily increasing vinylsilane concentration. The vinylsilane has reactivity with the glass surface (the silane end of the molecule) and with the polybutadiene (the vinyl end of the molecule), while the ethylsilane does not have reactivity with the polybutadiene. Figure 1 shows schematically the effect of vinylsilane concentration to the adhesion of polybutadiene to the treated glass.

Chemistry, though it is important, will not necessarily be adequate in explaining or improving adhesion. Stresses in the films making up the interface of interest are equally important since the stored strain energy in the film is the energy

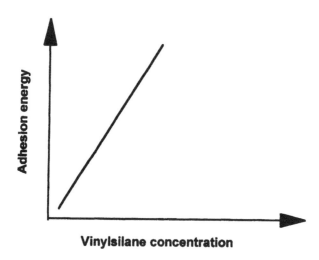

Figure 1. Schematic of polybutadiene adhesion to treated glass surface [20]. Adapted from Effect of Interfacial Bonding on the Strength of Adhesion, A.Ahagon and A.N.Gent, *J.Polym.Sci.: Polym.Phys.Ed.*, Copyright © 1975 John Wiley & Sons, Inc. Reprinted by permission of John Wiley & Sons, Inc.

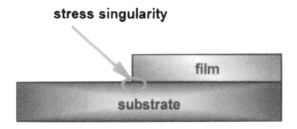

Figure 2. Schematic of a stress singularity at a step in a patterned bilayer structure.

source for delamination. As is well known, the crack will start at a flaw. This flaw may be particulate contamination, minute delamination due to surface uncleanliness, or a stress singularity (Figure 2) created during the build of the structure.

Adhesion improvement may, therefore, involve three components: 1) modification of interfacial chemistry, 2) change in the stresses in the structure, and 3) modification of structure stress singularities. If a near-interface (interphasial) or interface failure has been detected in the patterned bilayer structure, all three aspects should be looked at in order to improve adhesion. Interfacial chemistry involves substrate surface cleanliness and chemical interaction of the film with the substrate with or without an adhesion promoter. Reduction of the stresses in the structure may involve film thickness reduction, polymer film curing conditions change (ramp rate, cooling rate, peak temperature, etc), or change in metal film deposition conditions (temperature, energy of sputtered species, etc.). Elimination of strong stress singularities involves avoiding 90° angles between the sidewall and the bottom of the feature. The latter may be difficult to achieve as sloped walls will reduce pattern density, which is not a desirable result. Stress singularities are important for example with chip underfill polymers [21].

Often the only variable that can be changed to improve adhesion at a given interface is the interfacial chemistry. Film thickness and deposition conditions are usually set by other material properties requirements (such as degree of curing, crystallinity, dielectric or conductive properties) than adhesion. Modification of the interfacial chemistry will not always be the answer for adhesion improvement. If the locus of failure is in one of the materials making up the interface (a cohesive failure), attempting adhesion improvement via interface chemistry will be futile.

From the vast number of application-specific adhesion test methods, the 90° peel test and the MELT were chosen for the following reasons:
- applicability to thin film microelectronics adhesion testing with both brittle and ductile films
- simple and low cost test sample preparation
- simple and low cost test procedure
- large enough delaminated area can be created to allow locus of failure analysis with surface-sensitive analytical equipment
- both tests are being practiced in our laboratories

The rest of the paper is organized in the following manner:

2. Adhesion measurement: Purpose and definitions
3. Peel test and locus of failure analysis
4. Modified edge lift-off test (MELT) and locus of failure analysis
5. Summary

2. ADHESION MEASUREMENT: PURPOSE AND DEFINITIONS

Historically adhesion measurement had had three main purposes [1]:

1) To rank materials
2) To improve adhesion at poor interfaces
3) To determine the fundamental adhesion for the purpose of structure reliability modeling.

The more important one of these three is the second one, i.e. to improve adhesion at a poor interface. Adhesion does not usually carry as much importance in material ranking as do the other fundamental material properties such as thermal stability, crystallinity, dielectric behaviour, etc. The last one (#3) deals with **fundamental adhesion,** which is defined as the energy required to break the bonds at the weakest plane in the adhering system under adhesion measurement conditions used [1, 22, 23]. In theory this should correlate with the bond energies across the interface per unit area. The reliability modeling may, for example, assume that a film will stick to the substrate if (in the case of residual stress-driven delamination) [10]:

$$W_f \geq (h\sigma_R^2)/(E'\lambda) \qquad (1)$$

Where W_f [1] is fundamental adhesion, h is film thickness, σ_R is residual stress in the film, E' is the appropriate Young modulus (plane strain or plane stress) and λ is a cracking parameter [10]. Incidentally, in the original reference [10] the symbol γ has been used for the fundamental adhesion (W_f). Both the fundamental adhesion and the mechanical properties of the film as a function of stressing (thermal, temperature and humidity, etc.) are needed for the structure reliability modeling. **Interfacial adhesion** (which can be fundamental or practical) is measured when the locus of failure is at the very interface of the two materials making the interface. This is rare, but has been documented in the literature [23-26]. **Practical adhesion** is defined as the energy required to disrupt the adhering system irrespective of the locus of failure. It includes all or some of the other dissi-

---

[1] As per my conversation with Mittal, we have chosen W_f as the symbol for fundamental adhesion, because W_a is commonly used for the work of adhesion (or the thermodynamic adhesion). This is done to avoid confusion with surface tension, the symbol for which is commonly γ (see for example: A.W. Adamson, *Physical Chemistry of Surfaces*, 5th edition, John Wiley, New York (1990)).

pated energies due to the measurement process along with the fundamental adhesion. Practical adhesion is what is usually measured [1, 22-24, 27-28]:

Practical adhesion = F(Fundamental adhesion, all other energy dissipating processes)

The only way to understand what was actually measured during the adhesion testing is to characterize the locus of failure (LOF) [1]. In a bilayer structure, adhesion testing may result in the following (including, in some form, energies dissipated related to the film and substrate material properties) [1]:

1) Measurement of the cohesive strength of one of the materials making the interface
2) Measurement of the cohesive strength of the interphasial material
3) Measurement of cohesive strength of the reaction products layer
4) Measurement of practical adhesion
5) Measurement of some combination of 1-4

Crocker stated already in 1968: "Determination of the locus of failure is of first importance to interpretation of the results of any strength measurement on an adhesive joint" [26].

Figure 3 shows the possible loci of failure for a modified edge lift-off test (MELT) sample with two test materials.

The failure loci 2,3, 5,6, 8,9 are within 5-10nm of the respective interfaces, i.e. an interphasial failure in one of the materials making the interface. These failures are not considered cohesive, and the adhesion may be improved via surface modification schemes [1]. The definition of an interphasial failure is an arbitrary one: if some XPS signal (at $35°$ take-off angle) of the underlying material is detected the failure is an interphasial one. If no XPS signal is detected of the underlying material the failure is cohesive. With a cohesive LOF, surface treatments will not improve adhesion.

It is obvious that a careful analysis of the locus of failure is imperative. Surface sensitive analytical techniques (X-ray photoelectron spectroscopy (XPS), Auger electron spectroscopy (AES) with scanning electron microscopy (SEM), Atomic Force Microscopy (AFM) or other microscopy techniques) should be employed to determine the locus of failure.

It is not difficult to find papers in the literature dealing with methods to measure the fundamental adhesion of a given interface [8, 29-30]. Nor it is difficult to find papers claiming quantitative adhesion measurement results in the literature [5, 6, 31]. It is unlikely that an accurate or exact measurement of the desired property (in this case: fundamental adhesion) is achieved with precision and repeatability in any of the cited publications. More often than not it appears that the concept 'quantitative adhesion measurement' is used when what is actually meant is the precision and/or repeatability of the measurement method [32-33].

Since much of the study of the energy dissipation during adhesion testing has been done with the peel test, this discussion will be continued in the next section.

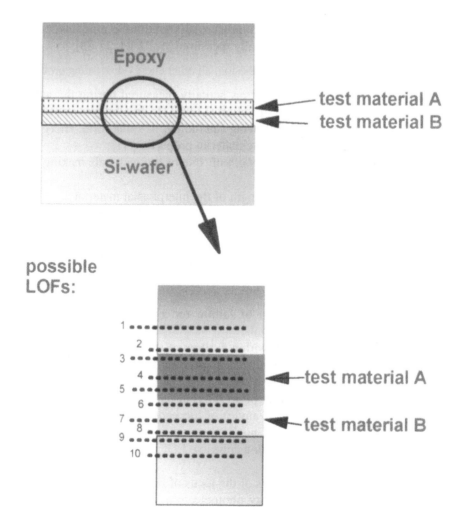

Figure 3. Schematic of the possible loci of failure in a two test material MELT sample. Notice that interfacial, mixed mode (crack path from one material through the interface into the other material and back) and reaction products layer LOFs are not included [1].

3. PEEL TEST AND LOCUS OF FAILURE ANALYSIS

An extensive amount of work has been done through the years on the analysis of the peel test [7, 8, 29, 34-49]. Peel test results have been found to be dependent on peel film thickness [28, 41-43, 50], film and substrate material properties [28-29, 41-43], peel rate [39-40, 51], peel test ambient [40, 50, 52], and macroscopic peel angle (Figure 4) [36, 44].

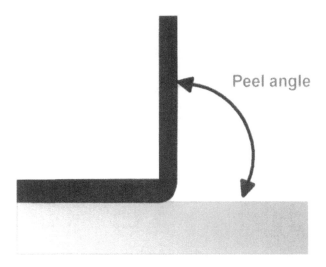

Figure 4. Schematic of peel test with peel angle defined.

Peel test sample preparation is shown schematically in Figure 5. In the particular case shown, the plated Cu film acts as a backing material (or superlayer [10]) to give adequate cohesive strength to the sample to allow peel testing to be performed. The backing material can be any material of choice as long as it and its thickness are kept constant throughout the testing [1]. The backing material in peel testing is there more for the purpose to give cohesive strength than to bring more available energy to the system.

The release layer can be copper or any other material that bonds poorly to one of the test materials. It must withstand the processing for sample build and not cause test area surface/interface contamination.

It is *impossible* to determine the fundamental adhesion from the peel test due to the inability to account for all energy dissipating processes [1]. Some of these (non-adhesion) energy dissipating processes are:

1) Heat [34]
2) Energy stored in the peel strip [49]
3) Acoustic emission [53]
4) Fractoemission [53]
5) Local deformation ahead of the crack tip [29, 44]
6) Macroscopic deformation of the peel strip and deformation of the substrate [29, 44].

The first two can be evaluated using calorimetry. Fractoemission is detected with insulator interfaces and acoustic emission with brittle materials. The acoustic and fractoemission have been assumed to be negligible [49], which may not always be true. Acoustic emission is used as a method in scratch testing to evaluate adhesion [54], and can be audibly detected during the peel of certain interfaces.

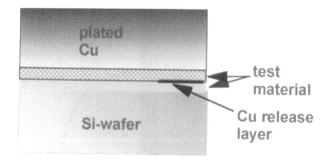

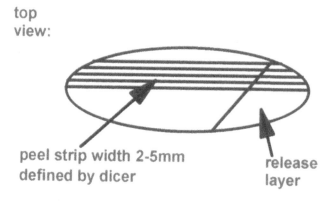

Figure 5. Schematic of peel test sample preparation [1].

An estimation of the energy dissipated in the deformation ahead of the crack tip and in the macroscopic deformation of the peeled film is difficult requiring careful fracture mechanics analysis [29, 44]. These energy dissipating processes, though often discussed in connection with the peel test, are not necessarily specific to peel test alone.

It is possible to use the peel force (=force required to separate the interface per unit width of the peel strip) directly as a relative measure of adhesion if the peel test and specimen variables mentioned above (peel film and its thickness, substrate, peel rate, peel ambient, and macroscopic peel angle) are kept constant [1]. This is because the peel force appears to be a function of fundamental adhesion so that when it increases so does the peel force [1]. It is not unreasonable to think that energy dissipated as heat, in acoustic or fractoemission, locally at the peel crack tip, in macroscopic deformation of adherend and the adherate as well as energy stored in the peeled film (which was omitted in the earlier discussion [1]) all are a function of the fundamental adhesion so that when it increases so would the energy dissipated in these processes. Small changes in the effective fundamental adhesion (a value derived from peel force corrected for work expenditure due to adherend plastic deformation [1, 29-28, 41-44]) seem to result in large changes in the peel force [28, 39, 41-43, 46]. This is advantageous as it offers higher level of distinction between samples [1].

Since the LOF characterization is of critical importance to adhesion testing, the following will focus on XPS and SEM data on peel test locus of failure analyses. Figure 6 shows the SEM locus of failure analysis of selected polyimide (PI)/oxide samples. The peeled PI film failure surface is shown [55].

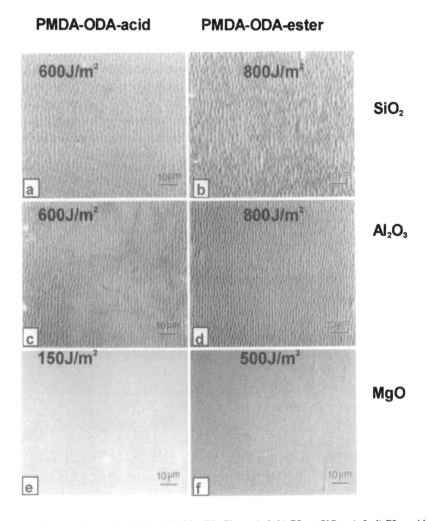

Figure 6. Failure surface of peeled polyimide (PI) films. a) & b) PI on SiO_2, c) & d) PI on Al_2O_3, e) & f) PI on MgO. The polyimide in each case is marked on the top of the column of pictures [55].

The peel direction is perpendicular to the stick-slip striations seen as vertical lines across the SEM pictures. As the peel force is reduced the spacing between the stick-slip caused striations decreases resulting finally in a featureless image, as shown in Figure 6e. It should be noted that these striations are not detected on the oxide surfaces [55]. The XPS locus of failure analysis shows polyimide on all failure surfaces (peel side or the substrate side) except in the case of PMDA-ODA-acid on MgO where Mg is detected also on the PI side of the peel locus of failure. The presence of Mg on the PI side indicates failure in the reaction product layer in this particular case. A salt formation between the acidic PMDA-ODA-acid and the basic MgO was proposed [55].

The effect of adhesion on the stick-slip behaviour is seen more clearly in Figure 7. In this case, both the polyimide and the substrate are the same, thereby removing any material property differences that could cloud the issue. The change in the peel force was brought about by exposing the samples to high temperature and humidity for differing lengths of time [56]. The XPS locus of failure analysis results show unequivocally that the failure in each case is in the polyimide close to the poly-imide/γ-APS (γ-aminopropyltriethoxysilane)/SiO_2 interface, but not at it [56]. It is clear from Figure 7 that the spacing between the stick-slip striations is a function of peel force; the higher the peel force the wider the spacing between striations.

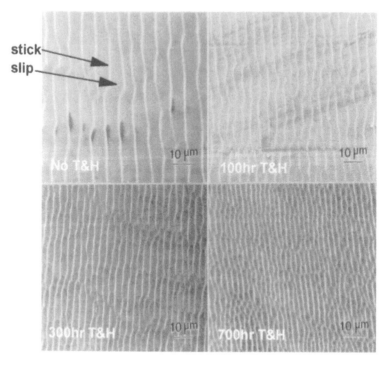

Figure 7. PMDA-ODA PI / γ-APS / SiO_2. Stick slip behaviour as a function of exposure time to elevated humidity and temperature conditions. The measured peel forces are $1200J/m^2$, $1050J/m^2$, $850J/m^2$ and $750J/m^2$ for no T&H, 100hr T&H, 300hr T&H and 700hr T&H, respectively [56].

Stick-slip behaviour has been discussed by Tsai and Kim [57]. However, their analysis considers poor adhesion cases only, i.e. the adherend would not deform plastically. It should also be pointed out that the peel rates used in their analysis of the stick-slip phenomenon were 0.1m/s and 2m/s. These peel rates are very high. In thin film adhesion testing the peel rates ordinarily vary between 0.5mm/min – 4mm/min [50-51, 55-56]. The stick-slip behaviour is a function of peel rate [57]. Therefore, another method should be developed to account for the energy dissipated in the stick-slip process in practical thin film adhesion measurement.

Figures 6 and 7 showing SEM micrographs present a rather uniform picture of the failure surface. This, however, may only be true locally over a small area. Over a wider area, the surface characteristics of the failure locus are more variable, as seen in Figure 8 showing SEM results on a Cr/PI (BPDA-PDA) failure locus.

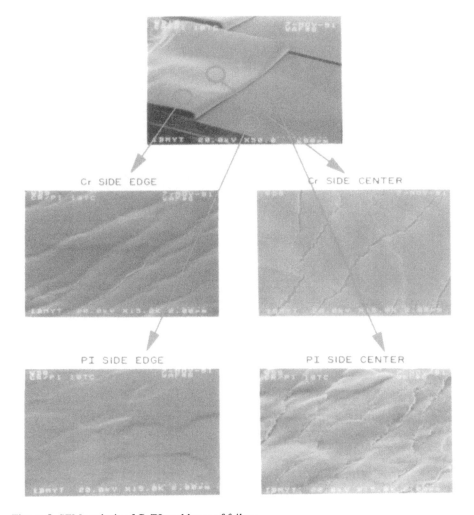

Figure 8. SEM analysis of Cr/PI peel locus of failure.

It is clear that there are differences in the appearance of the failure locus between the edge and the center portions. At the edge no stick-slip striations are detected on the PI coated substrate, while in the center portion they are clear. On the Cr side of the failure, the stick-slip behaviour can be detected both on the edge and in the center. With closer observation, there is a significant difference between the two locations: the center has what appears to be cracks, while the edge portion does not show them. This type of non-uniformity in the failure locus has not been considered to date to the best knowledge of this author in any calculation proposing to evaluate fundamental adhesion at a given interface. Figure 9 shows a clearer image of such cracking behaviour (crack branching) in Ta/PI test sample [58].

With careful study of the SEM picture below, one may be able to detect cracking of the Ta layer below the residual polyimide. A clearer picture of this type of secondary cracking (crack branching) of brittle films with strong adhesion is seen in Figure 10.

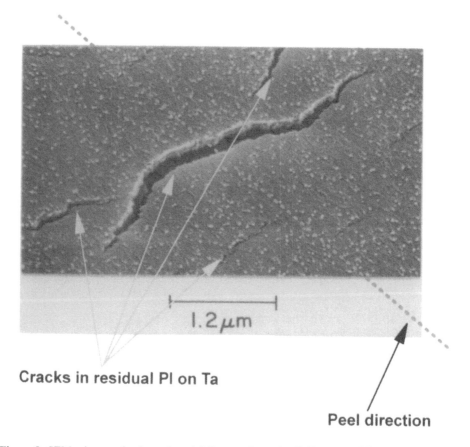

Cracks in residual PI on Ta

Peel direction

Figure 9. SEM micrograph of metal peel failure surface of a Ta/Ar-sputter/PI structure [58]. The cracks in the residual PI are perpendicular to the main peel crack path as well as perpendicular to the peel direction, which is marked in the figure.

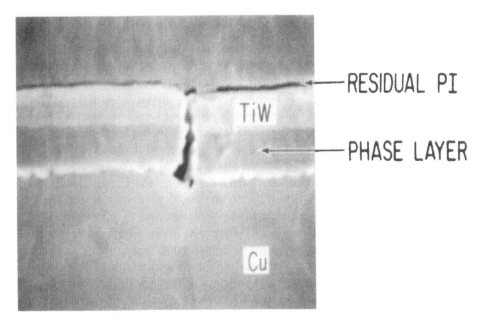

Figure 10. Cross-sectional SEM analysis of peeled metal film from a Cu/phase layer (adhesion promoting layer)/TiW/PI test sample. Note the secondary cracking (crack branching) extending through the more brittle metal films and blunting into the more ductile Cu [59].

There clearly are energy-dissipating processes that are not simple to account for in an adhesion test. One may not even know about them all, unless a very detailed analysis of the testing process is done. This detailed analysis does not only include the XPS and SEM analyses discussed here, but must include the calorimetric evaluation of the test, as well as acoustic and fractoemission detections. It is unlikely that one would go to such lengths in a practical situation to assure that all energy-dissipating processes were being taken into account in order to calculate the fundamental adhesion. As it has been stated before [1], for practical situations in the manufacturing environment, where quick response in problem solving is financially essential, a way has to be found where a simple straightforward adhesion testing can be used with reliable results. The 90° peel test, has proven to be an invaluable and relatively quick method in adhesion problem solution. The test is by no means perfect from the fracture mechanics point of view, and requires (as do all the present thin film adhesion tests) results verification with actual structure build and reliability testing [1]. This is because we do not usually know the mode of interface loading and/or the yielding (particularly important with tests that result in large strains, such as the peel test) conditions in the test or in the actual thin film product.

4. MODIFIED EDGE LIFT-OFF TEST AND LOCUS OF FAILURE ANALYSIS

The motivation for the development of the modified edge lift-off test (MELT) was to have an adhesion test that would not rely on mechanical energy for the delamination of an interface, but rather use the strain energy stored in the films making up the interface [60]. This may have been driven by Mittal's statement [23]: *"The choice of the test for measuring practical adhesion should be based upon the type of stresses the test specimen is going to encounter in practice."* This concept is exceedingly simple to state, but quite a different issue in practice. What it basically states is: A) the details of the stress state of the product interface as a function of build and use conditions must be known, and B) the details of the stress structure of the adhesion test specimen should be known to the same degree. Getting at the stresses in a particular structure will require some type of stress modeling, because one is not only interested in the global stress, but in the singularities (stress concentration sites) in the structure and in the test specimen. In order to model such stresses, free standing thin film mechanical properties (in-plane and out-of-plane) are needed as a function of process and use conditions. In today's microelectronic structures the film thicknesses are regularly less than 1μm. This causes nearly insurmountable problems with mechanical properties characterization as the techniques available today (nanoindentation, for example) are plagued with problems resulting in measurement inaccuracies. Besides, it is difficult, if not impossible, to make free standing films of submicrometer thick layers.

The stored strain energy in some films can be increased by cooling the sample down to liquid nitrogen temperature [12-13, 60]. The increase in the stored strain energy is dependent on the difference between the thermal coefficients of expansion (ΔCTE) of the film and the substrate. A schematic of this behaviour is shown in Figure 11.

BPDA-PDA polyimide on Si-wafer, for example, is a material pair that would fall onto the ΔCTE~0 line, while some epoxies are at the ΔCTE>>0, which is one reason why a silica filled epoxy was chosen as the backing material (or super-layer) for the MELT samples [60]. There is some concern about the need for cooling the samples down to liquid nitrogen temperature, as the lower temperature may be below the ductile-brittle failure transition with some materials. Also, these low temperatures are not experienced in the device build or use.

The sample preparation for MELT is simple. The wafer with the interface of interest is coated with a thick (150-250μm) epoxy coating using a doctor blade. Adhesion at the epoxy/test material interface is important. Oxygen plasma exposure to clean the test material surface works in most cases. With copper no treatment is necessary. The film is cured at 180°C for an hour. The samples are snap-cleaved to a sample size of about 2.5cm x 2.5cm. The samples are placed in a N_2 flushed enclosure, which is heated and cooled by N_2-gas. Convection cooling is preferred over conduction cooling since the specimens will bow during the test. The rate of cooling is well controlled but slow. The sample is first exposed to annealing past the glass transition temperature (T_g) of the epoxy to relax stresses. It is then cooled and the temperature of the onset of delamination (crack propagation) is recorded. The observation of delamination is visual. The temperature of delamination onset will be used for the determination of the stress in the epoxy film from a calibration curve (stress-temperature plot determined using the same rate of cooling as in the testing). The effective fracture toughness, K_{eff} (practical adhesion) is calculated using the following equation:

$$K_{eff} = \sigma_0(h/2)^{1/2} \tag{2}$$

Where σ_0 is the residual stress in the epoxy film and h is its thickness in meters. The above equation requires a precrack of length 'a' to be present between the substrate and the coating so that a/h >0.025 and assumes plane-strain conditions [60]. This analysis strictly speaking only accounts for crack-opening mode, which was a reasonable conclusion from the development of the test where experiments were run with epoxy/glass samples. The K_{eff} was found to correlate K_{IC} (fracture toughness) of the epoxy, and the failure was well into the epoxy film (cohesive failure) [60].

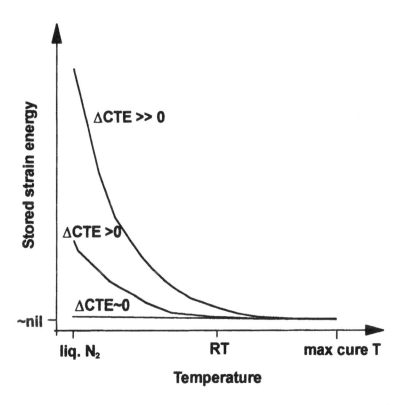

Figure 11. A schematic of the degree of stored strain energy in a film on a substrate as a function of the thermal coefficient of expansion difference between the film and the substrate.

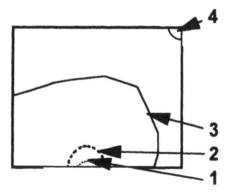

Figure 12. A schematic of crack initiation and growth during MELT testing of low k ILD/SiO₂ sample. See text for the number assignments.

In Figure 12 a general type of crack initiation and propagation is shown for low k ILD (=interlayer dielectric)/SiO₂ samples tested in our laboratories.

The crack initiation may start somewhere at the edge (1) or corner (4) of the sample. The crack growth may follow the path (1) to (2) to (3). Even though the cooling rate is monotonic, the crack growth often is not. The crack may arrest and then grow further as temperature is lowered sometimes ending in a catastrophic failure of the sample (sudden total delamination of the coating) and sometimes the total delamination is not experienced within the experimental temperature limits. It is obvious that in these cases the plane-strain condition (radius of crack-front curvature is infinite) does not hold [60]. The delamination in the MELT is not nearly as well controlled as it is with the EDT (edge delamination test) where a circular hole is cut into the coating and delamination distance from the edge is determined [12, 60], as shown in Figure 13.

In the EDT the onset of delamination is not so important as the delamination distance, which is a function of the hole diameter [12-13, 60]. The reason why one would want to use MELT as opposed to EDT is the ease of sample preparation (EDT requires lithography for the circular hole definition, though 'cookie cutters' have been used [61]).

Changes in the sample preparation for the MELT testing to control the delamination shape have been suggested. One may consider an etchback process to create a uniform precrack by etching into the film from the edges. Shaffer [60] has used this approach with polymer/metal interfaces. He chose FeCl₃ as the etchant to create the precrack. Cyclotene and Ultradel (Dow Chemical) interfaces with AlCu and Cu were extremely sensitive to the FeCl₃ etchant. Shaffer showed that the calculated K_{eff} was a function of time in the etchant. Even 5 second exposure of Cyclotene/Cu test samples showed etchant effect in the results.

Something akin to the etchback has been used for the study of photoresist adhesion to SiO₂ surface. The basis for the approach was the interfacial migration of ionic solutions [62]. It was found that neutral salts, such as LiCl or NaCl or acids such as HCl did not migrate to any appreciable extent. However, aqueous HF and basic ionic solutions did migrate. NaOH was chosen as the test solution, because the test (or probe) solution *should not attack the film or the substrate*.

We chose to look at SiN/Cu interface that was exposed to FeCl$_3$ precrack etching process.

Figure 14 shows a schematic of the etchback process results.

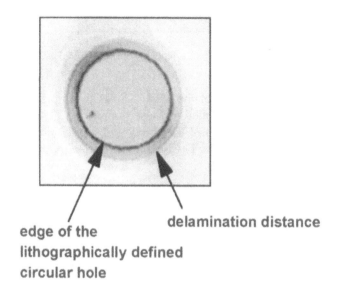

**edge of the
lithographically defined
circular hole**

delamination distance

Figure 13. Edge delamination test (EDT) results on Cyclotene/Al/Si-wafer test sample [13].

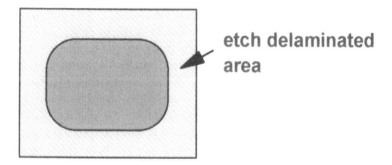

**etch delaminated
area**

Figure 14. A schematic of back etching to create a precrack for the MELT testing.

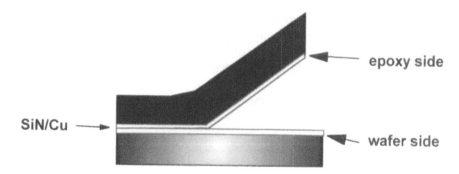

Figure 15. A schematic of the SiN/Cu MELT sample hypothetical failure path.

Figure 15 shows a schematic of the MELT SiN/Cu sample after test completion. No SEM analysis was performed on these samples.

The results of the XPS analysis of the epoxy and wafer sides of the failure are shown in Table 1. Samples with the $FeCl_3$ etchback and without it were analyzed.

Table 1.
Elemental composition (%) of SiN/Cu MELT sample LOF. XPS analysis results

Sample	C	O	N	Cl	F	Si	Cu
1A) SiN/Cu, no etch / epoxy side	10	11	30	ND	0.6	44	4
1B) SiN/Cu, no etch / wafer side	25	32	ND	ND	ND	ND	44
2A) SiN/Cu, FeCl₃ etch / epoxy side	60	17	11	0.7	ND	9	1
2B) SiN/Cu, FeCl₃ etch / wafer side	64	21	3	0.8	ND	ND	11
3) Epoxy surface	70	23	3	0.9	0.5	3	ND

Chlorine is only found at the interface of the sample that was etched with $FeCl_3$ (F1s found on sample 1A on the epoxy side of the delaminated sample may be due to SiN deposition chamber cleaning process residue). The etching has apparently altered the character of the LOF: the failure surfaces of the samples exposed to $FeCl_3$ etch have more carbon, and N1s is found on the substrate side. Since no Si is seen on sample 2B, the N1s peak detected cannot be due to SiN, but is more likely due to epoxy (amine curing agent). It could be that the interfacial chlorine is due to the epoxy (Cl and F are synthesis residues) penetration through the SiN film (<100nm thick), which is more significant with the etching than without it. The MELT failure locus with these samples is in the interphasial region near the SiN/Cu interface. The XPS Cu2p data show Cu on both sides of the failure thereby indicating that the actual LOF would be in the Cu near the SiN/Cu interface both with and without the $FeCl_3$ etch. This conclusion is corroborated by the lack of Si on the wafer side of the failure. XPS C1s data for samples 3, 2A, and 1A are compared with C1s due to ordinarily observed ambient contamination on a Si-wafer surface in Figure 16.

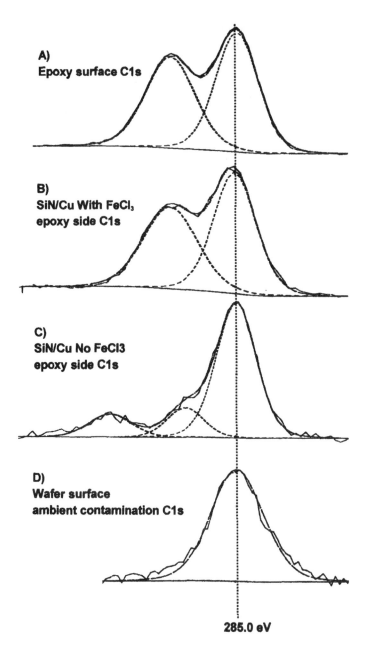

Figure 16. Comparison of C1s data of epoxy surface (A), SiN/Cu + FeCl₃; epoxy side of the failure (B), SiN/Cu no FeCl₃; epoxy side of the failure (C), and ambient contamination (D).

The following are obvious from the above figure:

1) The C1s found on the failure surfaces (Figures 16B and C) is different from the ambient contamination C1s (Figure 16D) pictured here. It is not unusual, however, to see the two higher binding energy (to the left of the main peak at 285eV) C1s peaks at a distinctly lower intensity than the main peak on a surface contaminated by ambient exposure (Figure 16C). What is highly unusual is to see the type of C1s shown in Figure 16B on an inorganic surface after exposure to ambient. Notice that Table 1 shows the presence of SiN on the epoxy side of the failure.

2) The C1s data of the sample with the etch vs the sample without it are different.

Since it has also been shown that an etchant has an effect on the measured adhesion [60], this approach should not be used for the definition of the precrack. Figure 17 shows an alternative scheme for MELT sample preparation using the release layer approach familiar from the peel test sample preparation. A crosshatched pattern of the 10-20nm thick release layer is deposited through a mask.

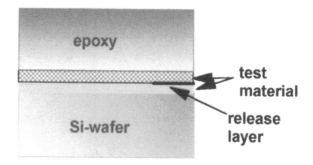

top view:

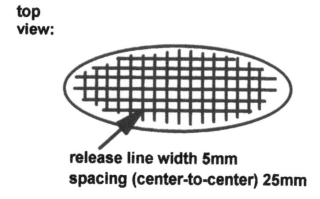

release line width 5mm
spacing (center-to-center) 25mm

Figure 17. MELT sample preparation using a release layer approach, much like that in the peel test [1].

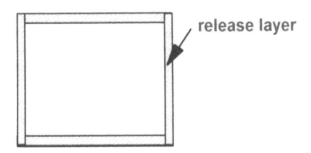

Figure 18. A schematic of MELT specimen with release layer for initial crack definition.

The top view of the sample after successful dicing along the middle of the 5mm wide release layer strips is shown in Figure 18. This type of sample fulfills the steady-state requirement for crack propagation (pre-crack length should be ~5 x film thickness) [60].

The rest of the paper will focus on low-k ILD/SiO$_2$ MELT results (Modern Metalcraft Cryostage MELT equipment) and locus of failure characterization. The low-k ILD thickness was about 0.75μm. The epoxy thickness was on the order of 200μm. No release layer or etchback was used with these samples. The samples were snap-cleaved with the assumption that this process created a precrack [60]. At times, the delamination at the edge of the sample was detected during the snap-cleave operation. The temperature of the onset of crack propagation was recorded. The practical adhesion calculated (equation 2) from the experimental data was about 0.24MPa(m)$^{1/2}$ for samples tested. The loci of failure were analyzed using XPS, optical microscopy, and SEM. XPS results for the as prepared, unstressed samples (identified as T-0 samples) are shown in Table 2.

Table 2.
XPS locus of failure characterization of low-k ILD/SiO$_2$ T-0 MELT samples. Elemental composition (%)

| Sample | Epoxy side | | | | | | Wafer side | | | | |
	C	O	N	Si	Cl	F	C	O	N	Si	F
1	97	3	-	-	-	~1	49	33	-	19	-
2	97	3	-	-	-	~1	46	31	-	21	2
Virgin low-k ILD surface	98	2									
Epoxy surface	70	23	3	3	0.9	0.5					

The data in Table 2 suggest that the locus of failure is in the low-k ILD close to the ILD/SiO$_2$ interface but not at it, i.e. the XPS data suggest interphasial failure. The presence of fluorine at the interface is not understood at this time, but it may be due to the SiO$_2$ deposition chamber cleaning process, which uses F-based chemistry.

The SEM analysis of the failure locus gives a new point of view to the situation. Figure 19 displays an SEM micrograph of the Si-wafer side of the failure locus.

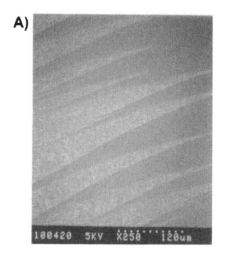

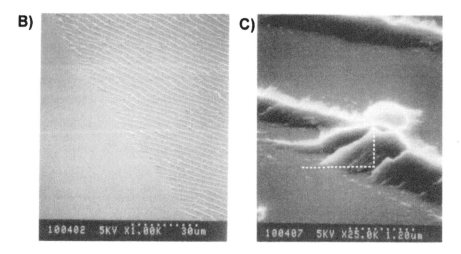

Figure 19. SEM micrographs of Si-wafer side of the MELT failure locus of low-k ILD/SiO$_2$ sample. A), B), and C) show different magnifications of the same general area of the sample. The dotted lines in micrograph C) are imaginary lines for the surface and feature heights. The SEM sample tilt angle was 30°.

The estimated feature height on the surface in Figure 19 C is about 1.6μm, which is about twice the height of the low-k ILD. This would, therefore, suggest failure into the epoxy through the thickness of the ILD. Notice also the significant difference in the failure characteristics within very small distances (Figures 19 A & B). It should be pointed out that the spot size used in the XPS analysis was an oval with 600μm and 1080μm axes. Therefore, the XPS results are an average over relatively large area covering different loci of failures. Figure 20 shows another aspect of the wafer side failure locus.

At this location the height of the feature at the vertical dotted line is about 3.2μm, which is about four times the low-k ILD thickness. Crack-branching is obvious in Figure 20B. The cracked-off part is about 2μm thick. The brittle nature of the failure is characteristic to epoxies [63]. The SEM micrograph series shown in Figure 21 is for the same sample but now from the epoxy side of the failure.

Based on the height of the features seen in Figures 19 and 20 the fibers shown in Figure 21 are likely epoxy. Figure 21A shows a particularly good example of the varied character of the MELT locus of failure in this sample. Within an area of analysis of about 450μm x 375μm, several obviously different loci of failure

A)   **B)**

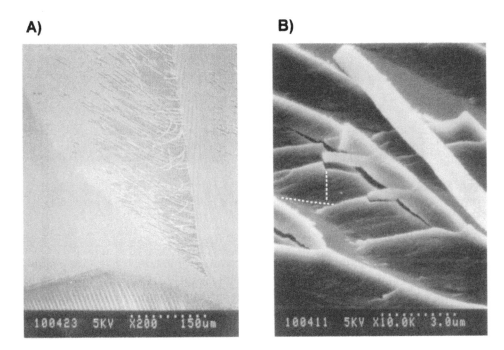

Figure 20. Another Si-wafer side failure area in the epoxy/low-k ILD/SiO₂/Si-wafer MELT sample. A) and B) are of the same general area but at different magnification. The dotted lines (B) again mark the imaginary lines for the surface and height of the image. The SEM sample tilt angle was 30°.

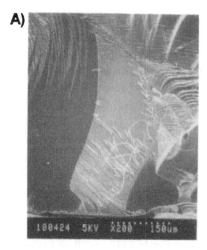

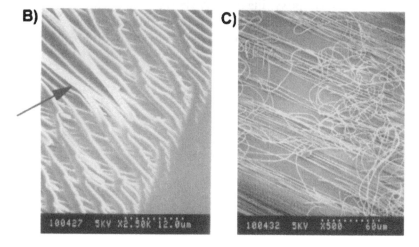

Figure 21. SEM micrographs of the epoxy/low-k ILD/SiO$_2$/Si-wafer MELT locus of failure. A), B), and C) are of the same general area of the epoxy side of the failure. The SEM sample tilt angle was 30°.

can be detected. It is quite possible that at the onset of crack propagation a multitude of failure loci can be experienced simultaneously (recall that the onset of crack propagation is judged visually). If such varied paths are present as shown in Figure 21A, it is impossible to determine what the onset of crack propagation actually measures.

Figure 22 shows a schematic of the failure paths in the MELT samples studied, most of which are shown in Figures 19-21. The one not shown is the final failure into the silicon substrate itself with the conchoidal nature characteristic to silicon fracture.

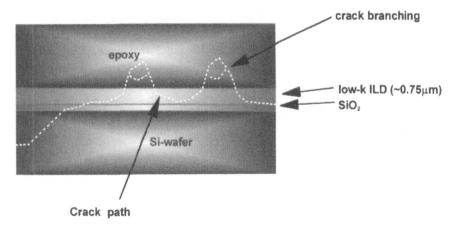

Figure 22. Schematic of low-k ILD/SiO$_2$ MELT test sample failure paths.

Fracture in the Si-wafer can happen at the onset of the crack propagation and move from there back to the test materials. Mode mixity influences the direction of interphasial cracks. When mode mixity is negative the crack may turn into the substrate [60, 64]. This combined with Figures 19-21 suggests significant changes in the mode mixity during the MELT process. Since the failure locus was not controlled into the region of interest in this case, it is very difficult to make statements about the measured practical adhesion values. Obviously they may represent cohesive strength of the epoxy, the cohesive strength of the low-k ILD etc., or some measure of adhesion. As opposed to the peel test, where the peel force required to separate the interface reaches a constant value (with stick-slip causing relatively consistent fluctuation across the mean peel value) and stays constant for a few centimeters of peeled interface, the MELT does not always allow separation of a large area at constant force (temperature). This is due to the mechanism used for delamination in the test. The energy in the peel test is supplied by external means with relatively controlled macroscopic crack path shape, while in the MELT it is internal and crack path shape can be varied.

After thermal cycling to elevated temperatures samples 1 & 2 show nearly identical locus of failure characteristics as determined by XPS (Table 3.).

Table 3.
XPS locus of failure characterization of low-k ILD/SiO$_2$ thermally stressed samples. Elemental composition (%)

	Epoxy side						Wafer side				
Sample	C	O	N	Si	Cl	F	C	O	N	Si	F
1	96	3	-	-	-	1	40	39	-	21	-
2	97	3	-	-	-	-	43	36	-	20	0.5

The practical adhesion calculated from the MELT data was about $0.27\text{MPa(m)}^{1/2}$ for the thermally stressed samples. This is within the experimental error from the practical adhesion calculated for the T-0 samples. Optical microscopy at 1500x, however, shows very different behaviour when comparing the T-0 results with the thermally stressed ones (Figure 23). The locus of failure characterized with microscopy shows a significantly smoother character than what was detected for the T-0 samples. The failure paths (mode of interface loading) are apparently not the same for these two sets of samples. Therefore, it is not reasonable to compare the results [1].

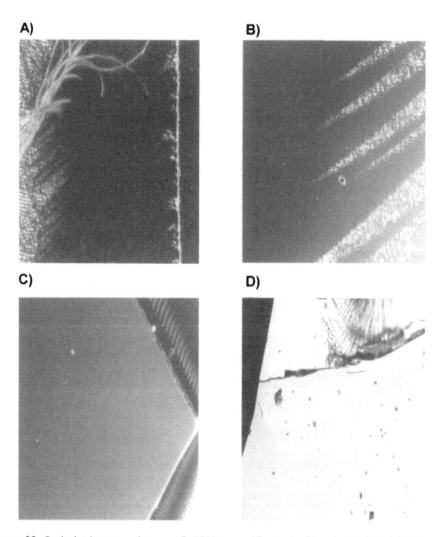

Figure 23. Optical microscopy images (@ 1500x magnification) of low-k ILD/SiO$_2$ MELT samples. Wafer side of the failure locus. A) sample #1 T-0, B) sample #2 T-0, C) sample #1 after thermal stressing, and D) sample #2 after thermal stressing.

5. SUMMARY

The measurement of practical adhesion, for example, with the $90°$ peel test can be done with reliable results if certain restrictions are obeyed in the sample preparation and measurement. A direct measurement or calculation of the fundamental adhesion from thin film adhesion tests is not possible due to the inability to account for all energy dissipating processes during the test. Modified edge lift-off test (MELT) or edge delamination test (EDT) may prove to be valuable tests for adhesion measurement, if precrack can be defined consistently and if the crack path can be controlled. Locus of failure analyses with surface sensitive analytical tools and high resolution microscopy are of primary importance in the interpretation of adhesion measurement data. Without such analysis it is not possible to know what was actually measured. Therefore, any adhesion test that does not allow locus of failure analysis is useless.

Acknowledgements

The author expresses her thanks to Dr. J.Hedrick for low k ILD/SiO$_2$ sample preparation, Dr. J.Hay and Mr. E.Liniger for the SiN/Cu etchback sample preparation.

REFERENCES

1. L.P. Buchwalter, *J.Adhesion*, **72**, 269 (2000).
2. D. Maugis and M. Barquins, *J.Phys.D: Appl.Phys.*, **11**, 1989 (1978).
3. D. Maugis, *J.Adhesion Sci.Technol.*, **10**, 161 (1996).
4. R.H. Dauskardt, S.Y. Kook, A. Kirtikar, and K.L. Ohashi, in: *High Cycle Fatigue of Structural Materials*, W.O. Soboyejo and T.S. Srivatsan (Eds.), p.479, The Minerals, Metals & Materials Society (1997).
5. X. Dai, M.V. Brillhart, and P.S. Ho, *Proc. IEEE Electr.Comp.Technol.Conf. 1998*, 132 (1998).
6. X. Dai, M.V. Brillhart, and P.S. Ho, *Mater.Res.Soc.Symp.Proc.*, **515**, 15 (1998).
7. A.K. Moidu, A.N. Sinclair, and J.K. Spelt, *J.Testing Eval.*, **23**, 241 (1995).
8. A.K. Moidu, A.N. Sinclair, and J.K. Spelt, *J. Testing Eval.*, **26**, 247 (1998).
9. A. Bagchi, G.E. Lucas, Z. Suo, and A.G. Evans, *J. Mater. Res.*, **9**, 1734 (1994).
10. A. Bagchi and A.G. Evans, *Interface Sci.*, **3**, 169 (1996).
11. B. Mirman, *Trans. ASME J.Electr.Packaging*, **117**, 340 (1995).
12. E.O. Shaffer, II, F.J. McGarry, and F. Trusell, *Mater.Res.Soc.Symp.Proc.*, **308**, 535 (1993).
13. F.J. McGarry and E.O. Shaffer, II, *Mater.Res.Soc.Symp. Proc.*, **356**, 515 (1995).
14. D.E. Packham, *Intl.J.Adhesion Adhesives*, **16**, 121 (1996).
15. A.A.Volinsky, N.I. Tymiak, M.D. Kriese, W.W. Gerberich, and J.W. Hutchinson, *Mater.Res.Soc.Symp.Proc.*, **539**, 277 (1999).
16. A.J. Kinloch, *Adhesion and Adhesives. Science and Technology*, Chapman and Hall, New York (1990).
17. A.N. Gent, *Plastics Rubber Intl.*, **6**, 151 (1981).
18. A.N. Gent, *Intl.J.Adhesion Adhesives*, 175 (April 1981).
19. A.N. Gent, *Adhesives Age*, 27 (February 1982).
20. A. Ahagon and A.N. Gent, *J.Polym.Sci: Polym.Chem.Ed.*, **13**, 1285 (1975).
21. S.L. Buchwalter, Personal communication (1999).
22. K.L. Mittal, *Pure Appl.Chem.*, **52**, 1295 (1980).

23. K.L. Mittal, in: *Adhesion Measurement of Thin Films, Thick Films, and Bulk Coatings*, K.L.Mittal (Ed.), pp.5-17, STP 640, ASTM, Philadelphia (1978).
24. K.L. Mittal, in: *Adhesion Measurement of Films and Coatings*, K.L. Mittal (Ed.), pp.1-13, VSP, Utrecht, The Netherlands (1995).
25. J.J. Bikerman, in: *Adhesion Measurement of Thin Films, Thick Films, and Bulk Coatings*, K.L.Mittal (Ed.), pp. 30-40, STP 640, ASTM, Philadelphia (1978).
26. G.J. Crocker, *Rubber Chem. Technol.*, **42**, 30 (1968).
27. A.N. Gent and G.R. Hamed, *Plast. Rubber Mater. Appl.*, 17 (February 1978).
28. J.Kim, K.S. Kim, and Y.H. Kim, *J. Adhesion Sci.Technol.*, **3**, 175 (1989).
29. K.S. Kim and N. Aravas, *Intl.J. Solids Structures*, **24**, 417 (1988).
30. A.J. Kinloch, in: *Fatigue of Advanced Materials*, R.O. Ritchie, R.H. Dauskardt, and B.N. Cox, (Eds.), p.439, Materials and Component Engineering Publications Ltd, Birmingham, UK (1991).
31. H.D. Goldberg, G.S. Cha, and R.B. Brown, *J.Appl.Polym.Sci.*, **43**, 1287 (1991).
32. L.P. Buchwalter, in: *Polyimides: Fundamentals and Applications*, M.K. Ghosh and K.L. Mittal (Eds.), p.587, Marcel Dekker, New York (1996).
33. L.P. Buchwalter, Polyimide Adhesion Characteristics to Selected Inorganic Surfaces, D.Tech. Thesis, Helsinki University of Technology, Helsinki (1997).
34. R.J. Farris and J.L. Goldfarb, in: *Adhesion Measurement of Films and Coatings*, K.L. Mittal (Ed.), p.265, VSP, Utrecht, The Netherlands (1995).
35. A. Hagemeyer, H. Hibst, J. Heitz, and D. Bauerle, *J.Adhesion Sci.Technol.*, **8**, 29 (1994).
36. A.N. Gent and S.Y. Kaang, *J. Adhesion*, **24**, 173 (1987).
37. A.N. Gent and G.R. Hamed, *J.Appl.Polym.Sci.*, **21**, 2817 (1977).
38. A.N. Gent, *Rubber Chem. Technol.*, **47**, 202 (1974).
39. A.N. Gent and A.J. Kinloch, *J.Polym.Sci., Part A-2*, **9**, 659 (1971).
40. A.N. Gent and R.P. Petrich, *Proc. Roy. Soc. A*, **310**, 433 (1969).
41. K.S. Kim, Elasto-Plastic Analysis of the Peel Test, T.&A.M. Report No. 472, University of Illinois, Urbana-Champaign (1985).
42. K.S. Kim, *Mater.Res.Soc. Symp.Proc.*, **119**, 31 (1988).
43. K.S. Kim and J. Kim, *ASME Trans.J.Eng.Mater.Technol.*, **110**, 266 (1988).
44. A.J. Kinloch, C.C. Lau, and J.G. Williams, *Intl.J.Fracture*, **66**, 45 (1994).
45. M.D. Thouless and H.M. Jensen, *J. Adhesion*, **38**, 185 (1992).
46. I.S. Park and J. Yu, *Acta mater.*, **46**, 2947 (1998).
47. Y.B. Park and J. Yu, *Mater.Sci.Eng.*, **A266**, 109 (1999).
48. Y.B. Park, I.S. Park, and J. Yu, *Mater.Sci.Eng.*, **A266**, 261 (1999).
49. J.L. Goldfarb, R.J. Farris, Z. Chai, and F.E. Karasz, *Mater.Res.Soc.Symp.Proc.*, **227**, 335 (1991).
50. L.P. Buchwalter and R.H. Lacombe, *J.Adhesion Sci.Technol.*, **2**, 463 (1988).
51. L.P. Buchwalter and R.H. Lacombe, *J.Adhesion Sci.Technol.*, **5**, 449 (1991).
52. A.N. Gent and J. Schultz, *J. Adhesion*, **3**, 281 (1972).
53. D.M. Mattox, in: *Handbook of Deposition Technologies for Films and Coatings*, R.F. Bunshah, (Ed.), p.643, Noyes Publications, Park Ridge, NJ (1994).
54. S.J. Bull, in: *Adhesion Measurement of Films and Coatings*, Vol.2., K.L. Mittal (Ed.), VSP, Utrecht, The Netherlands, in press.
55. T.S. Oh, L.P. Buchwalter, and J. Kim, *J.Adhesion Sci.Technol.*, **4**, 303 (1990).
56. L.P. Buchwalter, T.S. Oh, and J. Kim, *J.Adhesion Sci.Technol.*, **5**, 333 (1991).
57. K.H. Tsai and K.S. Kim, *Intl. J. Solids Structures*, **30**, 1789 (1993).
58. L.P. Buchwalter, *J.Adhesion Sci. Technol.*, **7**, 941 (1993).
59. L.P. Buchwalter, in: *Polymer Surfaces and Interfaces: Characterization, Modification and Application*, K.L. Mittal, and K.W. Lee (Eds.), p.147, VSP, Utrecht, The Netherlands (1997).
60. E.O. Shaffer, II, Measuring/Predicting the Adhesion of Polymeric Coatings, Ph.D Thesis, Massachusetts Institute of Technology (1995).

61. C.L. Bauer, The Determination of the Mechanical Behavior of Polyamic Acid / Polyimide Coatings, Ph.D Thesis, University of Massachusetts, Amherst (1988).
62. K.L. Mittal and R.O. Lussow, in: *Adhesion and Adsorption of Polymers, Part B*, L.H. Lee (Ed.), p.503, Plenum, New York (1980).
63. C. Feger, Personal Communication (1999).
64. A.G. Evans, H.D. Drory, and M.S. Hu, *J.Mater.Res.*, **3**, 1043 (1998).

Adhesion Measurement of Films and Coatings, Vol. 2, pp. 49–77
Ed. K.L. Mittal
© VSP 2001

Testing the adhesion of hard coatings including the non-destructive technique of surface acoustic waves

HEINER OLLENDORF,[1] THOMAS SCHÜLKE[2, *] and DIETER SCHNEIDER

Fraunhofer Institut für Werkstoff- und Strahltechnik, Winterbergstrasse 28, 01277 Dresden, Germany
Present address:
[1] *White Oak Semiconductor, 6000 Technology Blvd., Sandston, Virginia 23150, USA*
[2] *Fraunhofer USA, Center for Surface and Laser Processing, 211 Fulton Street Suite 101, Peoria, Illinois 61602, USA*

Abstract—Direct measurement of the adhesion of surface coatings is still a challenge. Surface engineers, however, desperately need practical quantities to evaluate the adhesion of their coatings, since this property essentially determines the applicability of their products in industrial practice.

Several adhesion test methods came into use for this reason such as the scratch test, bending test, Rockwell test, cavitation test, and impact test. These techniques gained acceptance due to their easy application.

The conflicting results often reported by such tests gave reason to perform a comparison of these techniques by utilizing them to test the adhesion of TiN hard coatings on steel (film thickness 1.2 ...2.4 µm). The films were deposited by ion plating. Adjusting the duration of an in-situ cleaning process (argon ion pre-sputtering) modified the interface conditions for the subsequent film deposition.

The measured results from the various standard tests were investigated with regard to their correlation with the pre-sputtering time. Several test parameters of these methods were used to evaluate the adhesion: the friction work, acoustic emission activity and critical load of the scratch test, the critical strain and the defect density of the four-point bending test, the proportional damage area of cavitation test, the critical number of loading cycles of the impact test, and Young's modulus of the film as measured with the laser-acoustic method. For the majority of these parameters, a significant correlation with the pre-sputtering time was found, but while some of them indicated the expected improvement of the adhesion with increasing pre-sputtering, other parameters suggested an opposite effect. Summarizing these surprising results revealed that the evaluation of the film quality differed for local and global adhesion behaviors. An interesting aspect of the non-destructive adhesion testing was the utilization of laser induced surface acoustic waves. The laser-acoustic technique is gaining acceptance as a quick and robust technique for determining Young's modulus of thin films with thickness down to nanometers. The results demonstrate the promising possibility of laser acoustic method for fast, non-destructive and inexpensive evaluation of the adhesion behavior.

Keywords: Adhesion; non-destructive testing; Young's modulus; acoustic waves.

---

* To whom correspondence should be addressed. Phone: +1 (309) 999-5887, Fax: +1 (309) 999-5889, E-mail: tschuelke@fraunhofer.org

1. INTRODUCTION

The adhesion of the film to the substrate determines essentially the quality of coated materials. Therefore, many efforts have been made to describe theoretically the adhesion phenomenon and to develop relevant adhesion test methods.

It is generally accepted to distinguish two terms, "fundamental adhesion" and "practical adhesion" [1,2]. The term "fundamental adhesion" is used to describe the fundamental aspect that two materials adhere by the nature of the chemical bonds at the interface between them. This suggests quantifying the adhesion by the force necessary to break the interface bonds or by the specific work to separate the two materials.

The use of the engineering term "practical adhesion" expresses that the fundamental aspect does not satisfy the needs of the industrial practice for two reasons: the fundamental adhesion cannot be deduced for most of the materials of interest and does not alone determine the mechanical behavior of a coated material. The combination of bulk properties of the adhered materials such as strength, stiffness, and toughness influence the failure of coated materials as well as specific effects such as the loading regime, the residual stress, the defect density and the defect distribution.

The wide variety of coated and surface modified materials and their application in semiconductor, optical and machining industries have been motivating the development of many adhesion test methods, discussed by Mittal [1b, 2]. Some of the well-known methods are the pull-off method, the topple method, the Scotch tape test, the scribe test, the blister method, the peel test, etc. whose strengths and limitations have been discussed in several publications [1,2,3].

The application of hard coatings with a thickness of some micrometers or even less for wear protection introduced new challenges for testing the adhesion. Conventional techniques such as the pull-off method or the peel test fail due to the high loading capacity of hard coated materials. In detail, the following reasons limit the application of the mentioned methods for adhesion testing of thin hard coatings:
– The interface strength of the hard coating is much higher than the strength of the adhesive bond needed to attach the sample to the test jig.
– The high hardness and stiffness of the film affect the test device.
– The brittle failure behavior of the film material prevents a reproducible and quantifiable detachment of the films.
– The low thickness reduces the critical volume available for the test.

More appropriate techniques to study adhesion of thin hard coatings are the scratch test, the bending test, the impact test, the cavitation test, and the Rockwell indenter test. Although more appropriate, it must be accepted that these methods cannot provide generally comparable results that are independent of the specific test conditions. However, the techniques are employed due to the lack of alternative approaches.

A test program has been performed to critically study the results of those techniques on a well-defined test system consisting of a titanium nitride (TiN) coated steel sample. TiN was chosen because of its dominant role in wear resistant hard coating applications such as cutting tools and components. TiN films with different levels of adhesion to the steel substrate were prepared by ion plating. Varying the pre-sputtering time with argon ions enabled to adjust different levels of film adhesion. It is known that pre-sputtering the steel surface for 15 minutes guaranties a good adhesion of a TiN film [4]. Seven test series with a total number of 130 samples were prepared with pre-sputtering times, t_S, from 0.5 to 15 minutes.

The investigation focused on the following aspects:
- Deducing suitable test parameters,
- Evaluating the sensitivity of the test parameters to distinguish the varying adhesion quality,
- Testing the correlation of the test parameters with the pre-sputtering time t_S which was assumed to represent the adhesion quality,
- Comparing the results of the different methods.

Special attention was paid to the possibility of non-destructively testing the mechanical film quality. Although non-destructive test methods are highly desirable, they cannot yield direct information about the failure mechanism because they do not physically separate the film from the substrate. However, empirical correlations between the non-destructive test parameters and the failure characteristics are very helpful to interpret the failure mechanism. This knowledge can be applied in industrial quality control. A promising technique to "non-destructively predict" a possible practical adhesion failure is the method of surface acoustic waves because:
- Surface acoustic waves are mechanical vibrations that propagate along the surface of the material and are very sensitive to thin films.
- This acoustic wave mode penetrates to a depth of the material that depends on the frequency. The penetration can be adjusted so that wave motion reaches through the film-substrate interface and can provide information about the interface quality.
- The method enables to measure the stiffness (Young's modulus) of the film material. The elastic stiffness of the film compared to the substrate essentially determines how the loading stresses are induced and distributed in the coated material. Therefore, the stiffness has an effect on the wear and adhesion properties of the film.
- Defects in the film or the film-substrate interface are sources of potential film failure. These defects also reduce Young's modulus of the material. This well-known effect implies the deployment of Young's modulus measurements to predict the performance of a thin hard coating in terms of adhesion.

Short laser pulses focused on the sample surface can effectively induce wide band acoustic waves. Measuring the dispersion of the wave's phase velocity along the sample surface allows deducing Young's modulus of the TiN films with an uncertainty of less than ± 2.5 % [6,7,8].

An alternative non-destructive approach is acoustic microscopy. This technique can detect single interface imperfections and microscopic flaws [9] and was included in the test program.

2. TEST MATERIALS, PREPARATION AND CHARACTERIZATION

Substrate samples:
Substrate material: Steel 42CrMo4,
Sample dimension: 6 mm × 8 mm × 60 mm,
Treatment: Quenched and tempered to a hardness of 350 HV,
Surface finishing: Polished to a roughness R_a = 0.05 μm, stress free annealed at a temperature of 560 °C in vacuum.

Depositing TiN films of different levels of adhesion:
Deposition technique: ion plating
Film thickness: 1.2 μm to 2.4 μm
Pre-sputtering time t_S: 0.5, 1, 2, 3, 5 and 15 minutes

The adhesion was varied by using different pre-sputtering times t_S (plasma gas: argon). The best film adhesion was expected for t_S = 15 min, the worst adhesion for t_S = 0.5 min.

The technique for depositing TiN films with different levels of adhesion is schematically shown in Figure 1. Two reference specimens with t_S = 15 min were additionally prepared in each deposition run to identify variations of the stoichiometry of the film which could influence the adhesion effect. This was done by covering the samples intended for poor adhesion with an aluminum foil in the first phase t_P of the pre-sputtering process. Therefore, only the surfaces of the two reference samples were cleaned by argon etching during t_P. Afterwards, the aluminum foils were removed from the protected samples and both the reference samples and the test samples were argon-sputtered. The pre-sputtering time of the test samples was determined as t_S = 15 min - t_P. After finishing the pre-sputtering process, the TiN coating was deposited on all samples with the same process parameters. This sequence ensured that films with good and poor adhesion had been deposited simultaneously. The consistency of the test series was monitored by measuring Young's modulus of the reference coatings.

2.1. Film texture

The texture of twelve TiN films was investigated by x-ray diffraction. A weak [111] texture was found, which is known for TiN films deposited with moderate ion physical vapor deposition processes like ion plating [10].

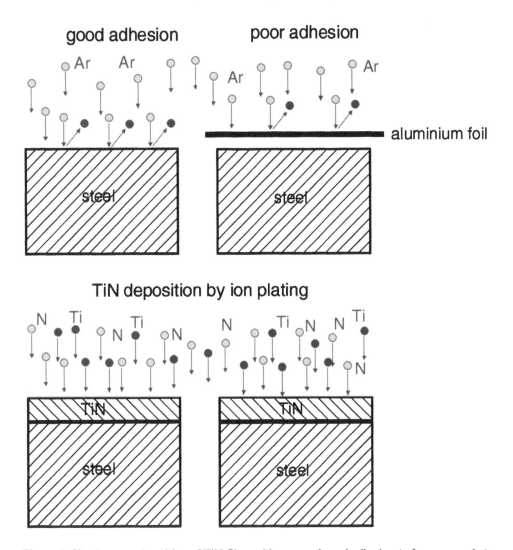

Figure 1. Simultaneous deposition of TiN films with poor and good adhesion (reference samples) with different times t_S of pre-sputtering with argon ions.

2.2. Stoichiometry

Eleven samples with pre-sputtering times $t_S = 0.5, 2, 5$ and 15 min were investigated by Auger Electron Spectroscopy (AES). The results revealed that the TiN films had a stoichiometric composition. No enhanced carbon and oxygen contents in the interface region of samples with $t_S = 0.5$ min could be found in the reference samples with $t_S = 15$ min. This result implies that the sputtering treatment has the expected influence on the interface build-up and subsequently on the adhesion.

2.3. Film hardness

The hardness of the film material was measured by the micro-indentation method (Shimadzu DUH 202) with a load of 1 g. The values of 2100 ± 300 HV are in agreement with numbers published in the literature. The micro-indentation tests were not sensitive enough to distinguish the quality of the different test series.

3. NON-DESTRUCTIVE TESTING OF THIN FILMS BY LASER-ACOUSTICS

3.1. Surface acoustic waves

The properties of the surface acoustic waves are shown in Figure 2. The amplitude of the wave motion exponentially decays within the material. The penetration depth can be estimated by the wavelength λ. The penetration depth of the surface wave depends on the frequency f due to the relation $\lambda = c/f$ with c as phase velocity of the surface wave. The higher the frequency, the smaller is the penetration depth. Therefore, thinner films can be investigated by analyzing waves with higher frequencies. Two aspects limit the highest detectable frequency: the bandwidth of the test equipment and the sound attenuation of the test materials.

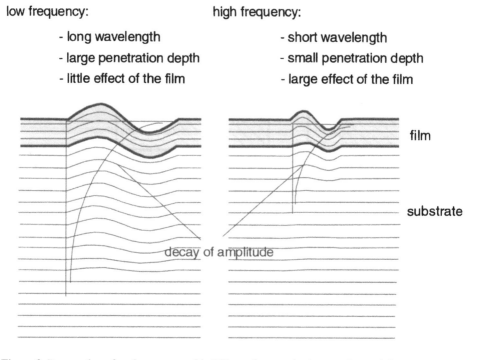

low frequency:
- long wavelength
- large penetration depth
- little effect of the film

high frequency:
- short wavelength
- small penetration depth
- large effect of the film

film

substrate

decay of amplitude

Figure 2. Propagation of surface waves with different frequencies in coated material.

Since the wave energy is concentrated near the surface, these waves are very sensitive to surface films, even when the film is much thinner than the penetration depth of the wave.

The phase velocity c is the wave parameter that can sensitively indicate the effect of a film. It depends on the elastic parameters and the density of the material. For a homogeneous and isotropic material described by Young's modulus E, Poisson's ratio v and density ρ, the following relation gives a good approach for the phase velocity c of the surface acoustic wave:

$$c = \frac{0.87 + 1.12v}{1 + v} \sqrt{\frac{E}{2\rho(1 + v)}} \tag{1}$$

A film whose properties differ from those of the substrate alters the effective elastic modulus and the density within the near surface region. Therefore, the effect of the film varies the propagation velocity of the surface acoustic wave. For coated materials, the phase velocity c depends on the elastic parameters, the density of both the film and substrate materials and the film thickness. This behavior is described by a more complex relation than equation (1). The theory behind was proposed by Farnell and Adler [11], and is implemented in the software of the laser acoustic test equipment.

Since the penetration depth of the surface acoustic wave decreases with increasing frequency (Figure 2), the film influences acoustic waves at higher frequencies more than the lower frequency waves, which penetrate deeper into the substrate. The result is that the phase velocity of the surface acoustic wave depends on frequency, a phenomenon called dispersion. The laser-acoustic technique measures this dispersion as a spectrum of the phase velocity versus the frequency (dispersion curve) and analyzes the curve by applying the Farnell-Adler theory to deduce the film modulus. A film with higher sound velocity than the substrate causes the dispersion curve to ascend, whereas a film with lower sound velocity causes a descending dispersion curve.

3.2. Laser-acoustic equipment

The laser-acoustic technique has proven advantageous for testing material surfaces with surface acoustic waves. The surface wave dispersion can be measured for a wide frequency bandwidth (up to 250 MHz) on comparably small samples surfaces (5 × 8 mm²) with a remarkable accuracy of $\Delta c/c < 10^{-3}$. The schematic representation of the test equipment is shown in Figure 3. Short laser pulses (pulse duration: 0.5 ns, energy: 0.4 mJ) of a nitrogen laser are focused through a cylindrical lens on the surface of the sample and generate wide-band surface waves. The acoustic waves are detected by a wide-band piezoelectric transducer [12] and recorded by a digital oscilloscope with a sampling rate of 1 GSa/s. The error in the signal acquisition is $\Delta t \leq \pm 0.25$ ns. Specimen and transducer are mounted on a translation stage that moves perpendicular to the position of the la-

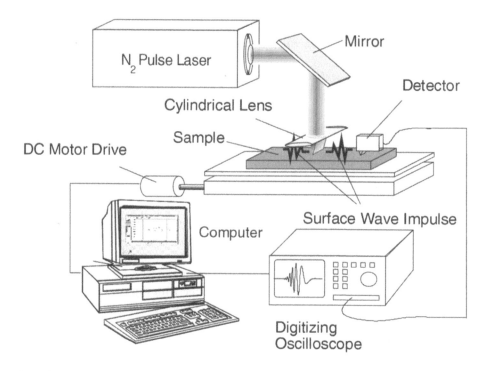

Figure 3. Schematic representation of the laser-acoustic equipment.

ser focus line to vary the distance x between acoustic wave generation and transducer. The absolute accuracy of the positioning of the translation stage was $\Delta x \leq \pm 3$ µm, but its repeatability was $\Delta x \leq \pm 1$ µm. The surface acoustic waveform is detected at different distances x_1 and x_2. Fourier transformation of the waveforms yields their phase spectra $\phi_1(f)$ and $\phi_2(f)$. The following relation (2) gives the phase velocity c depending on frequency f (dispersion curve).

$$c(f) = \frac{(x_2 - x_1)\omega}{\phi_2(f) - \phi_1(f)} \tag{2}$$

ω denotes the angular frequency. The measurement result represents a spectrum of phase velocities, the dispersion curve.

The theoretical curve is fitted to the measured dispersion curve to deduce Young's modulus of the film. To fit the film modulus, thickness, density and Poisson's ratio must be known. The film thickness was determined by the x-ray fluorescence technique (Fischerscope X-RAY 1020). A measurement uncertainty of $\Delta d/d < \pm 0.02$ was achieved by careful calibration and a counting time of 5 minutes. Because no deviation from the stoichiometry was found by the Auger analysis, a density of $\rho = 5.4$ g/cm³ was used for TiN films [10]. The Poisson's ratio was set to 0.2 [13].

The method yields in addition to the film modulus also Young's modulus of the substrate.

A description of the laser acoustic method is given in more details elsewhere [6].

3.3. Results of the surface acoustic wave technique

For illustrating typical results, Figure 4 shows dispersion curves measured for TiN films of thickness d = 1.3 and 1.6 μm deposited on steel and for the pure substrate as well.

The upper frequency limit is about 70 MHz. This is considerably lower than the bandwidth of the test equipment (250 MHz). This is caused by the effect of the ultrasonic attenuation due to scattering of the high frequency waves at the grain boundaries in the polycrystalline steel substrate. The full bandwidth can be achieved for single-crystal materials such as silicon wafers.

Comparing the curves of the coated samples with the non-coated sample reveals the way that the film influences the surface wave dispersion. For the non-coated sample, the phase velocity c is nearly constant. All surface waves, independent of their frequency and therefore independent of their penetration depth, propagate within the same material and have consequently the same propagation velocity. This is not true for the coated samples, for which the dispersion curves

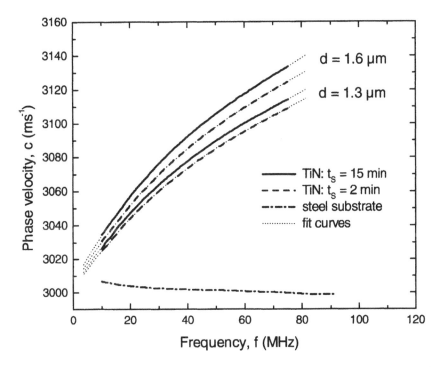

Figure 4. Dispersion curves of the surface acoustic wave propagation, measured for TiN coated steel with different film thicknesses d and pre-sputtering times t_S, and for the pure steel substrate as well.

increase with frequency. This reveals the higher sound velocity for the films compared to the substrate, due to the higher Young's modulus and the lower density of the film material. The TiN film can be said to accelerate the surface acoustic wave with increasing frequency, since the higher frequency waves are more sensitive to the film than the lower frequency waves, as shown in Figure 2. The slope of the dispersion curve increases with increasing film thickness. However, for the same film thickness the slope reduces with reducing pre-sputtering time t_S. This indicates that the film material lost stiffness, probably due to micro-defects in the film and at the interface that are expected to occur in the short-time sputtered samples.

Pores and defects may considerably reduce Young's modulus of materials [14]. For plasma-sprayed ZrO_2 coatings, a reduction of Young's modulus to 20 % compared to the bulk material was measured. Such drastic loss of stiffness was caused by the pores with high aspect ratio. In the same way the studies of damage layers in Al_2O_3 and GaAs suggest the surface waves to be a sensitive tool for testing enhanced defect density in regions near the material surface [15,16].

Young's moduli of the TiN films in Figure 4 were deduced by fitting the theoretical dispersion curves. The results in Table 1 show the expected reduction of the film modulus with reducing pre-sputtering time t_S.

Extrapolating the measured curves to zero frequency shows that they meet at nearly the same point, the velocity in the steel substrate (3010 m/s). It depends on the elastic constants and the density of steel. The fit procedure enabled the phase velocity at zero frequency to be determined and yielded Young's modulus of E = 212.5 ± 0.5 GPa for steel substrate.

3.4. Comparability

The comparability of the laser-acoustic results with alternative techniques such as micro-indentation and membrane deflection tests for measuring the elastic modulus of thin films was shown elsewhere [17]. These comparative studies were performed for hard coatings such as TiN, TiCN, CrN of thickness more than 1 μm. Comparing Young's modules of these films with results obtained with the laser-acoustic microscope and a Fischerscope H100VP-B (microindenter) revealed a linear correlation coefficient R = 0.995 and a slope of 1.01. A similar agreement was obtained with the membrane deflection test for measuring the modulus of 460 nm thick polysilicon films. The film moduli (150 to 170 GPa) measured with the membrane deflection deviate less than 5 GPa from the results of the laser-acoustic technique, which is within the error of both methods.

Table 1.
Young's moduli of TiN films of different thicknesses and pre-sputtering times t_S obtained from the dispersion curves in Figure 4

Film thickness d	pre-sputtering time t_s: 2 min	pre-sputtering time t_s: 15 min
1.3 μm	E = 433 GPa	E = 456 GPa
1.6 μm	E = 435 GPa	E = 454 GPa

4. ADHESION TEST METHODS

4.1. Scratch test

The tests were performed with a Rockwell-shaped diamond indentor continuously loaded with a normal force F_N from 0 to 70 N [7]. The transverse velocity of the sample to the diamond was 10 mm/min and the loading rate $dF_N/dx = 10$ N/mm. The scratches were 7 mm long. The friction force F_F and the acoustic emission signals were recorded. A typical data record is shown in Figure 5.

The following test parameters were deduced from the scratch test:
– Critical load deduced from microscopic evaluation of the crack pattern around the scratch,
– Friction work W_F along the scratch,
– Total number of the acoustic emission signals N_{AE} during the scratch test,
– Critical load L_C deduced from the Weibull statistical distribution of the first acoustic emission signals in a series of 20 scratch tests.

4.1.1. Evaluation of the Crack Pattern. The test criteria proposed by Burnett and Rickerby [18] were used to determine the critical load for the film detachment from microscopic observation of the scratch traces. The method revealed a lower adhesion for the test series with $t_S = 0.5$ min, but was not sensitive enough to distinguish the other adhesion levels. A significant correlation between the critical load and the pre-sputtering time was not found.

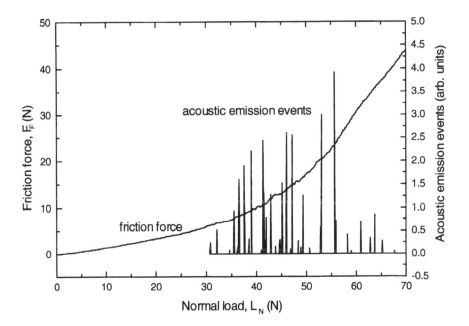

Figure 5. Data record of the Scratch test showing the friction force F_F and acoustic emission events with increasing load L_N for a film with pre-sputtering time $t_S = 1$ min.

4.1.2. Friction Work. The friction of the Rockwell diamond against the sample increased during the scratch test when the diamond removed the TiN film from the steel substrate. As long as the Rockwell diamond was scratching the TiN film, the friction force F_F was identical for all samples. In the range of the breakthrough of the film, F_F increased more for the samples with lower pre-sputtering time t_S. This suggested using the area under the curve as a test parameter for the adhesion [19]. It is defined as the friction work:

$$W_F = \int F_F \cdot dx = \int F_F / \frac{dF_N}{dx} \cdot dF_N .$$ (3)

Figure 6 shows correlations of the friction work W_F with the pre-sputtering time t_S for three ranges of film thickness: 1.2 to 1.6 µm, 1.6 to 2.0 µm, 2.0 to 2.4 µm. W_F reduces with increasing pre-sputtering time t_S. The films were detached at higher loads and the diamond scratched the film more than the substrate along the scratch trace of 7 mm. The correlation coefficients of R = -0.57 and -0.88 are significant.

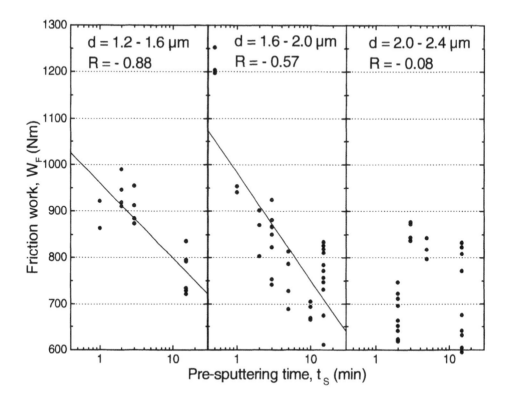

Figure 6. Friction work W_F versus pre-sputtering time t_S for three ranges of film thickness d, R is the correlation coefficient.

4.1.3. Acoustic Emission Activity. The acoustic emission signals detected during the scratch test were ascribed to brittle micro-fractures in the TiN film or the interface. The frequency of these events seems to be related to the defect density. Therefore, the number of the acoustic emission events N_{AE} is used as a test parameter for the adhesion. Figure 7 shows the correlations of N_{AE} with the pre-sputtering time t_S for the three different ranges of film thickness. The acoustic emission activity N_{AE} reduces with increasing pre-sputtering time t_S and shows the same behavior as the friction work W_F in Figure 6. This corresponds to the assumption that the defect density is lower in samples of higher pre-sputtering time. Significant correlation coefficients R from -0.56 to -0.62 were determined.

4.1.4. Statistical Evaluation of the Acoustic Emission. The Weibull statistical distribution function is applied for evaluating the fracture strength of hard and brittle ceramics and has also been proposed for the scratch test of hard coatings by Bull and Rickerby [20]. The fracture tests on a series of samples yield a distribution of strength values that can be described by the Weibull distribution function. This distribution function F(L) allows to deduce the critical load L_C for the fracture with a given probability and the parameter m depending on the shape of distribution [5]:

$$F(L) = 1 - \exp\left[-\left(\frac{L}{L_C}\right)^m\right] \tag{4}$$

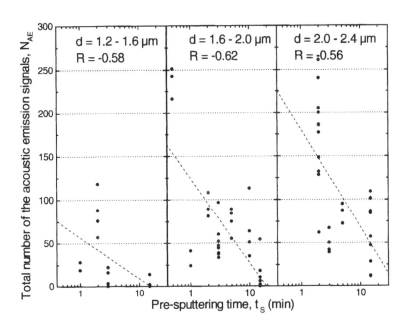

Figure 7. Total number of acoustic emission signals N_{AE} during the scratch test versus pre-sputtering time t_S for three classes of film thickness d, R is the correlation coefficient.

The statistical evaluation was applied to the acoustic emission detected during the scratch. To study the types of failure indicated by the signals, some tests were stopped after the first acoustic emission event. Microscopic investigations showed that this acoustic emission signal always correlated with the first detachment of the film. Therefore, the Weibull statistical distribution was applied to the first acoustic emission event detected in the scratch test. Twenty scratch tests were performed for one distribution, five tests on each one of four samples of a test series. The results are plotted in a double-logarithmic diagram as shown in Figure 8 for two sample series with a pre-sputtering time $t_S = 2$ min and 15 min, respectively. It represents the Weibull statistical distribution function in the form:

$$\ln(\ln(1/(1 - F(L)) = m\ln L - m\ln L_c \qquad (5)$$

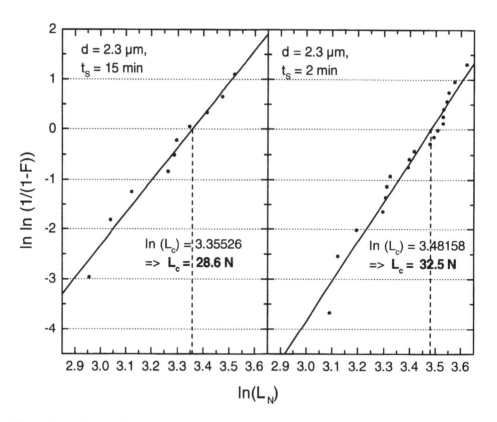

Figure 8. Double-logarithmic Weibull diagram for the probability F of the first acoustic emission event in the scratch test at the load L_N for two sample series with different pre-sputtering times t_S, the deduction of the critical load L_C is shown.

F(L) was deduced from the normalized number of the scratch tests having shown an acoustic emission at the load L. The critical load L_C can be determined from relation (5) for $\ln(\ln(1/(1-F(L)))) = 0$ which is illustrated in Figure 8. This is the load L up to that 63.2 % of all scratch tests have released an acoustic emission signal. The parameter m can be deduced from the slope of $\ln(\ln(1/(1-F(L)))$ versus $\ln(L)$ in relation (5). The nearly linear dependence in Figure 8 allows the conclusion that the first signals of the acoustic emission are Weibull distributed.

The correlations of the critical load L_C with the pre-sputtering time t_S are shown for the three ranges of film thickness in Figure 9. Correlation coefficients R from - 0.53 to -0.76 were determined. This suggests that the pre-sputtering time t_S influences the acoustic emission. However, it is emphasized that L_C reduces with increasing t_S. This means that the films with a better-expected adhesion show their first brittle micro-fracture at lower loads. Obviously, the L_C does not reflect the effect of the pre-sputtering process in the expected manner.

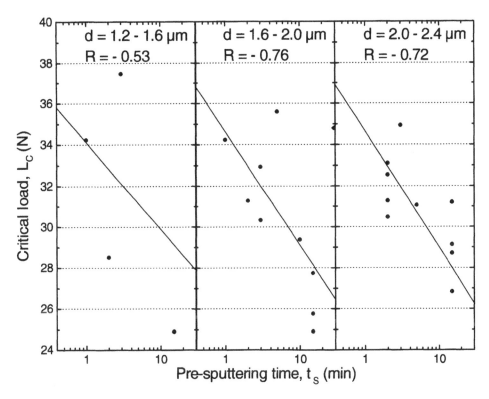

Figure 9. Correlations of the critical load L_C with the pre-sputtering time t_S for three classes of film thickness d, R is the correlation coefficient.

4.2. Four-point bending test

The four-point bending test is schematically shown in Figure 10. The sample was fixed in the jig in a way that the TiN film was loaded by compressive stress to make the results comparable to those of the scratch test. It was connected with an acoustic emission transducer to detect micro-fractures during the bending test. The samples were not coated within the area near the loading points to avoid the detection of acoustic signals other than those generated by the bending of the film. More details of the test procedure are described elsewhere [7,21]. The data record of the bending test (Figure 11) shows the stress σ versus the strain ε in the upper layer of the sample and the cumulative number of the acoustic emission signals. Two test parameters were determined. The critical strain ε_C was determined as the strain where 10 % of the total number of the acoustic emission signals has been detected. The crack density D_C was obtained from an optical micrograph of the coated area after the bending test. The stress-strain diagram did not yield information about the film, since the effect of the 2 μm thick film is too small compared to the 8 mm thick steel substrate.

The film on the sample with lower pre-sputtering time (t_S = 2 min) showed more and longer cracks in the film. The visible cracks were perpendicular to the surface. Detachment of the films was not observed. The crack density D_C is plotted versus the sputtering time t_S in Figure 12 a. D_C is significantly lower for t_S = 15 min than for t_S = 2 min. This behavior is in agreement with the one observed for the friction work W_F and the acoustic emission activity N_{AE} of the scratch test (Figures 6 and 7). However, the critical strain ε_C shows an unexpected behavior. Figure 12 b reveals that ε_C decreases with increasing t_S. One tenth of all microfractures occurred at a lower strain in films whose adhesion is expected to be better. This corresponds to the results obtained for the critical load L_C of the scratch test (Figure 9). It seems that the acoustic emission signals detected in bending and scratch tests reveal the same fracture mechanism.

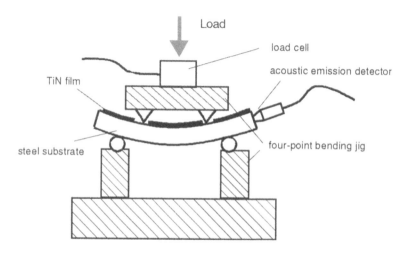

Figure 10. Schematic representation of the four-point bending test.

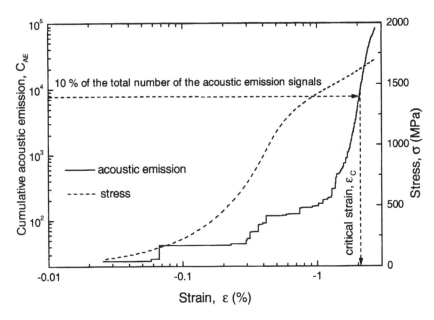

Figure 11. Data record of a four-point bending test on a TiN coated sample with pre-sputtering time $t_S = 2$ min, showing the cumulative number of the acoustic emission events C_{AE} and the stress σ versus strain ε, the deduction of the critical strain ε_C is shown.

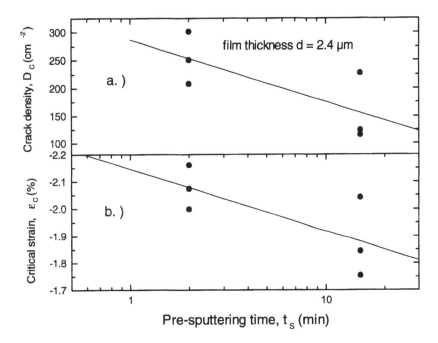

Figure 12. Results of four-point bending test. a.) Crack density D_C in the film after the test versus the pre-sputtering time t_S; b.) Critical strain ε_C versus the pre-sputtering time t_S.

4.3. Cavitation test

The cavitation test is performed in a water tank where the sample surface is exposed to high power ultrasound (Figure 13) [22]. An ultrasonic transducer with a diameter of 16 mm was arranged at a distance of about 0.5 mm above the sample. The high power ultrasound generates blisters on the surface of the sample. Having reached a critical dimension, the blisters collapse is accompanied by the formation of a microscopic water jet that impacts the surface with high pressure and generates a mechanical pulse. Since blisters are developing and collapsing permanently, this test acts like many local impact tests on a large area and with a high repetition rate.

The cavitation test attacks the surface at many spots and can lead to destruction and detachment of the film material [19]. An image processing is applied to determine the proportional area of the damage A_D. Figure 14 shows A_D versus the pre-sputtering time t_S. The damaged area A_D reduces with increasing pre-sputtering time t_S. This is an expected result. The cavitation test yields the same behavior of the adhesion of the TiN films as the friction work W_F and the acoustic emission activity N_{AE} of the scratch test and the crack density D_F of the four-point bending test.

4.4. Impact test

The impact test allows testing the local fatigue strength of the film-substrate system [23]. A schematic representation of the test apparatus is shown in Figure 15. A ball of cemented carbide repetitively pushed onto the sample surface cyclically loads the sample. A device that consists of a spring and a magnetic coil applies the cyclical loading. The impact frequency can be varied from 0 to 50 Hz and the

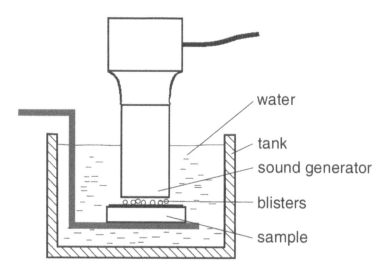

Figure 13. Schematic representation of the cavitation test.

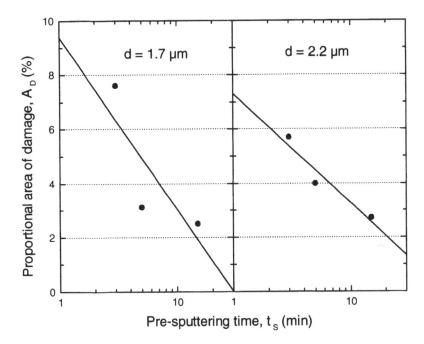

Figure 14. Proportional area of damage A_D versus pre-sputtering time t_S.

force of the impact can be varied from 0 to 1500 N. The impact force is controlled by a strain gauge. The damage of the sample is evaluated by an optical micro-scope. The test parameter is the critical number of loading cycles N_C up to that the surface does not show any damage. The result has the form of a Woehler diagram (Figure 16 a).

N_C versus t_S is presented in Figure 16 b. It shows that the critical number of cy-cles N_C decreases with increasing pre-sputtering time t_S. The impact test indicated that pre-sputtering makes the film adhesion worse. This corresponds to the results of the critical load L_C of the scratch test (Figure 9) and the critical strain ε_C of the four-point bending test (Figure 12 b).

4.5. Rockwell test

The advantage of this method is that it is easy to use even in an industrial envi-ronment. A conventional Rockwell hardness test is performed and the damage pattern of the hard coating around the indent is evaluated microscopically at a magnification of 100x. Standard damage pictures serve to classify the adhesion into six classes HF1 to HF6 [24, 25].

The Rockwell test was not sensitive enough to distinguish TiN films with dif-ferent pre-sputtering times. Even for test samples with a pre-sputtering time of only $t_S = 1$ min, this test indicates a good quality of the adhesion, which means it is not sensitive enough.

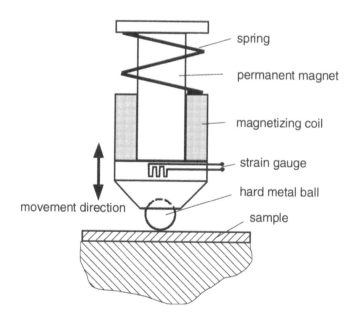

Figure 15. Schematic representation of the impact test.

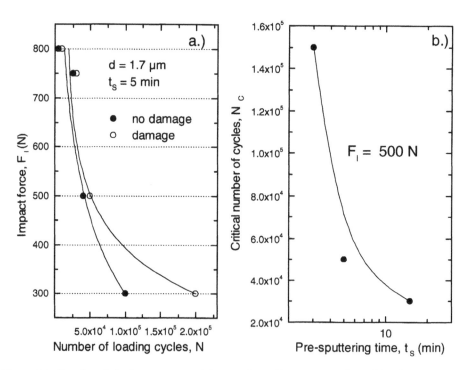

Figure 16. Results of the impact tests. a) Impact force F_I versus number of loading cycles N; b) Critical number of cycles N_C versus the pre-sputtering time t_S.

4.6. Acoustic microscopy

A scanning acoustic microscope (ELSAM, Leica Wetzlar) was used to search for defects in the films and interfaces of samples pre-sputtered with $t_S = 2$ and 15 min. The frequency of the acoustic waves was 1 GHz. The sample surface was scanned over regions of 1mm × 1mm with different focus depths of the acoustic field. This allowed studying the interface region.

The acoustic imaging did not show significant clues for defects in both samples [26]. This suggests that many of the effective defects are smaller than the resolution capability of the acoustic microscope (0.5 μm).

5. DISCUSSION

5.1. Non-destructive evaluation of the adhesion by surface acoustic waves

Figures 17 to 19 show correlations of Young's modulus E of the TiN films with test parameters indicating an improved adhesion with increasing pre-sputtering time t_S in terms of acoustic emission activity N_{AE} of the scratch test, crack density D_C of the four-point bending test and damage area A_D of the cavitation test. The film modulus E reduces by up to about 10 %, if N_{AE}, D_C and A_D increase. This suggests that N_{AE}, D_C and A_D express mechanical properties that are influenced by the defect density in the interface and the film. The film Young's modulus E as well as the test parameters for the adhesion N_{AE}, D_C and A_D represent the properties of large areas of the material surface. These can be said to indicate the "global" adhesion behavior.

It is remarkable that Young's modulus E of the TiN films also correlates with those adhesion test parameters indicating an adhesion trend opposite to the expected trend from the pre-sputtering process: the critical load L_C of the scratch test (Figure 20), the critical strain ε_C of the four-point bending test (Figure 21) and the critical number of cycles N_C of the impact test (Figure 22). Figures 20 to 22 show that Young's modulus E reduces, if the critical parameters L_C, ε_C and N_C increase. This is surprising at first sight, but corresponds to the observation that the pre-sputtering process assumed to improve the adhesion reduces the critical parameters L_C, ε_C and N_C. It is obvious that they do not reflect the global adhesion behavior such as N_{AE}, D_C and A_D do, but indicate instead a more locally reduced strength of the film and interface that does not depend on the quantity of the defects.

Young's modulus E of the TiN films varies only about 10 % with the effect of the pre-sputtering time t_S. Such small variations can be measured by the laser-acoustic technique. Careful film preparation and controlling the film quality by means of reference samples guaranteed that the stoichiometry of the TiN films did not vary more than acceptable for these investigations. Auger Electron Spectroscopy also confirmed this.

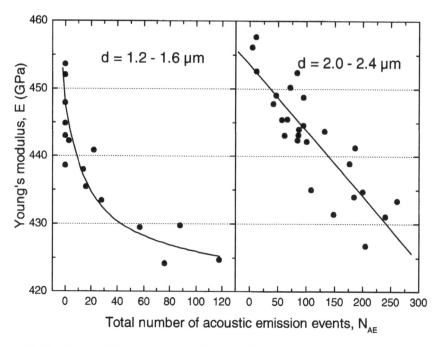

Figure 17. Correlation of Young's modulus E of the TiN films with the total number of acoustic emission events N_{AE} in the scratch test.

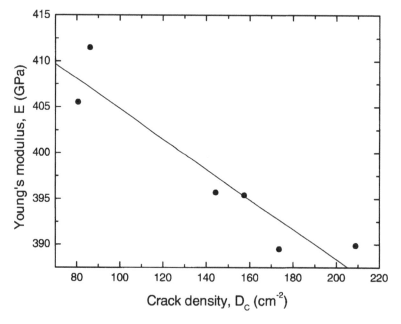

Figure 18. Correlation of Young's modulus E of the TiN films with the crack density D_C in the four-point bending test.

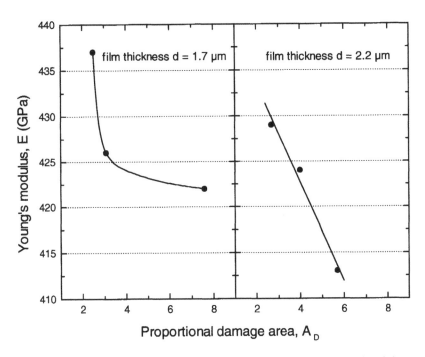

Figure 19. Correlation of Young's modulus E of the TiN films with the proportional damage area A_D in the cavitation test for different film thicknesses d.

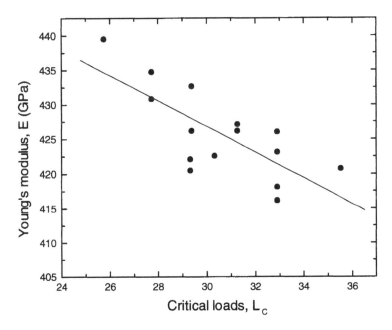

Figure 20. Correlation of the Young's modulus E of the TiN films with the critical load L_C of the Weibull distribution of the acoustic emission in independent scratch tests.

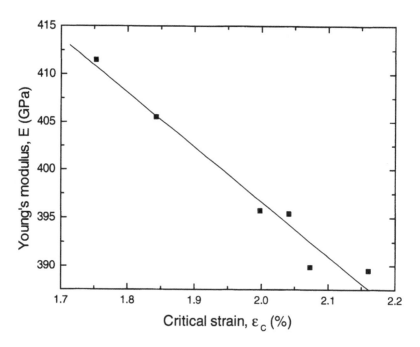

Figure 21. Correlation of the Young's modulus E of the TiN films with the critical strain ε_C of the acoustic emission in the four-point bending test.

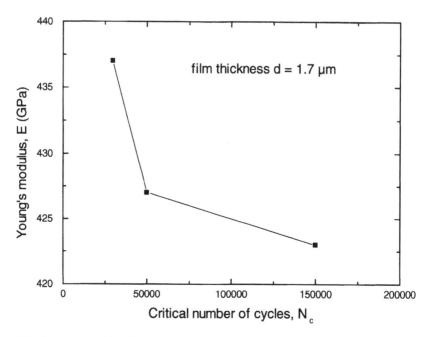

Figure 22. Correlation of the Young's modulus E of the TiN films with the critical number of loading cycles N_C of the impact test.

The present studies confirmed that laser-acoustics was a promising non-destructive method for testing thin films. The adhesion test parameters correlate with the film modulus E in a similar way as with the pre-sputtering time t_S. These correlations can be explained by the influence of micro-defects on the elastic modulus. In the present studies, the laser-acoustic technique could not distinguish between defects in the film and in the interface. It yields an effective modulus that contains the effect of both film and interface.

5.2. Evaluation of the adhesion test methods

Table 2 gives a summary of the test results and allows comparing the test methods. For each test method, the test parameters deduced for characterizing the adhesion are presented. The following criteria were applied to evaluate these parameters:
- Do they correlate with the duration of the pre-sputtering time t_S that is expected to improve the adhesion?
- Do they indicate an improvement or reduction of the adhesion with increasing pre-sputtering time t_S?
- Do they characterize global or local damages?

The sign "0" in the column "Correlation with pre-sputtering time t_S" denotes that this parameter does not significantly correlate with the pre-sputtering time t_S, "+" denotes that the parameter indicates an improvement of the adhesion with increasing t_S and "-" means that the parameter indicates a reduction of the adhesion with increasing t_S.

Table 2 enables the adhesion test methods employed in the present investigation to be evaluated as follows:
I. The quality of the adhesion of the TiN films could not be significantly distinguished by the following test methods:
    - Optical evaluation of the damage around the scratch trace,
    - Rockwell test
    - Acoustic microscopy.
II. The following test parameters (type 1) indicate an improvement of the adhesion with increasing pre-sputtering time t_S and are marked with "+":
    - Friction work W_F of the scratch test,
    - Total number of the acoustic emission events N_{AE} of the scratch test,
    - Crack density D_C of the four-point bending test,
    - Proportional area of damage A_D of the cavitation test,
    - Young's modulus of the film E.
III. The following test parameters (type 2) indicate a reduction of the adhesion with increasing pre-sputtering time t_S and are marked with "-":
    - Critical load L_C deduced from the acoustic emission in the scratch test,
    - Critical strain ε_C deduced from the acoustic emission in the four-point bending test,
    - Critical number of cycles N_C of the impact test.

Table 2.

Overview and evaluation of the results of all adhesion tests [19, 27]

Test method	Test parameter	Correlation with pre-sputtering time t_S	Global damage	Local damage
Scratch test	Microscopic evaluation of the scratch trace	0		
	Friction work, W_F	+	√	
	Total number of acoustic emission signals, N_{AE}	+	√	
	Critical load, L_C from the first acoustic emission event of 20 tests	-		√
Four-point bending test	Crack density, D_C	+	√	
	Critical strain, ε_C from acoustic emission	-		√
Cavitation test	Proportional area of damage, A_D	+	√	
Impact test	Critical number of cycles, N_C	-		√
Laser-acoustics	Young's modulus of the TiN film	+	√	
Scanning acoustic microscope	Imaging of the defects	0		
Rockwell test	Classification of damage pattern	0		

"0": no significant correlation with pre-sputtering time, t_S

"+": significant correlation with pre-sputtering time, t_S, parameters indicate an improvement of the adhesion

"-": significant correlation with pre-sputtering time, t_S, parameters indicate a reduction of the adhesion

"√": sign for global or local failure of the film

Taking into account the loading condition for which these parameters were deduced leads to the results:

1. The test parameters of type 1 ("+" correlation) characterize a global damage of the TiN-coated steel.
2. The test parameters of type 2 ("-" correlation) characterize a local damage of the TiN-coated steel.

The opposite behavior of local and global adhesion properties is at first sight surprising, but can be explained by a correlation of higher defect density in the film with lower residual stresses. The laser acoustic technique confirmed the higher defect density in the film with low global adhesion by yielding a lower Young's modulus. It can be assumed that the higher defect density enables a relaxation of residual stresses. Subsequently, a higher external load must be applied to trigger the first micro-fracture. This led to the experimental finding that "local" test parameters such as the critical load L_C deduced from the first acoustic emission signal in the scratch test or the critical strain ε_C of the four-point bending test indicated a better film quality for films which showed lower global adhesion in agreement with the pre-sputtering treatment.

However, it should be mentioned that several attempts have failed so far to detect significant differences in the residual stress by x-ray diffraction.

6. CONCLUSIONS

1. The comparative studies showed that the seven methods: scratch test, bending test, impact test, cavitation test, Rockwell imprint test, acoustic microscopy and surface acoustic wave technique used for testing the adhesion of TiN film did not provide coherent results. Whereas some of the test parameters deduced from the experimental data records indicated an improvement of the adhesion, other parameters showed a reduction of the adhesion. However, the results of all methods do not show a diffuse picture. Considering the character of the damages allows distinguishing the test parameters in types 1 and 2, which characterize global and local failures, respectively.
2. The application of only one test method can lead to a false evaluation of the practical adhesion. Application of several test methods is recommended to evaluate the film quality. However, classifying the test parameters by global and local failures of the film is helpful to avoid errors.
3. Although the scratch test has been often criticized, it has proven to be still the most practical test method. Compared to other techniques, it is relatively simple to use and does not need special specimen shape and preparation. Different test parameters can be deduced which characterize both the global and local adhesion behaviors. However, the light microscopic evaluation of the scratch trace normally used for evaluating the scratch traces did not allow a differentiation of the test samples. Acoustic emission proved more sensitive, if sufficient tests were performed for a reliable statistical analysis.
4. The four-point bending test provided similar results as the scratch test, but needs much more efforts for sample preparation and is an expensive test technique.
5. Cavitation and impact tests provided results that correlated significantly with the pre-sputtering time. However, whereas the cavitation test correlated in the expected manner, the impact test led to an opposite evaluation of the practical adhesion.
6. The laser acoustic method based on surface acoustic waves proved to be a useful non-destructive method for evaluating the adhesion properties of the films. It yields the elastic modulus that correlates with significant test parameters of the destructive adhesion tests in the same way as they correlate with the pre-sputtering time. This can be explained by the sensitivity of the method to the density of microscopic defects in the film and the film-substrate interface.

7. The significance of the results of all adhesion test methods is closely limited by the test material and film thickness. Keeping the film thickness constant is an important requirement to obtain meaningful results.

Acknowledgment

The authors would like to thank Dr. A. Mucha from MAT GmbH Dresden for the possibility of preparing well-defined samples, Mrs. Dr. R. Wiedemann from the Freiberg University of Mining and Technology for the x-ray diffraction investigations, Dr. G. Kirchhoff form FhG IWS Dresden for assisting in the four-point bending test, Dipl-Phys. V. Scherer from FhG -IzfP Saarbrücken for the investigations with the scanning acoustic microscope, Mrs. Dr. A. Schmidt-zum Berge from NUTECH Analytik- und Prüfzentrum Neumünster for the cavitation tests, and Dipl.-Ing. C. Wolff from RWTH Aachen for the impact tests.

REFERENCES

1. (a) K.L. Mittal in: *Adhesion Measurement of Thin Films, Thick Films and Bulk Coatings*, K.L. Mittal (Ed.), p. 5, STP No. 640, ASTM, Philadelphia, 1978.
   (b) K.L. Mittal in: *Adhesion Measurement of Films and Coatings*, K.L. Mittal (Ed.), pp. 1-13, VSP, Zeist, The Netherlands, 1995.
2. K.L. Mittal, Electrocomponent Sci. Technol. **3**, 21 (1976).
3. S.D. Brown in: *Adhesion Measurement of Films and Coatings*, K.L. Mittal (Ed.), pp. 15-39, VSP, Zeist, The Netherlands, 1995.
4. G.K. Wolf, Mater.-wiss. u. Werkstofftech. **24**,109 (1993).
5. W. Weibull, J. Appl. Mech. **18**, 293 (1951).
6. D. Schneider and Th. Schwarz, Surface Coatings Technol. **91**,136 (1997).
7. H. Ollendorf, D. Schneider, Th. Schwarz, G. Kirchhoff, A. Mucha, Surface Coatings Technol. **74-75**, 246 (1995).
8. D. Schneider, H. Ollendorf, and Th. Schwarz, Appl. Phys. **A61**, 277 (1995).
9. Y. Wang, Y. Jin and S. Wen, Wear **134**, 399 (1989).
10. J.E. Sundgren, Thin Solid Films **128**, 21 (1985).
11. G.W. Farnell and E.L. Adler in: *Physical Acoustics.*, W.P. Mason and R.N. Thurston (Eds.) p. 35, Vol. IX, Academic Press, New York, 1972.
12. H. Coufal, R. Grygier, P. Hess, and A. Neubrand, J. Acoust. Soc. Am. **92**, 2980 (1992).
13. X. Jiang, M. Wang, K. Schmidt, E. Dunlop, J. Haupt and W. Glisser, J. Appl. Phys. **65**, 3053 (1991).
14. D. Schneider, Th. Schwarz, H.-P. Buchkremer and D. Stover, Thin Solid Films **224**, 177 (1993).
15. D. Schneider, A. Krell, T. Weiss, and Th. Reich: Acta metall. mater. **40**, 971 (1992).
16. D. Schneider, R. Hammer, and M. Jurisch: Semiconductor Sci. Technol. **14**, 93 (1999).
17. D. Schneider, B. Schultrich, H.-J. Scheibe, H. Ziegele, and M. Griepentrog, Thin Solid Films **332**, 157 (1998).
18. P.J. Burnett and D.S. Rickerby, Thin Solid Films **154**, 403 (1987).
19. H. Ollendorf and D. Schneider, Surface Coatings Technol. **113**, 86 (1999).
20. S J. Bull and D.S. Rickerby, Surface Coatings Technol. **42**, 149 (1990).
21. W. Pompe, H.-A. Bahr, G. Gille, W. Kreher, B. Schultrich and H.-J. Weiss, Current Topics Mater. Sci. **12**, 267-325 (1985).
22. M. Pohl and M. Feyer, Tribologie + Schmierungstechnik **1**, 29 (1992).
23. O. Knotek, B. Bosserhoff, A. Schrey, T. Leyendecker, O. Lemmer, and S. Esser, Surface Coatings Technol. **54/55**, 102 (1992).

24. W. Heinke, A. Leyland, A. Matthews, G. Berg, C. Friedrich, and E. Broszeit, Thin Solid Films **270**, 431 (1995).
25. VDI guideline 3198, *VDI-Handbuch für Betriebstechnik*, Teil 3, Verein Deutscher Ingenieure, Düsseldorf, August 1992, p. 7.
26. V. Scherer, private communication, 1997.
27. H. Ollendorf, "Bewertung von Methoden zur Charakterisierung von Verschleißschutzschichten", Thesis, Heidelberg, 1998.

Adhesion Measurement of Films and Coatings, Vol. 2, pp. 79–106
Ed. K.L. Mittal
© VSP 2001

Scratch adhesion testing of coated surfaces – Challenges and new directions

J. MENEVE,[1] H. RONKAINEN,[2,][*] P. ANDERSSON,[2] K. VERCAMMEN,[1]
D. CAMINO,[3] D.G. TEER,[3] J. VON STEBUT,[4] M.G. GEE,[5] N.M. JENNETT,[5]
J. BANKS,[5] B. BELLATON,[6] E. MATTHAEI-SCHULZ[7] and H. VETTERS[7]

[1]*Vlaamse Instelling voor Technologisch Onderzoek (VITO), Materials Technology, Boeretang 200,
B-2400 Mol, Belgium*
[2]*VTT Manufacturing Technology (VTT), P.O. Box 1702, FIN-02044 VTT, Finland*
[3]*Teer Coatings Ltd. (TCL), 290 Hartlebury Trading Estate, Hartlebury, Kidderminster,
Worcestershire, DY10 4JB, UK*
[4]*Laboratoire de Science et Génie des Surfaces (LSGS), Ecole des Mines, Parc de Saurupt,
F-54042 Nancy Cedex, France*
[5]*National Physical Laboratory (NPL), Centre for Materials Measurement and Technology,
Queens Road, Teddington, Middlesex, TW11 0LW, UK*
[6]*Centre Suisse d'Electronique et de Microtechnique (CSEM), Rue Jaquet-Droz 1,
CH-2007 Neuchâtel, Switzerland*
[7]*Stiftung Institut für Werkstofftechnik (IWT), Badgasteiner Str. 3, D-28359 Bremen, Germany*

Abstract—The scratch test is extensively used, in industry as well as in research laboratories, to assess the adhesion strength of coated surfaces. The test consists of drawing a diamond stylus across the specimen surface under increasing normal load until a failure event occurs. The load at which the failure event occurs is called the critical normal load.

Only some of the observed failure events, however, are related to coating detachment at the coating/substrate interface and are therefore relevant as a measure of adhesion. Other failure events such as cracks and cohesive damage within the coating or substrate clearly cannot be used to assess adhesion properties, but may be equally important to determine the tribological behaviour of a coated component. Indeed, the scratch test is increasingly being regarded as a repeatable tribological test to assess the mechanical integrity of a coated surface.

In spite of its widespread use, however, the results, *i.e.* critical load values, obtained from scratch testing may thus reflect a variety of failure events. Since these are not always clearly defined when explaining results, the scratch test is often held in low esteem. Much of this confusion could be avoided and the value of the scratch test would be greatly enhanced, if the scratch test procedures were standardised.

In order to gain an increased understanding of the possibilities and shortcomings of the scratch test method a European EC SMT project on "Scratch Testing of Coated Surfaces" was established in 1995. The present paper reviews the main results of the project and discusses possible future directions for the scratch test method.

The sensitivity of the scratch test method to various test parameters (ambient test conditions, cleaning procedures, styli properties, and transfer layers originating from preceding scratches) has

---

[*] To whom correspondence should be addressed. Phone: +358 9 456 4485, Fax: +358 9 460 627,
E-mail: Helena.Ronkainen@vtt.fi

been assessed. As a major step forward, an atlas of scratch test failure modes was prepared to enable users of the scratch test to assign critical load values to agreed failure events. In addition, an extended round robin exercise has been conducted to evaluate the statistical scatter in scratch test results. The main conclusion of this work, however, is that uncertainties in the Rockwell C stylus tip shape constitute the major error source for the scratch test method.

Keywords: Scratch test; coated surface; adhesion measurement; Round Robin.

1. INTRODUCTION

Currently surface coatings are deposited in numerous ways and used for a variety of purposes: to provide resistance to wear and corrosion, to reduce friction, to provide special magnetic and optical properties, etc. The coatings possess different properties to fulfill the necessary requirements, but the functional characteristics and performance depend on the adhesion between the coating and the substrate. In many cases the adhesion between the coating and the base material is the limiting factor for a wider use of surface coatings in different applications. Practical adhesion, which is a function of the fundamental adhesion (interfacial bond strength) and the method-dependent factors, is determined as "the force or the work required to detach the coating from the substrate" [1]. It can be measured with various engineering coating adhesion test methods, the applicability and limitations of which have been reviewed by several authors [1-6]. The scratch test is a very widely and routinely used method for quick and simple monitoring of the practical adhesion of a coating to a substrate. The coatings studied by scratch testing span a wide range of applications from wear resistant coatings on cutting tools to optical coatings on glass. The scratch test has been successfully used in the investigation of the effects of deposition variables, interface layer structures, ageing, etc. [7-10]. It is used by industry for quality control purposes and by research laboratories for studying the mechanical strength of coatings on machine components. For example, over 200 industrial enterprises and research laboratories worldwide are currently using a single type among the commercially available scratch testers.

The conventional scratch test consists in drawing a diamond stylus across the coated surface under increasing normal load, either stepwise or continuous. As some well defined failure event is observed in a regular way along the scratch channel, the normal load corresponding to the failure event is called the critical normal load (L_C). The diamond stylus is usually a Rockwell C diamond indenter with a 200 μm tip radius, which is suited for coating thickness in the range 0.1 to 20 μm. This covers most of the applications of thin, hard engineering coatings. The driving forces for scratch induced failure are a combination of elastic-plastic indentation stresses and frictional stresses imposed by the scratch test and the residual stress present in the coating [11,12].

Similar to other engineering adhesion test methods, "adhesion values" (critical loads) obtained by scratch testing depend not only on the fundamental adhesion between the coating and the substrate, *i.e.* the interfacial bond strength, but also on extraneous factors, which are related both to the test itself (scratching velocity, stylus properties, etc.) and to other coating/substrate composite properties (hardness, modulus of elasticity, surface roughness, etc.) [13-19]. As a result, the scratch test can be at its best a semi-quantitative assessment method for coating/substrate adhesion; if tests are performed with care, the technique can be used to rank repeatably the adhesion of a number of similar coating/substrate systems. The test-specific influence factors, however, may lead to a large scatter in scratch test results even when the tests are performed on a single test sample with homogeneous properties. For example, a Round Robin on scratch testing conducted in 1988 showed that the reproducibility of scratch test results for different stylus-instrument combinations was poor (variations up to 70 % were observed) [20]. However, the test-specific parameters were not held constant during this Round Robin. Another Round Robin in 1990 again demonstrated the poor repeatability and reproducibility of test results [21]. In the latter case, the loading rate and traverse speed were kept constant and the diamond styli were cleaned prior to testing, but neither were the diamond stylus properties controlled nor were the instruments traceably calibrated.

The influence of normal loading rate and scratching speed, which are relatively easy to control, is well documented in the literature [4, 14, 15]. Even though the influence of friction between the scratch stylus and the coating on the measured critical loads has been noted by many researchers [15-19], friction between the scratch stylus and the specimen is not well understood at present. Some authors suggest that different frictional behaviours for different styli might be attributed to variable impurity levels in the (single crystalline) diamonds [15, 20], while other authors believe that well-adhering transfer layers originating from preceding scratch tests are responsible for the poor reproducibility of scratch test results with different diamond styli [4, 18, 22]. Moreover, it is known that friction between the stylus and the sample may depend on the ambient atmosphere and the procedure adopted for cleaning the diamond and the sample [18, 19].

The critical normal load is generally defined by a recognisable failure event along the scratch track which is observed *post facto* by reflected light microscopy. Only some of the observed failure modes, however, are related to detachment at the coating/substrate interface and are thus relevant as a measure of adhesion. Other failure events, such as cracking or cohesive failure within the coating or substrate may occur but clearly cannot be used to assess the coating/substrate adhesion strength. The latter, however, can be equally important to determine the behaviour of a coated component in a particular application. Indeed, the scratch test is increasingly being regarded as a tribological test to assess the mechanical integrity of a coated surface.

The critical normal load can also be determined by *in situ* techniques such as acoustic emission (AE) detection [23, 24], friction force (FF) [23, 25] and scratch depth measurement [26], but these methods cannot discriminate between adhesional and cohesive failure events nor can they always detect the very first failure event. Therefore, the use of microscopy remains as the only means of associating a failure mode with a measured critical normal load.

Several authors have classified the scratch test failure events [13, 27-30] typically for Physical Vapour Deposited (PVD) titanium nitride (TiN) on a variety of substrates, mostly engineering steels. Since the scratch test is widely used for different types of coatings, it reflects a variety of failure processes, which should be clearly defined when explaining the results for a specific coating/substrate system.

In spite of its widespread use the results, *i.e.* critical load values, obtained from scratch testing may thus reflect a variety of failure events. Since these events are not always clearly defined when explaining the scratch test results, the scratch test is often held in low esteem. Much of this could be avoided if test procedures were standardised, and the value of the scratch test would then be greatly enhanced. In order to gain an increased understanding of the possibilities and shortcomings of the scratch test method a European EC SMT project on "Scratch Testing of Coated Surfaces" was established in 1995 [31]. The present paper reviews the main results of the project and discusses possible future directions for the scratch test method.

The scratch test has been the subject of a European Pre-standard ENV 1071-3:1994 established by the European Standards Committee CEN TC184 "Advanced Technical Ceramics" WG5 'Test Methods for Ceramic Coatings'. This ENV is now ready for approval by CEN TC 184, hence it is expected that it will become a European Standard (EN) in the year 2000. The results of the above-mentioned European project were used to improve the quality of this standard.

In the present paper the selected parts with results of the experimental scratch testing work in the project "Scratch Testing of Coated Surfaces" are described.

2. MATERIALS AND EXPERIMENTAL PROCEDURE

2.1. Characterisation of diamond styli

Since the quality of the diamond stylus was assumed to have a great influence on the scratch testing, the diamond styli used in the research work were specially prepared for the project. Three sets of five similar diamond styli were produced out of two type Ia motherdiamonds of different qualities and a Ib synthetic motherdiamond, to enable the assessment of the possible effects of diamond impurities on scratch test results. Secondary ion mass spectrometry (SIMS) and Raman spectroscopic analyses confirmed that the diamonds were of significantly different quality. The measurements using 3-D profilometry were carried out to determine the tip radii of the styli. The results represented in Table 1 clearly show that the radii quite often deviated considerably from the standard radius of Rockwell C indenters, even though the tip radius was specified to be 200 µm.

Table 1.
Stylus radii and distance from the tip up to where it is spherical. The reported errors correspond to standard deviations of five measurements

Stylus number	A (Ia)				B (Ia)				E (Ib)			
	Tip radius	Tip radius error	Spherical distance	Spherical distance error	Tip radius	Tip radius error	Spherical distance	Spherical distance error	Tip radius	Tip radius error	Spherical Distance	Spherical distance error
	(μm)	(μm)	(μm)	(μm)	(μm)	(μm)	(μm)	(μm)	(μm)	(μm)	(μm)	(μm)
1	202	10	14.0	2	202	5	13.6	2	199	8	11.8	2
2	201	7	12.6	1	220	11	13.6	2	201	7	13.0	2
3	205	7	16.9	2	209	13	16.0	2	251	8	12.0	2
4	217	9	14.0	1	233	11	13.0	2	226	10	13.0	2
5	216	11	13.7	2	211	7	14.6	2	287	8	13.0	2

At diamond indenter manufacturers workshops, the <100> crystallographic orientations were determined empirically by sensing the most wear resistant axes during the grinding process. Laue XRD analyses, however, revealed that this method was not infallible, since deviations up to a maximum of 45° were observed. A true <100> direction in the plane perpendicular to the diamond holder axis was labelled on the indenter holder, to define a standard procedure for the scratches to be carried out in the project, *i.e.* parallel with this direction and with the mark preceding the scratch (Fig. 1). A start load of 5 N was used to ensure the identification of the start of the scratch track. The identification of the crystalline orientation in styli also enabled the investigation of the possible effects of the orientation of the diamonds.

2.2. Coated samples

The substrate material used in the study was M2 hardened high speed steel (HV1=827-858 ± 11). The typical surface flatness of the substrates was 2.2 ± 0.6 μm over 30 mm and the substrates were lapped and polished to surface roughness R_a=10 ± 5 nm. All uncertainties are quoted at the 95% confidence level.

The TiN coatings of constant composition, mechanical properties, surface roughness, thickness and coating-substrate adhesion were deposited in a UHV compatible reactive PVD equipment. The equipment was coupled directly to an Auger Electron Spectrometer (AES) in order to transfer the cleaned substrates in-vacuum, for verification of the cleanliness by differential AES. To maintain a constant coating/substrate interface, the substrate surface was sputter cleaned until

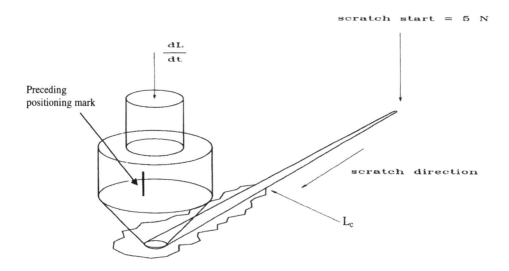

Figure 1. Convention of scratch test procedure: start load of 5 N, scratch parallel to the preceding indenter positioning mark on the indenter holder (<100> crystallographic direction). dL/dt refers to the loading rate and L_C to the critical load connected to a specific failure mode.

the oxygen content fell below the detection limit. Thus the surface was cleaned of contaminants as well as oxide layers. A Ti adhesion promoting interlayer was deposited immediately prior to reactive sputtering of the TiN coating. The objective was to standardise the adhesion, which was best judged to be carried out by using an atomically clean surface and which was justified by the excellent reproducibility of scratch test performance for the coatings. The thickness of the TiN coatings used ranged from 2.10 ± 0.15 µm to 2.59 ± 0.14 µm and a separate travelling set was also produced with a TiN thickness of 1.24 ± 0.09 µm (expected uncertainties quoted at 95% confidence level).

2.3. Test procedure

The scratch testing was carried out with different commercially available scratch test instruments. The operating parameters used in the study were loading rate of 100 N/min and a traverse speed of 10 mm/minute. A start load of 5 N was used in order to enable the identification of the start of the scratch track (Fig. 1). The ultimate critical load (UCL), which is defined as the critical load corresponding to a failure event beyond which no additional information on the coating/substrate mechanical behaviour under scratch testing is obtained, was determined. The maximum load was then limited to 60 N (= UCL + 10 N) to prevent unnecessary wear of the stylus during the course of the study. Prior to testing and during the testing the samples and the diamond styli were cleaned according to instructions produced in the project.

The critical loads were assessed by optical microscopic examination with a magnification of 200. Four clearly distinguishable failure events were identified, as shown in Fig. 2, and the associated critical load values were determined according to the definitions described in Table 2. Coating chipping at the scratch channel edges was associated to L_{C2} irrespective of whether or not the substrate was exposed, since the results of this study indicated that this was rather a matter of coincidence. Therefore, three critical load values, L_{C1}, L_{C2} and L_{C3}, were used for failure determination. The acoustic emission signal and the friction force were monitored during the scratching and the critical loads (L_{CAE} and L_{CFF}) were determined accordingly.

2.4. Instrument calibration

It is clearly important to ensure that the force and displacement measurements, which are carried out with scratch test instruments, are correct if relevant values for critical loads are to be quoted. The instrument stiffness characteristics may also influence scratch test results, *e.g.* by triggering the toughness related failure events through the instrument mechanical response. Therefore, traceable calibration procedures for the displacement and force measurements, as well as for the determination of the static and dynamic stiffness characteristics of a scratch testing system, were developed in the project. These procedures were split into two categories, namely mandatory and optional. The mandatory procedures must be

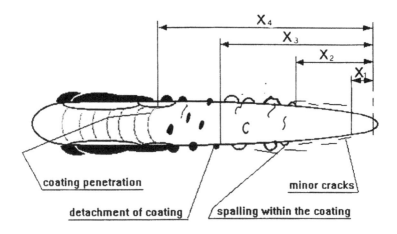

Figure 2. Schematic representation of the different failure events used in the scratch testing.

Table 2.
Description of the failure events corresponding to the critical loads

Distance from start	Description of failure event	Critical load
X_1	Longitudinal coating cracks at the scratch track edges	L_{C1}
X_2	Chipping at the scratch track edges, substrate not exposed	L_{C2}
X_3	Chipping at the scratch track edges, substrate exposed	L_{C2}
X_4	Ploughing of the indenter through the coating	L_{C3}

performed if results obtained from a scratch test are to be quoted, and a value for the uncertainty in measurements is to be derived.

The mandatory procedures in the scratch testing was considered to comprise:
- The procedure to estimate sample planarity prior to testing.
- The calibration of the load cell which measures applied load.
- The calibration of the measurement of horizontal displacement of the indenter across the sample.

The optional procedures in the scratch testing was considered to comprise:
- The calibration of the load cell which measures frictional load.
- The calibration of the horizontal static stiffness parameters.
- The calibration of the vertical static stiffness parameters.
- The calibration of the measurement of vertical displacement of the indenter into the sample.

- A simple procedure which yields qualitative information on the dynamic stiffness response of the scratch test system.
- A procedure for the evaluation of the mechanical impedance response of the scratch tester.

These procedures were used by the participants in the reproducibility study to quantify the behaviour of their test systems, and formed the basis for the analysis of the results. A full description of the calibration procedures can be found in reference [32]. Figs. 3 to 6 show examples of the experimental set-ups used for carrying out the calibration measurements. The accuracy of the calibration was generally better than 1% of the measuring range.

Figure 3. Instrumental set-up for the calibration of the normal force measuring system, using a calibrated load cell.

Figure 4. Instrumental set-up for the calibration of the horizontal displacement measuring system, using a calibrated long-stroke LVDT.

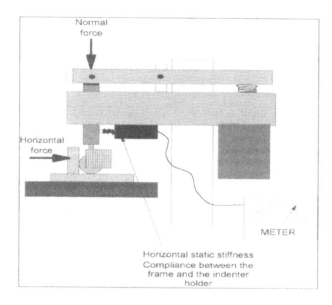

Figure 5. Instrumental set-up to determine the horizontal compliance of the indenter holder, using a calibrated short-stroke LVDT.

Figure 6. Instrumental set-up for the calibration of the friction force measuring system, using a thumb screw device and a calibrated load cell.

3. SCRATCH TEST SENSITIVITY STUDY

3.1. *Effect of cleaning procedures*

In order to increase the reliability of scratch testing it is necessary to clean the diamond stylus and the coated surface prior to testing. The effect of different cleaning procedures and equilibrium (stabilisation period between cleaning and testing) times was assessed by comparing the conventional cleaning procedure with a simple procedure carried out with less toxic liquids. The cleaning procedures are summarised in Table 3. The effect of equilibrium times prior to scratch testing was assessed in the range 3 to 60 minutes.

The results showed only minor changes in the L_C values for different cleaning procedures and equilibrium times as represented in Fig. 7. This was confirmed also by the overall mean values and standard deviations obtained in the study for L_{C2} =24.7±4.6 N and L_{C3}=39.7± 4.1 N, which indicated good repeatability of the scratch test even though a range of different cleaning procedures were used. This observation demonstrated that the simple cleaning procedures, for the specimen and the stylus, making use of petroleum ether, was as good as more sophisticated procedure using toxic solutions of acetone, toluene and n-hexane. The stabilisation time after the cleaning did not appear to be critical, and the test could hence be performed practically immediately after the cleaning. This study yielded the following recommendations for cleaning:

Coated specimen:
Before each series of scratches, at least every day:
- Ultrasonic bath for 5 minutes in clean pro analyse (99.1 % purity) petroleum ether.
- At least 3 minutes equilibration time before testing.
If drying stains are observed:
- Wiping with a soft tissue soaked in petroleum ether.
- At least 3 minutes equilibration time before testing.
- During testing, the specimen surface must be kept free of fingerprints.

Stylus:
Before each scratch:
- Wipe with a soft tissue soaked in petroleum ether.
If adhering debris is still observed under an optical microscope (magnification: 200 X):
- Mechanical cleaning with #2400 and #1200 SiC emery paper, and wiping with a soft tissue soaked in petroleum ether.
- At least 1 minute equilibrium time before testing.
- During testing, the specimen surface and stylus tip must be kept free of fingerprints.

Table 3.
The conventional and the simpler and less toxic cleaning procedures used in the study

Conventional cleaning		Less toxic cleaning	
Diamond stylus	Coated samples	Diamond stylus	Coated samples
- rubbing with SiC paper to remove adhered residues - wiping with acetone soaked tissue, drying in warm air flow - microscopic examination	- 5 min. ultrasonic bath in toluene - 5 min. bath in n-hexane - drying in oven (30 min. 110-120 °C)	- rubbing with SiC paper to remove adhered residues - wiping with petroleum-ether soaked tissue - drying in warm air flow - microscopic examination	- 5 min. ultrasonic bath in petroleum ether - drying in warm air flow

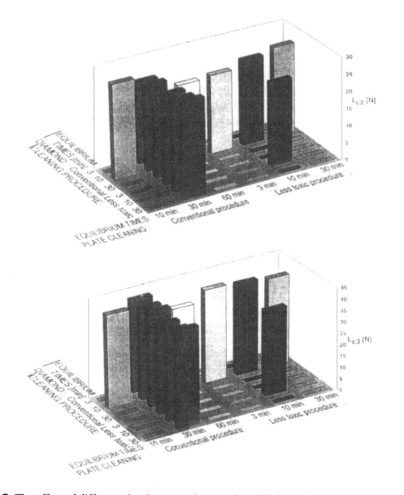

Figure 7. The effect of different cleaning procedures and equilibrium times on critical load values L_{C2} and L_{C3} in scratch testing.

3.2. *Effect of ambient atmosphere*

The effect of variations in the temperature and relative humidity of the ambient conditions was evaluated in the range 20 to 30 °C and 20 to 80 % relative humidity. No significant effect of the ambient temperature was observed between 20 and 30 °C. The relative humidity of the ambient air, in some cases, seemed to influence slightly the scratch test results in the range studied, but not dramatically (Fig. 8). As a consequence, it was concluded that the conditioning of the ambient atmosphere was not a prerequisite for scratch testing, but tribology laboratory conditions (22±2 °C temperature and 50±10 % relative humidity) should nevertheless be recommended whenever possible to adopt.

3.3. *Effect of diamond quality*

When the influence of the diamond quality was assessed, the results indicated that the possible effect of diamond impurities was of only secondary importance when compared to the effect of the indenter geometry and the presence of defects at the stylus tip. Figure 9, for example, shows the critical load values corresponding to the easily recognisable failure event of "chipping of the coating at the scratch track edges" obtained on the same TiN coated high speed steel (HZ) specimen, using three different fresh and clean styli, A5, B5 and E5. The stylus E5 yielded high critical load values due to its large tip radius of 287 µm. Styli A5 and B5 had tip radii which were closer to 200 µm, but stylus A5 showed a surface cavity near its tip (Fig. 10), which may explain the lower L_C values due to local stress concentration caused by the cavity.

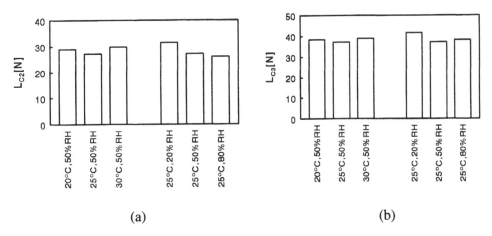

Figure 8. Influence of ambient temperature and relative humidity on critical loads corresponding to the 'chipping with substrate exposure' (L_{C2}) failure event (a), and the 'ploughing through the coating' (L_{C3}) failure event (b).

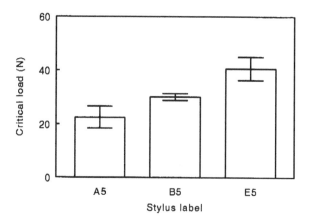

Figure 9. Influence of stylus properties: critical loads L_{C2} with 95% confidence error bars.

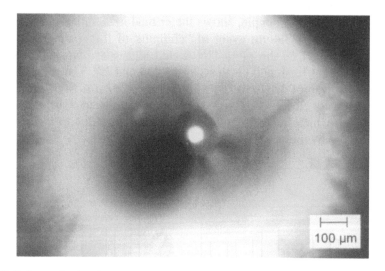

Figure 10. Defect at the tip of stylus A5 (scratch direction is from right to left).

Significantly different critical loads were observed when performing scratch tests by rotating the E5 stylus in its holder. This discrepancy can be explained by the deviation from the spherical shape of the particular indenter (Fig. 11 (b,d)). No such differences as a function of stylus orientation were observed when a stylus with a better spherical tip shape, such as the B5 indenter was used (Fig. 11 (a,c)).

The results of this work demonstrate that the availability of a scratch test reference specimen imposes itself to enable the control and follow-up (wear during usage) of deviations from the required tip shape. Besides the control of stylus quality, the reference specimen would also allow the detection of errors in the load and displacement parameters as well as other malfunctioning of the instrument.

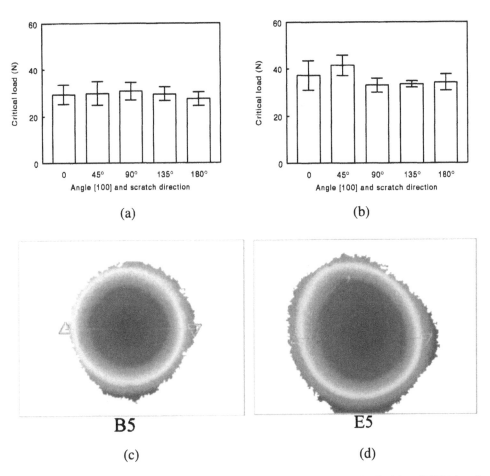

Figure 11. Critical loads corresponding to the coating chipping (L_{C2}) obtained with styli B5 (a) and E5 (b) as a function of stylus orientation, and the corresponding contour mappings of the styli tips (c,d; zero degree scratch direction is from right to left).

3.4. Effect of transfer layers

The effect of transfer layers was investigated by deliberately contaminating a scratch stylus with titanium (Ti). Heavy contamination, by sliding the stylus against a worn Ti sputter target, resulted in high friction against the TiN specimen (stick-slip motion), and the chipping failure modes at the scratch track edges were obscured due to Ti transfer to the TiN coating (Fig. 12). A slightly contaminated stylus, obtained by sliding the stylus against a thin Ti film on top of a TiN coating, resulted in increased friction during the initial part of the scratch track as well. These findings stress the importance of inspection and cleaning of the stylus before each experiment. It was demonstrated that the cleaning procedure developed in this study performed adequately.

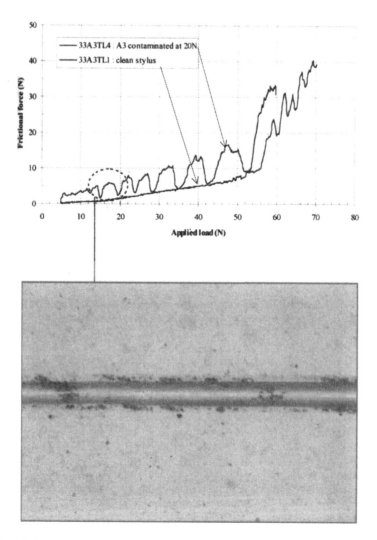

Figure 12. Friction graphs of a Ti-contaminated and a clean styli. Examination of the wear track demonstrated that the increase in friction corresponded to transfer of Ti onto the TiN coating.

4. REPRODUCIBILITY OF SCRATCH TEST

4.1. Influence of test equipment

The reproducibility of the scratch test was assessed by carrying out scratch testing with four different test devices. The diamond stylus set A1/B1/E1 and TiN coated specimens were exchanged between the participants and different instruments in the following chronological order: (1) VTT (VTT Scratch Tester), (2) VITO (CSEM Revetest instrument), (3) NPL (modified CSEM Revetest) and (4) TCL (TCL ST-2200 Scratch Tester).

The results were analysed according to the conventional metrological procedures. Based on the calibration results, the following generalised relationship between the true normal load, L, and the measured distance between the scratch track start and the failure event, x_m, was established (neglecting surface roughness effects and penetration by elastic or plastic deformation of the surface):

$$L = \frac{L_r}{x_r}(x_m + x_{corr}) + S_{spr}(h - h_{start}) + L_{corr} + L_{start} \tag{1}$$

where L_r = normal loading rate; x_r = horizontal displacement rate; x_{corr} = systematic error in the displacement reading; h-h_{start} = vertical indenter tip displacement relative to the start position; S_{spr} = the spring stiffness; L_{corr} = systematic error in the normal load reading; L_{start} = start load (5 N). This equation was used to calculate the systematic deviation from the critical load, which is usually determined as $10 \cdot x_m + 5$ (L_r = 100 N/min.; x_r = 10 mm/min.; L_{start} = 5 N).

Random calibration errors were calculated according to:

$$\delta L = \sqrt{\sum_{i=1}^{8}\left(\frac{\partial L}{\partial e_i}\delta e_i\right)^2} \tag{2}$$

where $e_1 = L_r$, $e_2 = x_r$, $e_3 = x_{corr}$, $e_4 = x_m$, $e_5 = h$, $e_6 = S_{spr}$, $e_7 = L_{corr}$ and $e_8 = L_{start}$. The δe_i (i = 1 to 8) values corresponded to the 95% confidence level of each parameter.

The total uncertainty at the 95% confidence level, including the measurement error, is thus given by:

$$L_c \pm \Delta L_c = L_c \pm \sqrt{(\delta L)^2 + \left(\frac{ts}{\sqrt{n}}\right)^2} \tag{3}$$

where n is the number of measurements, t is the so-called Students' t-value which depends on the number of measurements (*e.g.* the t-value using 10 measurements is 2.262), and s represents the unbiased standard deviation of the measured critical load values based on the sample.

A representative result of the reproducibility study is shown in Fig. 13, where the individual participant bars are displayed in chronological order. It can be seen that for a stylus that remained intact during the exercise (stylus A1), the critical loads obtained by the four participants agreed within 15%. The stylus B1 was damaged during the experiments at TCL and the stylus E1 showed progressive wear during the course of the study (Fig. 14), which resulted in lower critical load values towards the end of the exercise. The typical measurement uncertainty at the 95% confidence level was 20%. The results demonstrated that for similarly shaped styli, which were not damaged, the critical load values agreed within the error margins (solid and clear bars in Fig. 13), indicating that the possible effect of diamond impurities was negligible.

It was checked whether the observed tendencies could have been influenced by ageing of the specimens, the location of the scratch on the specimen, or by possible temperature fluctuations during aircraft transportation, but no significant effects were observed.

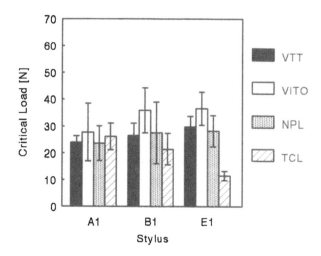

Figure 13. Reproducibility exercise: critical load L_{C2} with 95% confidence error bars.

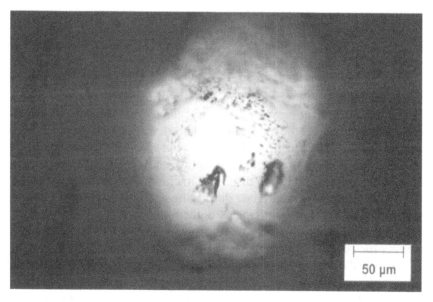

Figure 14. Wear defects at the tip of stylus E1 after the reproducibility study (scratch direction is from top to bottom).

4.2. Influence of operator

To assess the variability introduced by different operators, the same scratched specimen was inspected by the four participants using the same microscope. A representative result is shown in Figure 15, which demonstrates that measurement errors introduced by different trained operators lies within 5 to 10%.

4.3. Acoustic emission (AE) and friction force (FF) measurements

Critical loads obtained by monitoring the acoustic emission (AE) and friction force (FF) signals have been correlated with the L_C values determined by microscopic inspection of the scratch track. Figure 16 shows that the values agree quite well with particular failure events, although no absolute correlation could be obtained. Furthermore, this work revealed that interpretations used to derive critical load values from AE and FF signals varied strongly from one laboratory to another. In addition, it is known that the AE signal amplitude strongly depends on the procedures (hard- and software) used for capturing, transmitting and processing this signal. Perturbations in the FF signal may also be caused by machine artefacts, e.g. by distortions in the translation mechanisms. Hence, it was concluded that AE and FF signals could not be used as a reliable means for determining scratch test critical loads. Only after a large series of experiments on the same coating type, in order to establish the statistics of correlation with a certain failure mode, these techniques can at their best be used as a warning system in the quality control of coated components. Inspection of the scratch track by microscopic observation remains the only reliable means of associating a failure event with a measured critical normal load.

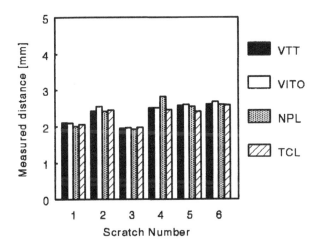

Figure 15. Operator influence: Distance between the start of the scratch track and the 'chipping without substrate exposure' (L_{C2}) failure event.

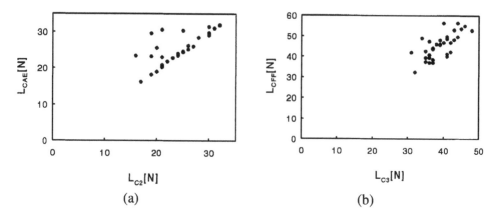

Figure 16. Correlation between critical loads corresponding to (a) the 'chipping at the scratch track edges' failure event and critical loads determined by an increase in the AE signal, and (b) the 'ploughing through the coating' failure event and critical loads determined by the first significant perturbations in the FF signal.

5. SCRATCH TEST FAILURE MODES

In order to assist users of the scratch test in the standardised reporting of scratch test results, an atlas of scratch test failure modes was produced. In addition of TiN coatings that were specially produced for the project [31], other coatings, like SiC, Al_2O_3, and a-C:H, were also used for the cataloguing work. These coatings were deposited with various techniques presently used for thin film deposition. Different failure mechanisms were classified in terms of plastic deformation, cracking, spallation (where the coating flakes off) and perforation. It was decided to include the widest possible series of coated specimens available in the participating laboratories, even though some of the failure modes reported in the literature may be irrelevant from a practical point of view, e.g. those originating from brittle coatings on unrealistically soft substrates.

The pictures of the failure modes, twenty in total, have been assembled in a poster format (1.7x0.9 m²), and each figure is accompanied by a concise description of the observed failure mode as well as the coating/substrate system. The critical load values at which the failure events occurred are reported. Two examples of the figures in the poster of scratch test failure modes are shown in Figs. 17 and 18.

The present catalogue of scratch test failure modes cannot be claimed to be comprehensive. However, it should be regarded as an important step towards the standardised reporting of critical loads in scratch testing.

One must bear in mind though that for assessing the quality of a coated component, the magnitude of and subtle differences in the observed failure event may often be equally important as the failure mode itself. In addition, there is still much research work to be carried out in order to understand the mechanisms of each failure mode.

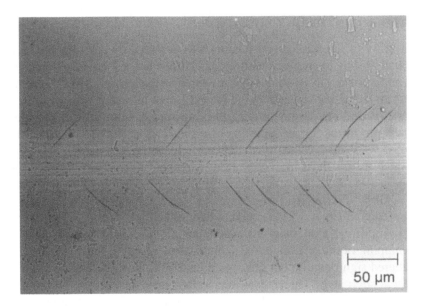

Figure 17. Forward chevron cracks at the borders of the scratch track. PACVD DLC (6.6 µm) on hardened and polished M2 steel (64 HR$_C$); L$_C$ = 36 N.

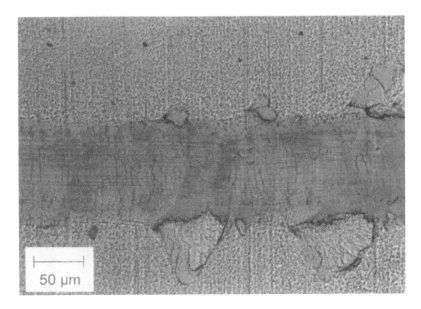

Figure 18. Spallation along the scratch track borders, both cohesive and interfacial. PVD TiN (3.8 µm) on hardened and ground M2 steel (64 HR$_C$); L$_C$ = 42 N.

6. ROUND ROBIN INTERCOMPARISON

The reliability of the scratch test method was assessed by carrying out an extensive Round Robin intercomparison study. Altogether 13 organisations, five industrial enterprises (IND) and eight research organisations (RES), from eight European countries participated in the Round Robin intercomparison.

Calibration and measurement guidelines, instructions for critical load determination, and pro forma for the reporting of results and a set of TiN coated AISI M2 specimens, were distributed to all the participants. Each participant calibrated their scratch testers by using calibration equipment exchanged between the participants. In testing, each participant used their own scratch stylus and each performed at least 10 scratches.

By the end of 1997, 13 participants had submitted results, which were analysed. Raw critical load data corresponding to the easily recognisable failure event 'chipping of the coating at the scratch track edges' is shown in Fig. 19. The data scattered significantly, but this was primarily believed to be due to the use of imperfect scratch styli. The critical load values may increase in the case of flattening of the scratch tip (gradual wearing, resulting in a larger apparent tip radius and a lower contact pressure), or decrease in the case of cracked regions (chipping) at the stylus tip, which result in local stress concentrations in the contact. The very low values obtained by RES3, for example, are undoubtedly caused by the high roughness of the indenter (Fig. 20). After the Round Robin, control measurements were carried out on all test samples at VITO, by using a single

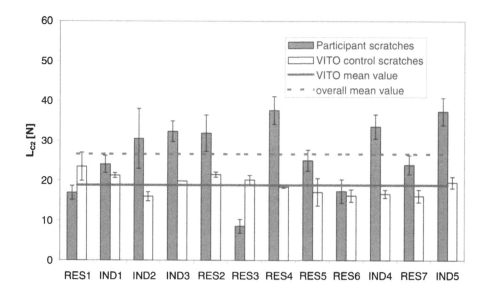

Figure 19. Round Robin: Critical loads L_{C2} with 95% confidence error bars for TiN coated AISI M2 steel specimens as measured by the participating research (RES) and industrial (IND) organizations.

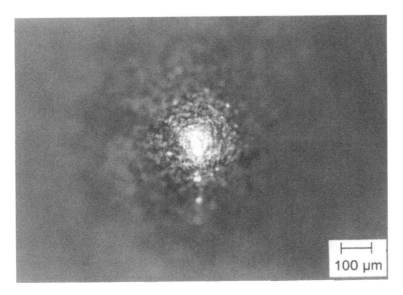

Figure 20. An optical micrograph of the diamond stylus used at RES3 in the Round Robin (scratch direction is from top to bottom).

diamond stylus. In this case the critical load values agreed better, as can be seen in Fig. 19. The results obtained by RES8, where a constant load operation mode instrument was employed, were not included in Figure 19. In that case, chipping of the coating was observed for normal loads of 30 N and higher.

7. DISCUSSION

The sensitivity investigation of the scratch test showed that cleaning of both the coated specimen and the stylus was necessary, and practical and adequate cleaning procedures were defined. The use of contaminated styli resulted in a different friction behaviours and a loss of information. Hence, it is essential that the stylus is inspected and/or cleaned before each scratch. The conditioning of the ambient atmosphere did not seem to be a prerequisite for scratch testing, but tribology laboratory conditions (22±2 °C temperature and 50±10 % relative humidity) are nevertheless recommended. The main conclusion of the sensitivity study, however, is that the uncertainties in the Rockwell C stylus tip shape are the major source of error for the scratch test method.

The reproducibility was assessed between four different instrument types. The instruments were calibrated according to traceable calibration procedures developed in the project, and errors were quantified using the conventional metrological protocol. Wear of the styli, which were exchanged between the participants, resulted in significant differences in the critical load values. It was also demonstrated that acoustic emission and friction force monitoring could not be used as stand-alone methods for determining the critical load values in scratch testing. Inspection of the scratch track by microscopic observation remains the most reliable means of associating a failure event with a measured critical normal load.

A wide range of typical scratch test failure modes was catalogued. These failure events were classified into plastic deformation and different forms of cracking, spallation and coating perforation events, and were assembled in a poster format. This poster – Atlas of Failure Modes – represents an important step forward to the standardised reporting of the critical load values in scratch testing.

An extended Round Robin intercomparison exercise included 13 participants from eight European countries. In this case, each participant used their own scratch stylus, and the critical load values scattered markedly. In this study there were strong indications that the scatter in test results was primarily due to the use of imperfect scratch styli.

The results of this study will contribute to the standardisation work on the scratch test. The scratch test has been the subject of a European Pre-standard ENV 1071-3:1999 [32] established by the European Standards Committee CEN TC184 'Advanced Technical Ceramics' WG5 'Test Methods for Ceramic Coatings'. This ENV is now ready for approval by CEN TC 184, so it is expected that it will become a European Standard (EN) in the year 2000.

Through the present work, several shortcomings and/or possible improvements in the use and design of current scratch test instruments were identified:

- In general, the instrument manuals lack procedures for calibration of the most important parameters such as the normal load rate. Manufacturers should thus develop customised calibration procedures, as well as provide the necessary hardware, for their respective instruments in accordance with the procedures worked out under the FASTE project. Less evident parameters, which normally remain constant, such as the instrument stiffness characteristics, should be calibrated at the manufacturers' plants and the values should be supplied to the customers.

- Wherever possible, the stiffness of the instrument should be increased. This refers especially to the horizontal stiffness (indenter and specimen table), which results in a stick-slip motion between the indenter and the surface.

- In situ measurement of the specimen displacement would be desirable, to enable a direct correlation between normal load, displacement, and other signals such as the friction force and acoustic emission.

- Spring loaded rigs should have a (stiff) levelling mechanism to preclude the variation of load rate due to oblique surfaces. In addition, the operator should be warned that the load rate may vary considerably with the position of the spring, and the compliance of the test specimen. Ideally, mechanisms with in situ control of the load should be used.

The project revealed that uncertainties in the Rockwell C stylus tip shape are a major source of error for the scratch test method. There are, however, no readily adaptable methods to ensure the use of correctly shaped styli. As a consequence, a follow-up European project, "A Certified Reference Material for the Scratch Test" [33] was established. A certified reference material, presenting a repeatable failure event within a known critical load interval, is being developed, as a means for controlling the proper functioning of a scratch test rig. This will not only allow the control of deviations from the required tip shape, but also the detection of errors in the load and displacement calibrations as well as any other malfunctioning of the instrument. In addition, the manufacturing process of diamond Rockwell C indenters should be improved, which will be also addressed in the project.

The scratch test "Atlas of failure modes" needs to be extended and refined. For the scientifically sound classification of failure events, a good understanding of the mechanisms of failure is a necessity. In a current SMT project, 'Multimode Scratch Testing (MMST)', in situ video monitoring and enhanced signal capturing and processing techniques are being developed, which will greatly facilitate the understanding of scratch test failure mechanisms [34].

Only some of the observed failure events in scratch testing are related to detachment at the coating/substrate interface and are thus relevant as a measure of adhesion. Other failures, such as cracks and cohesive damage within the coating or substrate, may be equally important to determine the behaviour of a coated component in a particular application. The scratch test should indeed be regarded as a repeatable tribological test to assess the mechanical integrity of a coated surface, including that of bulk materials. Within this scope, it would be a big step forward if the operation modes were to include, for example, constant load (uniaxial and reciprocal) test procedures. The latter is one of the targets of the aforementioned MMST project. Another important issue would be the extension to smaller scale (micro-, nano-) contacts, to enable the probing of near-surface (coating-only) properties, much like the evolution from the conventional hardness tests to the current nano-indentation techniques. This would require, *inter alia*, the use of a sharper stylus and higher load resolutions and hence, enhanced calibration procedures and measurement skills.

8. CONCLUSIONS

The following conclusions can be made on the basis of the present research work reported:

- Cleaning of the specimen and indenter is essential, and realistic and adequate cleaning procedures have been defined.
- Tribology laboratory conditions (T = 22±2 °C; RH = 50±5 % relative humidity) are recommended for the scratch test method.
- The control of the stylus shape is imperative, in the as-received condition as well as during usage, in order to detect wear at the stylus tip.
- The calibration of the scratch tester is a necessity and traceable calibration procedures have been developed. The accuracy of the calibration is generally better than 1% of the measurement range.

- Failure events that are associated with scratch test critical load values must be clearly defined. The Atlas of failure modes that was established in the work represents an important step forward to the standardised reporting of scratch test results.
- The critical load values derived from perturbations in the acoustic emission or friction force cannot be used as stand-alone values to determine critical load values. The only reliable means of determining scratch test critical load values remains microscopic inspection of the scratch track.
- The typical measurement uncertainty at the 95% confidence level is 20%, and different operators introduce errors in the range 5 to 10 %.
- Under optimum conditions, the reproducibility between different scratch test instruments is better than 15 %.

Acknowledgements

The Standards, Measurements and Testing Programmes of the European Commission are acknowledged for the support to the present research activities. VTT Manufacturing Technology and VITO are indebted to Simo Varjus and to Danny Havermans, respectively, for carrying out most of the actual scratch testing.

LIST OF ACRONYMS, SYMBOLS AND ABBREVIATIONS

SIMS = secondary ion mass spectrometry.
3-D = three-dimensional.
Rockwell C = standard shape of diamond indenter (EN 10109): 120° cone with a 0.2 mm tip radius.
Laue XRD = X-ray diffraction technique to determine crystal structures.
$\mathbf{L_C}$ = scratch test critical normal load.
RH = relative humidity.
HSS = high speed steel.
Ti = titanium.
TiN = titanium-nitride.
LVDT = displacement transducer (linear variable differential transformer).
AISI M2 = USA standard grade high speed steel (corresponds to DIN S 6-5-2 (WN 1.3343)).
UCL = ultimate critical load
L = normal load.
$\mathbf{L_{corr}}$ = systematic correction in the normal load reading.
$\mathbf{L_{start}}$ = start load.
$\mathbf{L_r}$ = normal loading rate.
$\mathbf{x_m}$ = measured distance.
$\mathbf{x_{corr}}$ = systematic correction in the displacement reading.
$\mathbf{x_r}$ = horizontal displacement rate.

h - h$_{start}$ = vertical indenter tip displacement relative to the start position.

S$_{spr}$ = load spring stiffness.

n = number of measurements.

t = Students' t-value, to determine statistical confidence level.

s = unbiased standard deviation based on the sample.

δL = random calibration error in L.

ΔL$_C$ = uncertainty in L$_C$.

AE = acoustic emission.

FF = friction force.

SiC = silicon carbide.

Al$_2$O$_3$ = aluminium oxide.

a-C:H = amorphous hydrogenated carbon (form of diamond-like carbon).

DLC = diamond-like carbon.

PACVD = plasma assisted chemical vapour deposition.

PVD = physical vapour deposition.

HRC = hardness according to the Rockwell C scale.

SMT = EC Standards, Measurements and Testing Programme.

REFERENCES

1. K.L. Mittal, in *Adhesion Measurement of Films and Coatings*, K.L. Mittal (Ed.), pp. 1-13. VSP, Utrecht, The Netherlands, 1995.
2. K.L. Mittal, Electrocomponent Sci. Technol., **3**, 21 (1976).
3. A.J. Perry, Thin Solid Films, **107**, 167 (1983).
4. J. Valli, J.Vac. Sci. Technol., **A4**, 3007 (1986).
5. P.A. Steinmann and H.E. Hintermann, J. Vac. Sci. Technol., **A7**, 2267 (1989).
6. P.R. Chalker, S.J. Bull and D.S. Rickerby, Mater. Sci. Eng., **A140**, 583-592 (1991).
7. B. Olliver and A. Matthews, J. Adhesion Sci. Technol. **8**, 651-662 (1994).
8. C.H. Lin, H.L. Wang and M.H. Hon, Thin Solid Films **283**, 171-174 (1996).
9. H. Ronkainen, J. Vihersalo, S. Varjus, R. Zilliacus, U. Ehrnstén and P. Nenonen, Surf. Coat. Technol., **90**, 190-196 (1997).
10. S. Zhang and H. Xie, Surf. Coat. Technol., **113**, 120-125 (1999).
11. D.S. Rickerby, Surf. Coat. Technol., **36**, 541 (1988).
12. P.J. Burnett and D.S. Rickerby, Thin Solid Films, **154**, 403 (1987).
13. P.J. Burnett and D.S. Rickerby, Thin Solid Films, **157**, 233 (1988).
14. S.J. Bull and D.S. Rickerby, *in Plasma Surface Engineering*, E. Broszeit, W.D. Munz, H. Oechsner, K.T. Rie and G.K. Wolf (Eds.), p. 1227. DGM, Oberursel, 1989.
15. P.A. Steinmann, Y. Tardy and H.E. Hintermann, Thin Solid Films, **154**, 333 (1987).
16. P.C. Jindal, D.E. Quinto and G.J. Wolfe, Thin Solid Films, **154**, 361 (1987).
17. R. Rezakhanlou and J. von Stebut, in *Mechanics of Coatings*, D. Dowson, C.M. Taylor and M. Godet (Eds.), p. 183. Elsevier, Amsterdam, 1990.
18. S.J. Bull, D.S. Rickerby, A. Matthews, A. Leyland, A.R. Pace and J. Valli, Surf. Coat. Technol., **36**, 503 (1988).
19. J. Valli and U. Mäkelä, Wear, **115**, 215 (1987).
20. A.J. Perry, J. Valli and P.A. Steinmann, Surf. Coat. Technol., **36**, 559 (1988).
21. H. Ronkainen, S. Varjus, K. Holmberg, K.S. Fancey, A.R. Pace, A. Matthews, B. Matthes and E. Broszeit in *Mechanics of Coatings*, D. Dowson, C.M. Taylor and M. Godet (Eds.), p. 453. Elsevier, Amsterdam, 1990.

22. J. von Stebut, in *Plasma Surface Engineering*, E. Broszeit, W.D. Munz, H. Oechsner, K.T. Rie and G.K. Wolf (Eds.), p. 1215. DGM, Oberursel, 1989.

23. J. Sekler, P.A. Steinmann and H.E. Hintermann, Surf. Coat. Technol., **36**, 519 (1988).

24. S. Yamamoto and H. Ichimura, J. Mater. Res., **7**, 2240 (1992).

25. J. Valli, U. Mäkelä, A. Matthews and V. Murawa, J. Vac. Sci. Technol., **A3**, 2411 (1985).

26. T.W. Wu, R.A. Burn, M.M. Chen and P.S. Alexopoulos, Mater. Res. Soc. Symp. Proc., **130**, 117 (1989).

27. S.J. Bull, paper presented at ICMCTF-91, San Diego, April 22-26, 1991.

28. J. von Stebut, R. Rezakhanlou, K. Anoun, H. Michel and M. Gantois, Thin Solid Films, **181**, 555 (1989).

29. P. Hedenqvist, M. Olsson, S. Jacobson and S. Söderberg, Surf. Coat. Technol., **41**, 31 (1990).

30. S.J. Bull, Tribology Intl., **30**, 491-498 (1997).

31. European Commission - Measurements and Testing Programme, Project 'Development and Validation of Test Methods for Thin Hard Coatings - FASTE', contract MAT1-CT94/0045

32. prENV 1071-3:1999 Advanced technical ceramics - Methods of test for ceramic coatings-Part 3: Determination of adhesion and other mechanical failure modes by a scratch Test, CEN European Committee for Standardization, Central Secretariat, rue de Stassart 36, B-1000 Brussels, Belgium.

33. European Commission – Standards, Measurements and Testing Programme, Project 'A Certified Reference Material for the Scratch Test – REMAST', contract SMT4-CT98/2238.

34. European Commission – Standards, Measurements and Testing Programme, Project 'Multimode Scratch Testing (MMST): Extension of Operation Modes and Update of Instrumentation', contract SMT4-CT97/2150

Adhesion Measurement of Films and Coatings, Vol. 2, pp. 107–130
Ed. K.L. Mittal
© VSP 2001

Can the scratch adhesion test ever be quantitative?

S.J. BULL[*]

*Department of Mechanical, Materials and Manufacturing Engineering, University of Newcastle,
Newcastle-upon-Tyne, NE1 7RU, UK*

Abstract—The scratch test has been used to assess the adhesion of thin hard coatings for some time now and is a useful tool for coating development or quality assurance. However, the test is influenced by a number of intrinsic and extrinsic factors which are not always adhesion-related and the results of the test are usually regarded as only semi-quantitative. The stress state around a moving indenter scratching a coating/substrate system is very complex and it is difficult to determine the stresses which lead to detachment. Furthermore the interfacial defect state responsible for failure is unknown. However, by a careful analysis of the observed failure modes in the scratch test (not all of which are related to adhesion) it is possible to find some which occur in regions where the stress state is relatively simple and quantification can be attempted.

Ideally engineers would like a material parameter (such as work of adhesion or interfacial toughness) which can be used in an appropriate model of the coating-substrate system stress state to determine if detachment will occur under the loading conditions experienced in service. These data are not usually available and the development of such models must be seen as a long term goal. It must be questioned whether the scratch test can ever give quantitative information suitable for this modelling analysis. This paper highlights the main adhesion-related failure modes and the stresses responsible for them and indicates where quantification is possible illustrating this with results from hard coatings and thermally grown oxide scales. The use of empirical calibration studies and quantification by finite element methods is discussed.

Keywords: Scratch test; buckle failure; spallation failure; residual and applied stresses; adhesion.

1. INTRODUCTION

The scratch test has been used for some time to provide a measure of coating/substrate adhesion [1-6]. In the most commonly applied version of the test a diamond stylus is drawn across the coated surface under an increasing load until some well-defined failure occurs at a load which is often termed the critical load, L_c. Many types of failures are observed which include coating detachment, through-thickness cracking and plastic deformation or cracking in the coating or substrate [7-10]. In fact several different failure modes often occur at the same time and this can make results of the test difficult to interpret.

---

[*] Phone: +44 191 222 7913, Fax: +44 191 222 8563, E-mail: s.j.bull@ncl.ac.uk

The types of failures observed in the scratch test depend on many factors and are most easily characterised in terms of the hardness of both substrate and coating (Figure 1). For soft coatings and soft substrates the test is dominated by plastic deformation and groove formation and little or no cracking is observed. For hard coatings on soft substrates deformation of the substrate is predominantly plastic whilst the coating may plastically deform or fracture as it is bent into the track created by plastic deformation of the substrate. Soft coatings on a harder substrate tend to deform by plastic deformation and some extrusion of the coating from between the stylus and the substrate may occur. Considerable thinning of the coating by plastic deformation will occur before plastic deformation and fracture of the substrate becomes significant. For hard coatings on hard substrates plastic deformation is minimal and fracture dominates the scratch response.

The scratch test is not well-suited to measure the adhesion of soft coatings but can give some information if the interfacial shear strength is less than the shear strength of either the coating or substrate. In general, the scratch test is most effective if the substrate does not plastically deform to any great extent. In such cases the coating is effectively scraped from it and the uncovering of the substrate itself can be used as a guide to adhesion. However, it is difficult for this to be quantified.

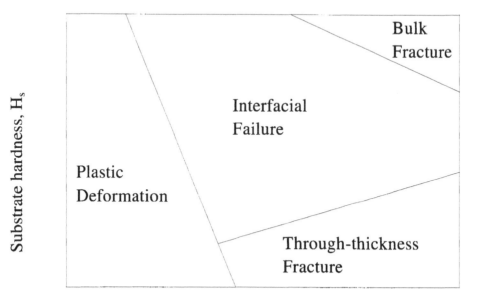

Figure 1. Schematic showing the various scratch test failure modes which dominate as a function of coating and substrate hardness (H_c and H_s, respectively).

The scratch adhesion test is much more useful for hard coatings, particularly when these are deposited on softer substrates. For a harder coating on a soft substrate the spallation and buckling failure modes arise from interfacial detachment [8,9,10] and can thus be used as the basis for an adhesion test. Both modes are amenable to quantification and are discussed in some detail in this paper. The origin of these failure modes and the theoretical basis for analysing them is introduced, together with finite element results aimed at improving quantification.

2. REQUIREMENTS FOR QUANTIFICATION OF THE SCRATCH TEST

If the scratch test is to be fully quantitative it must deliver a parameter which is representative of the state of adhesion of the interface but is not related to the other properties of the coating/substrate system such as hardness. The best parameter for this is work of adhesion which is a measure of the chemical bonding across the interface. However, most adhesion tests do not measure this fundamental adhesion but produce a practical adhesion measurement conflating fundamental adhesion with other factors which can be specific to a given material pair or test method [11-13]. Film adhesion is often characterised by the strain energy released per unit increase in delamination area, G, which is sometimes referred to as the interfacial fracture energy and can be used to generate an interfacial fracture toughness. For most practical purposes this measurement of practical adhesion is sufficient but it should be corrected for method-specific factors to ensure that the test is widely applicable and the data produced can be compared with that from other test methods.

The scratch test is usually only regarded as semiquantitative as there are a number of intrinsic and extrinsic parameters which are known to affect the measured critical load (Table 1). Many of these intrinsic factors are instrument-specific and require a careful calibration approach if results are to be compared between instruments. However, the extrinsic factors such as coating thickness and substrate hardness must also be known if the results of the test are to be understood. These parameters, together with the residual stress in the coating and its Young's modulus, are an important requirement for the models of the failure mode used to generate interfacial fracture toughness. There are thus three requirements for a quantitative scratch adhesion test:-
1) An adhesion-related failure mode,
2) A well-defined failure mechanism,
3) A method of determining the stresses which cause failure.
These will be discussed in more detail in the following sections.

Table 1.
Intrinsic and Extrinsic factors in the scratch test

Intrinsic	Extrinsic
Loading rate [3, 27, 28]	Substrate properties (hardness, elastic modulus) [28]
Scratching speed [3, 27, 28]	Coating properties (hardness, modulus, residual stress) [27, 28]
Indenter tip radius [3, 27]	Friction coefficient [3, 27]
Indenter wear [27]	Surface roughness [28]
Machine stiffness/design	

3. SCRATCH TEST FAILURE MODES

3.1. Soft coatings

In general soft coatings (hardness <5GPa) fail by plastic deformation whether deposited on softer or harder substrates. Coatings deposited with a porous microstructure (e.g. the open columnar structures produced by vapour deposition at low temperatures) may also show some evidence of fracture but this is not widespread. The scratch test is not very useful for assessing adhesion unless the interfacial shear stress is less than the shear strength of the softer component. In such cases stripping or flaking of the coating may occur if the adhesion is very poor but often there is little to see but a plastic groove after the test is complete.

When a soft coating is deposited on a very different, harder substrate, such as aluminium or gold on glass, the detection of interfacial failure is much easier. As the load is increased in the scratch test the soft coating is progressively plastically deformed until at the critical load the substrate is uncovered. This can be detected by a colour change or by the use of surface analysis techniques such as x-ray photoelectron spectroscopy (XPS) which are surface sensitive. However, XPS analysis is not always practical since failure does not occur exactly at the interface - in such cases x-ray mapping or backscattered imaging in the SEM can be used to determine a critical load but this does not represent the load for interfacial detachment (Figure 2). Since there is no sharp transition when the coating is stripped, selecting a critical load for coating detachment is almost impossible unless adhesion is poor.

The earliest attempts at scratch test quantification by Benjamin and Weaver [7] are most applicable when thin coatings are plastically deformed in the scratch test. According to these authors the critical shearing force for coating removal is a function of the scratch geometry, the substrate properties and the frictional force on the stylus. Thus, for a stylus of radius, R,

$$\tau = \frac{kAH}{\left(R^2 - A^2\right)^{\frac{1}{2}}} \tag{1}$$

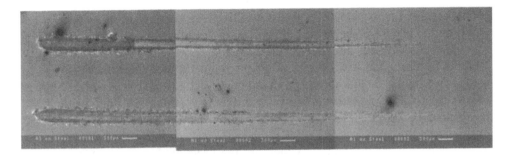

Figure 2. Scanning electron micrographs (backscattered images) of scratch tracks in a 100nm aluminium coating on 304 stainless steel showing the stripping of the coating at the critical load with a sharp change (top scratch) and a more gradual change as the aluminium coating is thinned (lower scratch).

where the radius of the contact $A=(L_c/\pi H)^{1/2}$, L_c is the critical load, τ is the critical shearing stress, H is the hardness of the substrate material and k is a constant varying between 0.2 and 1.0. The critical shear stress increases as the substrate hardness increases which agrees with experiment. This model assumes full plastic deformation (which is only applicable in a limited range of cases) and does not show the influence of coating thickness.

For soft polymeric coatings on harder metallic substrates the shear stress applied to the coating during the scratch test can lead to regions of delamination extending ahead of the stylus. In such cases a fracture mechanics model has been developed to assess adhesion based on the assumption that the stress field around a moving indenter can be given by the Boussinesq solution [14, 15] This is clearly not a complete solution as it does not deal with elastic mismatch at the coating-substrate interface but generates strain energy release rates comparable to those obtained by different adhesion test methods. However, the method requires a knowledge of the area and geometry of delamination which is not always easy to determine if the coating is not transparent and the same mechanism of failure is not often observed for other coating systems. For this reason the model is not widely applicable.

3.2. Hard coatings

The failure modes can be broadly split into three categories:
1. Through-thickness cracking - including tensile cracking behind the indenter [8,16], conformal cracking as the coating is bent into the scratch track [8,16] and Hertzian cracking [8]
2. Coating detachment - including compressive spallation ahead of the indenter [8,16], buckling spallation ahead of the indenter [8] or elastic recovery-induced spallation behind the indenter [8, 17]
3. Chipping within the coating.

The type of failure which is observed for a given coating/substrate system depends on the test load, the coating thickness, the residual stress in the coating and the substrate hardness and interfacial adhesion. Generally the critical load at which a failure mode first occurs, or occurs regularly along the track, is used to assess the coating though there is a distribution of flaws and hence of failures in most cases [6]. Comparisons between different coatings are only valid if the mechanism of failure is the same which requires careful post facto microscopical examination for confirmation.

The adhesion related failures which are the basis of the scratch adhesion test for hard coatings are buckling and spallation [9,10] and are described in more detail in the next section.

4. FAILURE MECHANISMS RELATED TO ADHESION FOR HARD COATINGS

4.1. Buckling

This failure mode is most common for thin coatings (t<10μm) which are able to bend in response to applied stresses. Failure occurs in response to the compressive stresses generated ahead of the moving indenter (Figures 3a and 3b). Localised regions containing interfacial defects allow the coating to buckle in response to the stresses and individual buckles will then spread laterally by the propagation of an interfacial crack. Spallation occurs when through-thickness cracks form in regions of high tensile stress within the coating. Once the buckle has occurred the scratch stylus passes over the failed region crushing the coating into the surface of the scratch crack formed in the substrate. Coating removal can be enhanced at this point or the failure may disappear completely depending on its size and the toughness of the coating.

Buckling failures typically appear as curved cracks extending to the edge of the scratch track or beyond. They are often delineated by considerable coating fragmentation and have major crack planes perpendicular to the coating/substrate interface. In most cases buckles form in the region of plastic pile-up ahead of the moving indenter (Figure 3c). The size of the buckle is typically less than or equal to the extent of pile-up. This would imply that the pile-up process controls the buckle failure mode to a great degree. This explains, to a large extent, the increase in critical load with substrate hardness for titanium nitride tool coatings on steel which is often reported [1] since in such coatings the buckle failure mode dominates. As the steel hardness increases plastic pile-up ahead of the indenter is reduced and the bending stresses induced in the coating by the pile-up are limited. A higher normal load is needed to develop equivalent pile-up and bending stresses and thus the critical load increases. The correlation between buckle diameter and pile-up diameter is very close for alumina scales on the oxide-dispersion strengthened alloy MA956 or TiN coating on stainless steel (Figure 4).

(a)

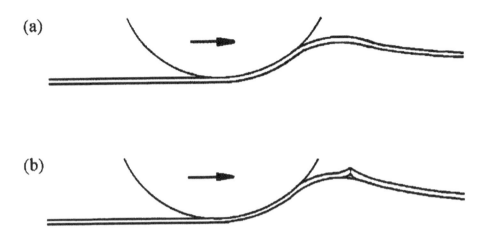

(b)

(c)

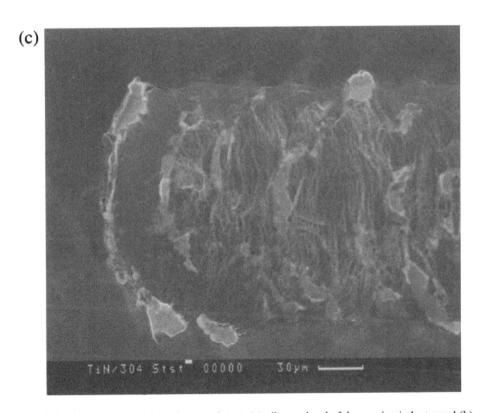

Figure 3. Buckling failure mode in the scratch test; (a) pile-up ahead of the moving indenter and (b) interfacial failure leading to buckling. Through-thickness cracking results in removal of coating material. Scanning electron micrograph (c) of buckle failures in TiN coated stainless steel.

(a)

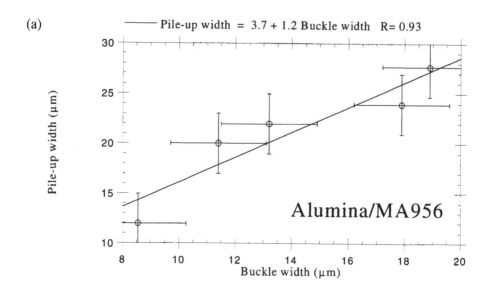

(b)

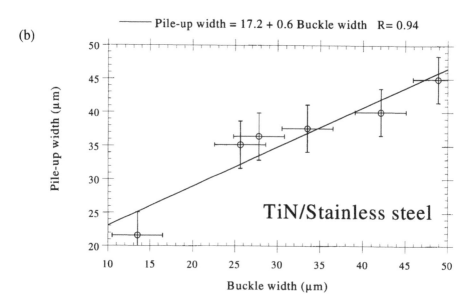

Figure 4. Relationship between buckle width and pile-up width in the scratch test (a) Alumina on MA956 (b) TiN on stainless steel.

For TiN coatings on steel, changes in buckle diameter can be produced by changes in interfacial structure and adhesion but within limits defined by the size of pile-up.

According to Evans [18] the critical buckling stress σ_b is given by:

$$\sigma_b = \frac{1.22E_c}{1-v_c^2}\left(\frac{t}{R}\right)^2 \tag{2}$$

where E_c and v_c are Young's modulus and Poisson's ratio of the coating, t is coating thickness and R is the radius of the buckled region. This predicts that the critical buckle stress increases with coating thickness as is mostly observed. However, this equation assumes a planar interface which is not the case for the buckle failures observed in this study since the buckle is associated with pile-up. A complex stress state is expected which is confirmed by the finite element modelling (see later).

For a curved interface, Strawbridge et al. [19] have shown that a tensile stress, σ_t, is generated normal to the interface by the action of the stress in the plane of the surface, σ_0, and the magnitude of this stress at the interface is given by:

$$\sigma_t = \sigma_0\frac{t}{R_i} \tag{3}$$

where R_i is the radius of curvature of the curved interface and t is the coating thickness. In the scratch test the applied stress (i.e. the sum of any residual stress and the stresses introduced by the scratch stylus) determines σ_0 and, in the case of a coating bent over the pile-up ahead of the moving indenter, R_i represents the radius of curvature of the pile-up. Since the amount of pile-up depends on the hardness of the substrate, the critical load in the scratch test should thus be proportional to R_i and inversely proportional to t. As the hardness of the substrate increases so R_i tends to increase and this behaviour is maintained for a fixed value of t but the critical load is not inversely proportional to thickness. This is due to the fact that R_i is actually a function of t - the extent of pile-up decreases as t increases as mentioned previously. In fact for thin TiN coatings on a range of steels experimental results indicate that R_i is proportional to t^2 which would imply that the critical load is in fact proportional to coating thickness which is close to what is observed. However, much more data are necessary to determine the validity of this observation.

4.2. Wedge spallation

For thicker (>10μm) coatings where bending is less common the buckling failure mode is not observed. In fact the coating can suppress the formation of a narrowly defined pile-up region (Figures 5d and 6) and the stresses ahead of the indenter are less complex. Adhesional failure now occurs by a different mechanism (Figure 5a-c). Initially compressive shear cracks form some distance ahead of the indenter through the thickness of the coating. These propagate to the surface and interface and generally have sloping sides which can act like a wedge. Continued forward motion of the indenter drives the coating up the wedge causing an interfacial crack to propagate. As the extent of interfacial failure increases the wedge lifts the coating further away from the substrate creating bending stresses within it. Large enough displacements will cause a region ahead of the indenter to be detached in response to the tensile bending stresses created. When this happens the scratch diamond can drop into the hole left by removal of the coating (Figures 5d and 6) and there is a dramatic increase in scratch width and scratch depth. Pile-up is then often seen beside the track until the stylus climbs up the wedge and out of the hole. Whereas such large failures are often observed for alumina scales on MA956, much smaller failures are often produced for vapour deposited TiN coatings and it is rare that the stylus drops into the hole left by the spalled coating. In this case the stylus passes over the edge of the spalled region creating considerable microfracture in the coating as it passes.

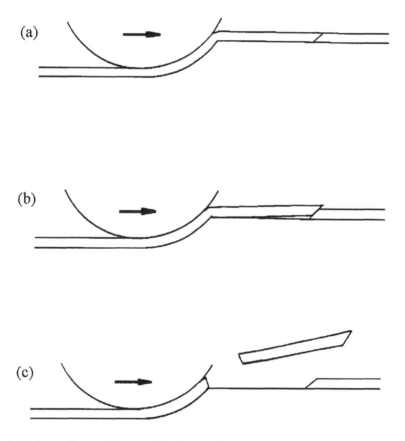

Figure 5. Wedge spallation failure mode in the scratch test; (a) wedge crack forms some way ahead of the moving indenter; (b) continued forward motion of the stylus drives the coating ahead of the indenter forward and it slides up the previously formed wedge crack opening up an interfacial crack; (c) through-thickness cracking close to the indenter leads to spallation. (d) Scanning electron micrograph of a wedge spallation failure in an alumina scale on MA956 oxidised at 1250°C for 100h.

(d)

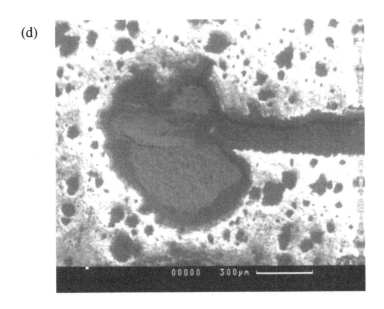

Figure 5. (Continued).

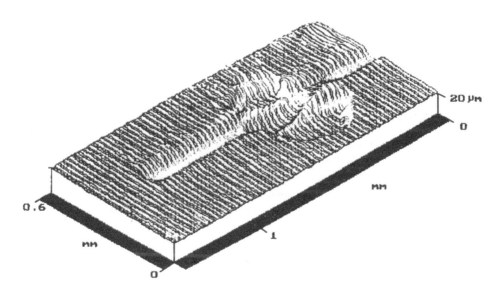

Figure 6. 3D stylus profilometry trace of a wedge spallation failure in 15μm alumina on MA956.

The wedge spallation failure mode depends on two distinct processes occurring [18]. Firstly a compressive shear crack must form and then interfacial detachment occurs. According to Evans [18] the biaxial stress necessary to cause the wedge crack, σ_w, is given by

$$\sigma_w = \left(\frac{4E_c G_f}{(1-v_c)\lambda} \right)^{1/2}$$ (4)

Where G_f is the coating fracture energy and λ is the width of the wedge spalled region. The biaxial stress to produce the spall, σ_{sp}, after shear cracking has occurred is given by

$$\sigma_{sp} = \left(\frac{E_c G_i}{(1-v_c)t} \right)^{1/2}$$ (5)

Where G_i is the strain energy release rate for a crack in the interfacial region (i.e. the interfacial fracture energy) which, in an ideal case, can be equated with the crack surface energy, γ, and hence fundamental adhesion. However, for all real interfaces other energy dissipation mechanisms are likely to be operating (e.g. plasticity, microfracture, heat generation, etc.) so G_i can be taken as a measure of practical adhesion. Since to get a visible wedge spall both the through-thickness and interfacial cracks must be formed, the total failure stress, σ_F, is given by the sum of equations (4) and (5). As Equation 5 has a $1/\sqrt{t}$ dependence the critical load for wedge spallation is expected to decrease as coating thickness increases. There is no requirement that the crack propagates exactly along the interface in this analysis though this is often the case if the adhesion is poor and the interface is sufficiently planar.

5. STRESSES RESPONSIBLE FOR FAILURE

The stresses around a moving indenter sliding across a coating/substrate system are complex and no analytical model exists which fully describes what is observed. Some progress has been made with finite element modelling, particularly in cases where both coating and substrate remain elastic, but this work is a simplification compared to what usually occurs in a real scratch test. A number of improvements to the approach are required and still need to be addressed:-

1. Realistic materials models are required which include the elastic properties of coatings and substrate, their yield and fracture strength and work hardening characteristics.
2. Cracking and the modification of the stress state by the presence of cracking needs to be implemented for hard coatings or substrates.
3. Modification of the stress field by changes in indenter/coating friction requires data for each coating/substrate system of interest.

Many materials properties of coatings are not well known and it seems highly unlikely that sufficiently good materials models will exist for all but a few model systems in the near future. Also, any failure mode which occurs close to the indenter is likely to be in a region where the stress field is changing rapidly and it will be difficult to determine exactly which stresses are responsible for failure. Currently finite element modelling can only be regarded as providing a guide to the stress fields around the moving indenter but can help to identify those failure modes which are generated by simple stress conditions.

6. EXPERIMENTAL

Samples of 304 stainless steel and several alloys which form a protective alumina scale on oxidation (FeCrAlY, MA956, FAL and Kanthal APM (compositions in Table 2)) were cut into 20 x 10 x 2mm sections, polished to a 1μm diamond finish, and cleaned and degreased in isopropyl alcohol prior to use. The stainless steel coupons were coated with TiN by sputter ion plating [20] at a temperature of 500°C and a bias voltage of -35V. Coatings with thicknesses in the range 1 to 15μm were deposited with a 120 nm titanium interlayer to promote adhesion. The alloy samples were isothermally oxidised in air at temperatures between 1150°C and 1300°C for times up to 1400h to produce alumina scales up to 20μm in thickness. The thickness of all scales or coatings was measured by either ball cratering or metallographic cross sections.

Scratch testing was performed using a CSEM manual scratch tester fitted with a Rockwell 'C' diamond (200 μm tip radius). This is a dead-loaded machine where a separate scratch is made for each applied load. A 3mm scratch was made at each load. For the tests reported here scratches were made at 2N intervals starting at 2N. Care was taken to place the scratches sufficiently far apart so that their deformation regions did not overlap. Critical loads for each failure mode were determined by post facto microscopic examination of the scratch tracks. The critical load criterion used was the lowest load at which the failure occurred more than twice along the scratch track. Since the total number of wedge cracks produced was low it was not possible to perform a full Weibull statistics analysis [6].

Table 2.
Composition of alloys and properties of the coatings investigated in this study

Alloy	Composition	Coating	Young's modulus (GPa)	Poisson's ratio
304 stainless	Fe-18Ni-9Cr-1Ti	TiN	450±35	0.28
MA956	Fe-20Cr-5Al-0.4Ti-0.5Y$_2$O$_3$	α-Al$_2$O$_3$	388±51	0.26
Kanthal APM	Fe-21Cr-6Al-0.3Si-0.1Zr	α-Al$_2$O$_3$	356±28	0.26
FeCrAlY	Fe-15Cr-4Al-0.9Y-0.4Ni	α-Al$_2$O$_3$	403±41	0.26
FAL	Fe-25Al-1Zr	α–Al$_2$O$_3$	216±88	0.26

Finite element modelling was used to assess the stress distribution for TiN coated stainless steel. A two-dimensional plane strain model, which models the indentation as a cylinder rather than a sphere, was implemented in the DYNA3D code in order to achieve reasonable run times. The mesh was chosen to be symmetric about the y axis. The coating thickness was set at 2μm and reasonable materials properties were used for both substrate and coating. For the stainless steel substrate plastic deformation and work hardening was allowed (yield stress, σ_y=500MPa; work hardening exponent, n=0.26) whereas for the TiN coating deformation is elastic up to fracture at 500MPa. The indenter/coating friction coefficient was fixed at 0.15. Two models were run for comparison: a static indentation where the maximum vertical indenter displacement was 2μm and a simulated scratch where the indenter was allowed to indent to 2μm and was then moved tangentially 10μm. Given the uncertainties about the difference between cylindrical and spherical indentations, as well as questions about the quality of the materials data, the absolute stress values generated must be questionable. However, the difference between static indentation and scratches will be instructive.

In order to calculate the practical adhesion in terms of the interfacial fracture energies, Young's modulus and Poisson's ratio of the coating are necessary. For all the scales and coatings investigated in this study Young's modulus was determined by nanoindentation testing on coatings that were at least 8μm thick. In order to reduce the scatter in the data the sample surface was polished prior to testing at a maximum load of 10mN; under these test conditions the contribution from the substrate is expected to be minimal. Young's modulus was extracted from the unloading curve by the method of Oliver and Pharr [21]. Quoted values in Table 2 are the average of ten measurements and are similar to what is expected for such coatings except in the case of the alumina scale on FAL which has properties very similar to the substrate. This oxide scale on this alloy is very friable and the measured indentation data are thus likely to be dominated by the substrate response. There was a small variation in Young's modulus of the oxide scales depending on the oxidation temperature but this was not significant in the temperature ranges investigated here. Handbook values have been used for Poisson's ratio in all cases.

7. RESULTS

7.1. Scratch test failure load regimes

For all coatings investigated the critical load for buckle formation increases as the coating thickness increases (Figures 7a and 7b). Wedge spallation does not occur until higher coating thicknesses and the critical load for wedge spallation decreases as thickness increases. This is exactly the same as has been observed previously [9,10] and is broadly in agreement with the theoretical predictions in Section 4.

(a)

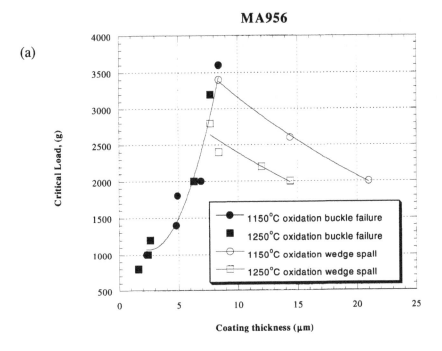

(b)

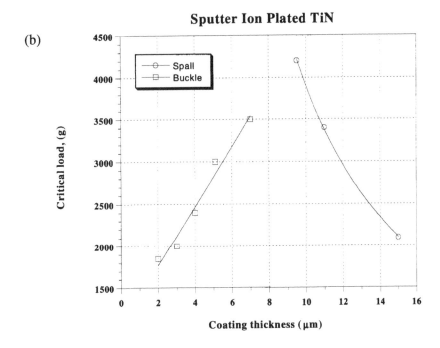

Figure 7. Variation of critical load for wedge or buckle formation as a function of coating thickness for (a) Alumina on MA956 and (b) TiN on stainless steel.

7.2. Finite element results

The main stress components in the coating at the coating/substrate interface have been extracted from the finite element data for both static indentation and scratching and are plotted in Figure 8. σ_{xx} (parallel to the surface) is tensile beneath the indenter in static indentation due to the bending of the coating as the substrate plastically deforms beneath it. At the edge of the contact the bending is in the opposite sense and compressive stresses are observed. This is exacerbated by pile-up. σ_{xx} quickly falls to zero outside the pile-up region. On moving the indenter the compressive stress is increased ahead of the indenter and reduced behind it, probably due to changes in the amount of bending in the coating. Well outside the contact region a compressive stress exists ahead of the indenter which approximates to a state of pure compression. In the region of bending at the edge of the contact the shear stress component τ_{xy} is also significant and the values at the leading edge of the indenter are increased by sliding. The stresses perpendicular to the interface σ_{yy} are compressive in the contact region as expected with tensile stress regions just outside in the bending zone. These tensile stresses are much reduced in the scratch case.

Clearly a very complicated stress state exists in the pile-up region close to the indenter where bending of the coating occurs. This is the region where buckling occurs and it makes the relationship between σ_b and L_c difficult to define. The stress state well ahead of the indenter where wedge cracking occurs is much simpler and a linear relationship between L_c and σ_F is expected for a given coating/substrate/indenter combination. This is amplified in the next section.

7.3. Quantification of failure stresses

The stresses responsible for coating detachment, σ_F, are a combination of the residual stresses remaining in the coating at room temperature, σ_R, and the stresses introduced by the scratch stylus, σ_s. Thus

$$\sigma_F = \sigma_R + \sigma_s \qquad (6)$$

σ_R can be measured for both TiN and alumina coatings by x-ray diffraction using the $\sin^2\psi$ method [22]. Table 3 contains the measurements made in this study.

The stresses induced by the indenter have been determined empirically. The critical load for coating detachment is known to decrease as the residual stress in the coating increases for a wide range of coatings such as TiN [23]. In the case of TiN coatings the residual stress can be increased by increasing the energy or flux of ion bombardment during deposition [24]. Equating the change in scratch test critical load with the difference in measured residual stress enables a calibration factor to be determined. In this study 1g normal load in the scratch test equates to a 2.3 MPa compressive stress ahead of the indenter. For the alumina scales grown on MA956 the residual stress can be changed by altering the oxidation temperature. Plotting the measured residual stress against critical load allows a

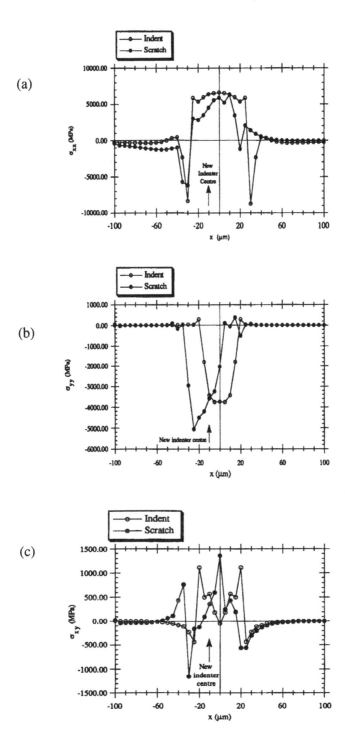

Figure 8. Stress components in the coating next to the coating/substrate interface determined for static indentation and scratching using the Finite Element code DYNA3D (a) σ_{xx}, (b) σ_{yy} and (c) σ_{xy}.

Table 3.

Residual stress at room temperature in the coatings measured by x-ray diffraction ($\sin^2\psi$ method)

Coating/Substrate System		Residual Stress (GPa)	
Al$_2$O$_3$ on MA956		8μm	20μm
Oxidised at	1150°C	3.75 ± 0.1	3.64 ± 0.1
	1200°C	4.04 ± 0.1	3.87 ± 0.1
	1250°C	4.11 ± 0.09	3.91 ± 0.05
	1300°C	4.24 ± 0.1	4.01 ± 0.1
TiN on 304 Stainless Steel		5μm	
	- 30V bias	2.10 ± 0.20	
	- 60V bias	6.03 ± 0.07	

calibration coefficient to be determined from the slope of the graph (Figure 9). Experiments have been performed at two different oxide thicknesses which represent the maximum and minimum values tested in this study. The calibration constant is almost the same in each case and is effectively constant within experimental error. Thus, for this material 1g normal load in the scratch test equates to an average value of 0.4 MPa. In both cases a linear relationship between critical load and stress is assumed which appears reasonable in these systems but further work is needed to determine the validity of the approach. Calculated failure stresses for alumina scales on MA956 and TiN on stainless steel are shown in Figure 10.

The wedge failure stresses (Figure 11) can then be used to determine the work of adhesion by plotting the calculated σ_F against the reciprocal of the square root of coating thickness (Figures 12a and 12b). It is then possible to separate the two components contributing to wedging failure (Equations 4 and 5); the slope of this curve can be used to determine the interfacial fracture energy, G_i, using the coating data in Table 2, whereas the intercept gives a measure of coating fracture strength, σ_w. These values are presented in Table 4.

It is important to determine if failure is really interfacial if the scratch test is to be used for adhesion assessment, so Auger Electron Spectroscopy (AES) was used to identify the locus of failure for both materials. In the case of the alumina scales there is always a thin layer of oxide on the uncovered substrate at the bottom of the wedge-spalled pit. However, this may have been formed after scratch testing due to the exposure of the bare metal substrate to the atmosphere. There is no evidence for substrate material on the underside of any spalled debris that was collected. It is, therefore, reasonably certain that failure occurs at or very near the interface once the wedge crack reaches the interfacial region. Assessing the failure locus of the TiN coating is more complex since a thin (~100nm) titanium interlayer was used to promote coating adhesion which dissolves a considerable amount of carbon and oxygen from the substrate surface in the early stages of

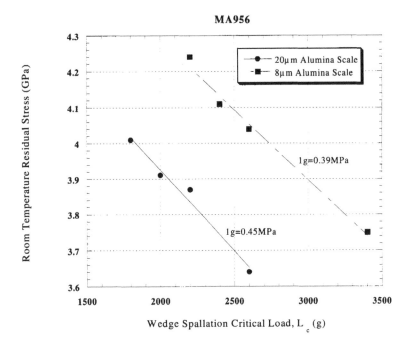

Figure 9. Variation of critical load with residual stress for 8μm and 20μm alumina scales grown on MA956 using a range of oxidation temperatures.

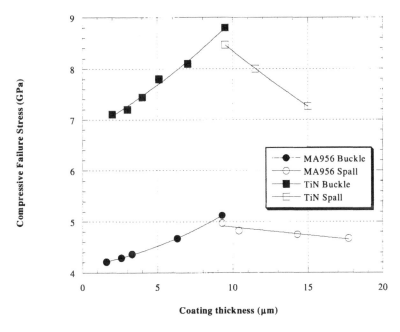

Figure 10. Variation of compressive failure stress with coating thickness for alumina on MA956 and TiN on 304 stainless steel.

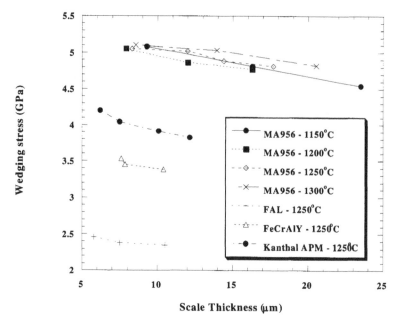

Figure 11. Variation of wedging stress with scale thickness for alumina scales on a range of alloys.

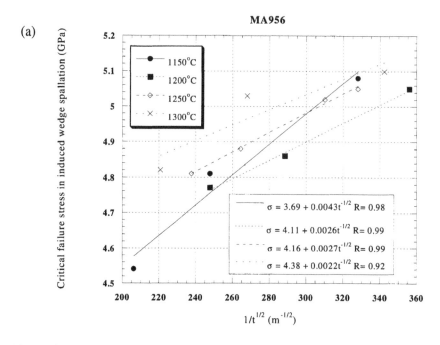

Figure 12. Variation of critical failure stress, σ_F with the reciprocal of the square root of coating thickness, t, for (a) alumina scales on MA956 at different oxidation temperatures, (b) alumina scales grown at 1250°C on a range of alloys and (c) TiN on stainless steel.

(b)

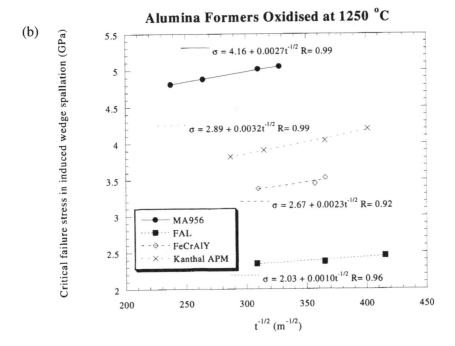

(c)

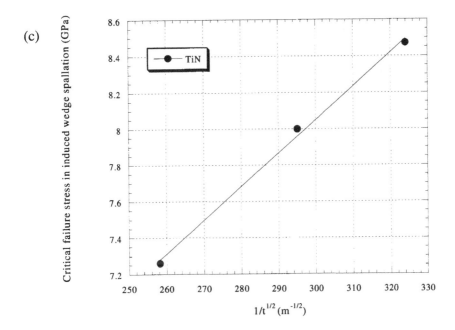

Figure 12. (Continued).

Table 4.
Wedge fracture stress and interfacial fracture energy determined from the scratch test. Unless otherwise indicated all oxide scales were grown at 1250°C

Coating/substrate	Interfacial fracture energy, G_i (J/m^2)	Wedge fracture stress (GPa)
TiN/304	451	2.52
Al_2O_3/MA956		
1150°C	35.0	3.69
1250°C	16.9	4.16
Al_2O_3 /Kanthal APM	21.3	2.9
Al_2O_3/ FeCrAlY	13.4	2.7
Al_2O_3/ FAL	3.3	2.0

deposition [25]. In the sputter ion plating process ion bombardment of the growing coating is used to promote adhesion by forming a pseudodiffusion zone in the interface region giving a metallurgical bond with no well-defined interface plane. By visual inspection using a light microscope the gold-coloured TiN was clearly removed at the bottom of wedge spalls but there was still considerable titanium present on the surface of the substrate. SIMS images of the surface showed that the nitrogen content of this surface layer was very low compared to the carbon and oxygen levels. It thus seems likely that failure has occurred within the titanium interlayer.

For both materials the interfacial fracture energies are higher than that expected from the fracture energy of the coating ($\sim 1J/m^2$) but lower than typical substrate values ($\sim 10^3J/m^2$). This also indicates that the failure crack is propagating at or near the interface with at least some crack tip plasticity occurring within the substrate. As G_i increases the effective coating/substrate adhesion increases so the results here indicate that the TiN/stainless adhesion is better than that for alumina/MA956 since the mechanical properties of the substrates, and hence the energy dissipated in crack-tip plasticity processes, are very similar. Since the TiN coated stainless steel has much smaller spalled regions and the coating is considerably more resistant to detachment during abrasion than the alumina scale the relative values of the fracture energies are as expected.

The interfacial fracture energy for alumina on MA956 is reduced as the oxidation temperature increases (Figure 13) but the fracture strength of the coating is actually increased. During the long exposures necessary to grow thick scales on the alloy at low temperatures void-like defects are known to grow in the scale. These will act as the crack nucleation sites that lead to failure [26]. If the scale has a constant toughness then the more defective low temperature scales would be expected to fail at a lower stress level. The better adhesion of the low temperature scales is more difficult to explain but may be due to different chemistry of the interfacial regions.

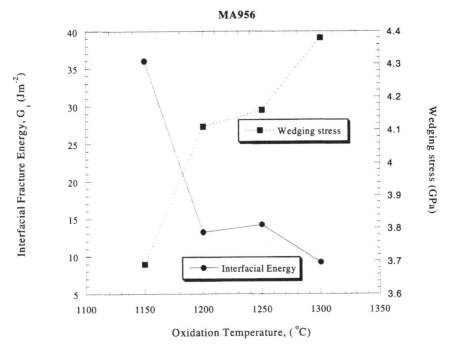

Figure 13. Variation of interfacial fracture energy and wedge fracture strength of alumina scales on MA956 as a function of oxidation temperature.

8. CONCLUSIONS

The scratch test is a good method for quality assurance/quality control testing of the adhesion of hard coatings and is useful in the development of new coatings for process optimisation. However, extracting quantitative adhesion data from the test which might be used as the basis of a performance model is almost impossible from the test data. In most cases, the stresses around the moving indenter are too complicated to be easily predicted and, therefore, the stresses driving coating failure are not known. However, some failure modes occur sufficiently far away from the indenter that the stress states are likely to be more simple and quantification is a possibility.

The two main adhesion related failure modes in the scratch test are wedge spallation and buckling. Buckling occurs for thin coatings which are able to bend in response to applied stresses. The stresses responsible for failure are complex due to the fact that buckling is confined within the region of pile-up close to the indenter. For thicker, stiffer coatings wedge spallation becomes the dominant failure mechanism. This occurs well ahead of the moving indenter and the stresses which are responsible for failure approximate to a state of pure compression. Wedge spallation stresses can, therefore, be quantified by calibration enabling an interfacial fracture energy to be determined.

To derive the maximum benefit from the scratch test better theoretical models for the stress fields associated with a moving indenter in a coating/substrate system are needed. These are most likely to be based on finite element analysis, but the modelling approach would need to include a large number of factors if the true stress state is to be predicted accurately enough. A considerable amount of development and validation work is thus required. The chances are that the scratch adhesion test will never be fully quantitative, but under some circumstances it can deliver useful information.

Acknowledgements

The author would like to thank Ian Gilbert for some SEM images and colleagues at Newcastle University for useful discussions.

REFERENCES

1. A J Perry, *Thin Solid Films*, **107**, 167 (1983).
2. P A Steinmann and H E Hintermann, *J. Vac. Sci. Technol.*, **A3**, 2394 (1985).
3. J Valli, *J. Vac., Sci. Technol.*, **A4**, 3001 (1986).
4. H E Hintermann, *Wear*, **100**, 381 (1984).
5. A J Perry, *Surf. Eng.*, **2**, 183 (1983).
6. S J Bull and D S Rickerby, *Surf. Coat. Technol.*, **42**, 149 (1990).
7. P Benjamin and C Weaver, *Proc. Roy. Soc. Lond.*, *Ser A*, **254**, 177 (1960).
8. S J Bull, *Surf. Coat. Technol.*, **50**, 25 (1991).
9. S J Bull, *Materials at High Temp.*, **13**, 169 (1995).
10. S J Bull, *Tribology Int.*, **30**, 491 (1997).
11. K.L. Mittal, in *Adhesion Measurement of Thin Films, Thick Films and Bulk Coatings*, K.L. Mittal (Ed.), STP No. 640, ASTM Philadelphia, (1978), pp5-17.
12. K.L. Mittal, *Electrocomponent Sci. Technol.*, **3**, 21 (1976).
13. K.L. Mittal, in *Adhesion Measurement of Films and Coatings*, K.L. Mittal (Ed.), VSP, Utrecht, The Netherlands, (1995), pp1-13.
14. S Venkataraman, D L Kohlstedt and W W Gerberich, *J. Mater. Res.*, **7**, 1126 (1992).
15. M D Kriese, D A Boismier, N R Moody and W W Gerberich, *Eng. Fracture Mechanics*, **61**, 1 (1998).
16. P J Burnett and D S Rickerby, *Thin Solid Films*, **154**, 403 (1987).
17. R D Arnell, *Surf. Coat. Technol.*, **43/44**, 674 (1990).
18. H E Evans, *Materials at High Temperature*, **12**, 219 (1994).
19. A Strawbridge, H E Evans and C.B. Ponton, *Mater. Sci. Forum*, **251/252**, 365 (1997).
20. D S Rickerby and R B Newbury, *Vacuum*, **38**, 161 (1988).
21. W C Oliver and G M Pharr, *J. Mater. Res.*, **7**, 1564 (1992).
22. D S Rickerby, A M Jones and B A Bellamy, *Surf. Coat. Technol.*, **37**, 111 (1989).
23. S J Bull, D S Rickerby, A Matthews, A Leyand, A R Pace and J Valli, *Surf. Coat. Technol.*, **36**, 503 (1988).
24. D S Rickerby and S J Bull, *Surf. Coat. Technol.*, **39/40**, 315 (1989).
25. P R Chalker, S J Bull, C F Ayres and D S Rickerby, *Mater. Sci. Eng.*, **A139**, 71 (1991).
26. J.P. Wilber, J.R. Nicholls, and M.J. Bennet, in *Microscopy of Oxidation 3*, S.B. Newcomb and J.A. Little (Eds), Institute of Materials Book 675, London, 1997, p207.
27. S J Bull, D S Rickerby, A Matthews, A R Pace and A Leyand, in *Plasma Surface Engineering Volume 2*, E. Broszeit, W D Munz, H Oechsner, K-T Rie and G K Wolf (Eds.), DGM Informationsgesellschaft, Oberursel, 1989, p1227.
28. P A Steinmann, Y Tardy and H E Hintermann, *Thin Solid Films*, **154**, 333 (1987).

Adhesion Measurement of Films and Coatings, Vol. 2, pp. 131–140
Ed. K.L. Mittal
© VSP 2001

Characterisation of thin film adhesion with the Nano-Scratch Tester (NST)

JAMES D. HOLBERY and RICHARD CONSIGLIO*

CSEM Instruments SA, Jaquet-Droz 1, CH-2007 Neuchâtel, Switzerland

Abstract—As the thickness of functional coatings decrease to satisfy structural and protective needs in thin film applications, quantitative instrumentation has become a necessity for adequate evaluation of system properties, particularly scratch resistance and adhesion at the film-substrate interface. The Nano-Scratch Tester (NST) is a new instrument overcoming the limitations of both the classical stylus scratch test (normal force range) and the scanning force microscope (SFM) techniques (short sliding distances) allowing for scratch lengths of up to 10 mm. Tangential force and penetration depths are simultaneously measured during the scratch process. To assist in the inspection of the deformed or damaged area, a scanning probe microscope may be integrated into the system. Experimental results are presented for a range of thin film adhesion applications including clear coatings and thin, multi-layer magnetic laminates. The results indicate very good reproducibility and confirm the application of this instrument to accurately characterization adhesion, elasticity, and mechanical integrity in coated systems where the film thickness is less than 1 μm.

Keywords: Adhesion; magnetic layers; polymer coatings; scratch test.

1. INTRODUCTION

Scratch testing represents a test form where some degree of controlled abrasive wear or adhesion between one material medium to another can be quantified under controlled conditions. In conventional scratch testing a diamond stylus is drawn across a sample under either a constant or progressively increasing normal load. The minimum or critical load point at which the onset of coating delamination occurs is used as a semi-quantitative measure of practical adhesion. The scratch test on a coated sample may be analyzed in terms of three contributions: (1) a ploughing contribution that depends on the indentation stress field and the effective flow stress in the surface region, (2) an interfacial friction contribution due to interactions at the indenter-sample interface, and (3) an internal stress contribution as any internal stress will oppose the passage of the indenter through the surface, thereby effectively modifying the surface flow stress [1-3].

---

* To whom correspondence should be addressed. Phone: 41-32-720-5560, Fax: 41-32-720-5730, E-mail: richard.consiglio@csem.ch

The critical failure point is occasionally difficult to distinguish by one method alone and thus is typically determined by one or more of the following techniques: (1) optical microscopy during or after the test [4,5], (2) chemical analysis at the bottom of the scratch channel (with electron microprobes), (3) acoustic emission [6], (4) frictional forces [7], and (5) Atomic Force Microscopy in post-testing analysis [8].

Recent advances in the development of sub-micrometer films and coatings has stimulated efforts to produce analytical test instruments with the ability to resolve adhesion, material response, and wear problems at the nanometer scale [9]. Specifically, several industrial and research applications such as thin magnetic layers, clear polymer coatings, thin diamond-like coatings (DLC's), etc., require an instrument that combines high resolution and the versatility to be applied to a wide variety of materials science problems [10,11].

In this paper, we outline the Nano-Scratch Tester (NST) designed to work within the normal force range of 10 μN and 1 N by utilizing a stylus mounted double-cantilever beam design (Fig. 1) of varying stiffness. This design reduces the torsional effect during the scratch test, increases the accuracy by maintaining the indenter in a vertical plane during measurement, and allows for the applied load ranges of 10 mN, 100 mN and 1 N with respective resolutions of 0.15, 1.5 and 15 μN (Table 1).

2. METHODS

The NST performs the scratch test on a sample mounted on a friction table that permits frictional variations to be characterised during the test. A precision motorized stage controls the X and Y-axes to allow the sample to be positioned with a lateral resolution of 1 μm that is critical for registration of a particular feature within a sample (Fig. 1). The load is applied by controlling the position of the NST Head using a displacement actuator.

A combination displacement actuator and displacement sensor, D_Z, monitors the surface profile of the sample under a constant load and controls the vertical displacement of the stylus. This technique can be utilized to measure the penetration depth (measured as the indenter displacement) during a scratching operation and to determine the viscoelastic response (residual depth) in the sample upon test completion. The force sensor, F_N, measures the deflection of the double cantilever beam and consists of a linear voltage differential transformer (LVDT). An adjustable feedback system allows the applied load to be maintained at the desired level. The specifications of the NST are summarized in Table 1.

Several loading modes are available: constant normal force mode and progressive normal force modes applied either as a single pass or a multi-pass scratch. In each case, prior to scratching, an initial surface profile of the sample is measured (pre-scan of the surface under a very low constant load). This data set is referred to as the "Profile" and provides the reference from which the penetration depth

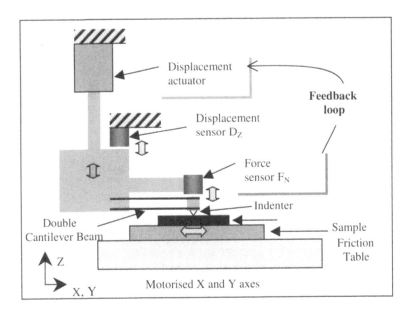

Figure 1. The principle of the Nano-Scratch Tester showing the double cantilever beam and independent force (F_N) and displacement (D_Z) sensors.

Table 1.
Specifications of the Nano-Scratch Test Instrument

Load	Standard Load	High Load	High Resolution
Maximum Load	100 mN	1 N	10 mN
Minimum Load	100 mN	1 mN	10 mN
Resolution	1.5 µN	15 µN	0.15 µN
Depth		Range A	Range B
	Max. Displacement	100 µm	1 mm
	Resolution	1.5 nm	15 nm
Friction		Range 1	Range 2
	Maximum Load	1 N	0.2 N
	Resolution	30 µN	6 µN
Loading Rate	Up to 5 N/min		
Scratch Speed	0.2 to 20.0 mm/min.		

(P_D) can be accurately calculated during scratching (Fig. 2). The residual depth (R_D) can also be measured (post-scan of the surface under a very low load) directly after a scratch test has been performed. In this mode, the NST serves as an accurate surface profilometer. In addition, the system can record the frictional force during scratching which allows one to compute the coefficient of friction

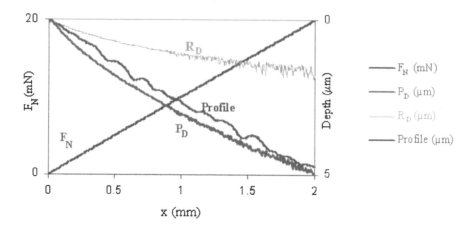

Figure 2. A sample output of a 2 mm NST test at a constantly increasing load from 0 to 20 mN (F_N), the sample Profile, the penetration depth (P_D), and the residual depth (R_D).

between the diamond tip and the tested material. For example, to concentrate the stress in thin hard films, smaller radii tips are typically utilized. In viscoelastic coatings, depending on the objectives of the investigator, larger radii tips are generally used for bulk properties or single layer coatings (Fig. 3 depicts a 10 μm diamond spherical indenter) and sharp tips are utilized for depth-specific properties in multi-layer polymer films.

The NST incorporates a standard optical microscope, but may be integrated with an SFM objective to provide additional capabilities [12]. This allows high resolution imaging of a scratched area in three dimensions and quantitative lateral and depth measurements. In addition, the surface morphology and roughness can be investigated before scratching.

3. RESULTS

3.1. Diamond-like carbon (DLC) deposited on tungsten

In the first example of the instrument capabilities we examined a thin hard coating of diamond-like carbon deposited on a tungsten substrate. The coating layer, nominally 100 nm, is loaded at a rate of 20 mN/minute at a scratch speed of 2 mm/minute. A diamond indenter with a radius of 2 μm has been used to create an overall scratch length of 1 mm with an initial scanning load of 0.05 mN (Fig. 4A). The friction sensor has been incorporated into the instrument and indicates an increase in events at the 4.2-mN-load level (F_F), the level that is considered the first point of critical failure. Additionally, the penetration depth (P_D) and residual depth (R_D) traces indicate that the primary failure has occurred at this level. The

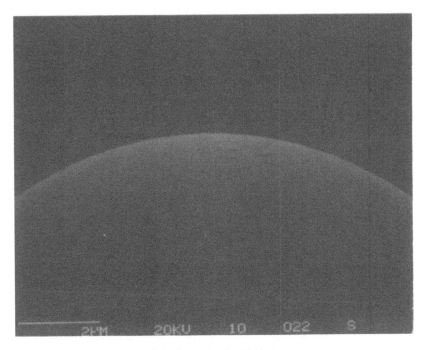

Figure 3. SEM image of a 10 μm radius diamond spherical indenter.

penetration depth is at approximately 300 nm at the point of critical load while a post-failure scan indicates the residual depth to be quite near to 200 nm. This indicates that the plastic deformation in the tungsten substrate was approximately 100 nm.

The optical micrograph (Fig. 4B) indicates two positions (as depicted by the white solid and white dashed lines): the first position the initial point of failure (solid line) and the second an arbitrary point within the failed region (dashed white line). The point within the failed region has been scanned using the objective-mounted SFM (Fig. 4C) that clearly indicates that non-uniform pile-up has occurred on both sides of the failed region.

3.2. Multi-layer magnetic films coated with DLC

Very thin hard films of amorphous carbon or DLC are widely used as protective overcoats for hard disks. In an effort to better understand the material behavior of these composite systems, several research groups have explored work of adhesion [13] and friction and wear properties [14]. Nanoscratch testing has been carried out on a typical hard disk sample made up of different multi-layers: these films consist of a 10-20 nm carbon overcoat covering a magnetic multi-layer resulting in a total thickness of approximately 120 nm (Fig. 5 left). In order to generate a maximum stress field at the multi-layer/substrate interface, a diamond was chosen with a small tip radius (2 μm) [15-18].

J.D. Holbery and R. Consiglio

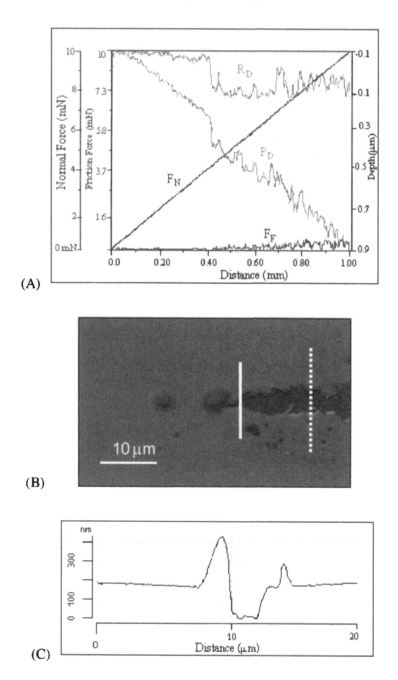

(A)

(B)

(C)

Figure 4. Scratch results on a thin, hard coating. The instrument (A) indicates load in the normal direction (F_N), frictional force (F_F), penetration depth (P_D), and residual depth (R_D). The critical load, as determined at the solid white line depicted in the optical micrograph (B), is 4.2 mN. A cross-sectional analysis using SFM at the location indicated by the white dashed line is depicted in the lower right hand view (C).

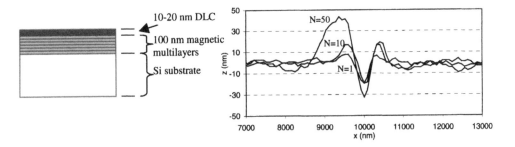

Figure 5. A schematic of a very thin hard film of amorphous carbon (or diamond-like carbon, DLC) on approximately 100 nm of magnetic multilayers supported by a silicon substrate is depicted on the left. SFM traces of the surface as a function of the number of fatigue passes are shown on the right. At 70 traces, the delamination was too great for this scale and is addressed in Figure 6.

Progressive normal load scratches were first performed at a sliding speed of 3 mm/min. For a 0 – 10 mN load range no evident damage was detected from optical microscopy at the end of the track. The loading rate was subsequently increased such that the load range was 0-40 mN and again no damage was evident.

In order to study the friction fatigue of this sample, a multi-pass, dual-direction constant load cycle was applied. By successively increasing the number of cycles, N, under a low load, fatigue failure assessment is possible. The first applied load used was 0.5 mN. Several scratches were made with N varying from 1 to 500 cycles, but no resulting track was detected during optical observations. A second set of measurements was carried out with an applied load of 2.0 mN. Scratches were made with N varying from 1 to 70.

Figure 5 (right) provides a series of SFM profiles of each scratch track after a specific number of cycles up to N = 50. We clearly see the evolution of deformation thought to be due to an increase in film layer strain with increasing N cycles. The beginning of the damage process appears at N=10 and rapidly increases until damage by delamination is evident at N = 50 implying that protective residual stresses have not developed in the near-surface layers [19-21]. At N = 70, blistering caused by delamination was too great for the scale depicted.

To verify the interfacial character of the damage, we chose to remove the blistered damaged area caused at N=70 passes. An additional purpose of this operation was to obtain information about multi-layer plastic deformation below the DLC coating. Transversal scratches were performed with the Y-translation table, the same tip and an applied load of 1.0 mN (4 scratches with 2-μm spacing). The result, shown in Fig. 6 (top), indicates coating failure and chip evacuation has occurred. SFM analysis of the transverse profile allows for an accurate study of the damage process; the SFM cross-sectional profile of the delaminated area confirms that the adhesion failure is located at the interface between the substrate and the first coating at approximately 120 nm, a result confirmed by comparing the known thickness of the multi-layer.

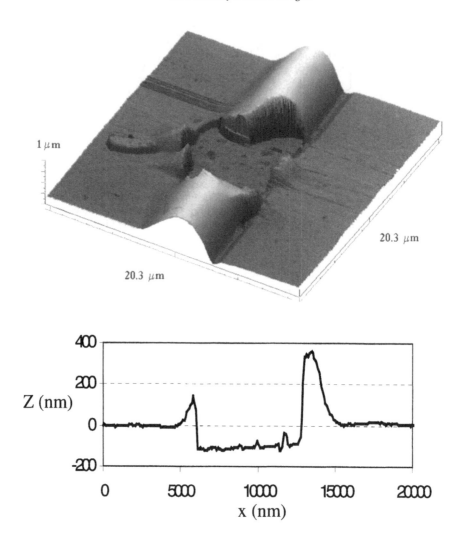

Figure 6. Detailed analysis of a multilayer coating on a thin hard film of amorphous carbon subjected to 70 multi-pass fatigue cycles. A 3-D AFM image of the region is shown on the top. Note the film has blistered in the direction of the scratch series; a subsequent transverse scratch has removed a portion of the blister to reveal the delaminated layer. The SFM trace on the bottom indicates that the film had delaminated at a depth of approximately 120 nm.

3.3. Comparison of TiO₂ coated substrates

A comparison of two TiO_2 coated glass substrates was made by performing a progressive normal load scratch of 0-50 mN with a 10 µm diamond spherical indenter to evaluate the effect of surface roughness on each sample. Thin coatings (< 100 nm) subjected to scratch damage fail by a combination of delamination and spallation [22-24]. These types of scratches offer the opportunity to study the adhesion between coatings that exhibit different friction coefficients under a progressive loading condition [11].

In our study, we compared two samples qualitatively by viewing the delamination area after scratching at 2.5 mm/min. over a length of 2.5 mm. As can be seen in Fig. 7, the area with high surface roughness exhibited a delamination area approximately one-third that of the sample with a low roughness indicating that the load is more highly concentrated with lower friction. While others have examined in detail the relationship between the stress field under loading and the bonding mechanism between film and substrate [3], in this case a strong factor influencing delamination area is the external surface friction coefficient and the resulting stress distribution under the 10 μm diamond spherical indenter. We acknowledge that in order to fully understand the mechanics of failure in this case further quantification of the experimental factors is necessary including interface characterization, surface friction factor determination, and a detailed stress analysis. However, as is common for industrial quality control evaluation, the scratch test for these coatings has been shown to be a very useful technique for direct comparison of processing and application variations.

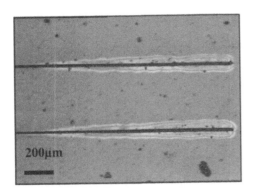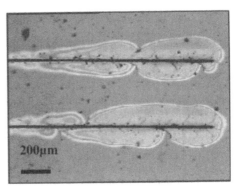

Figure 7. A comparison of two TiO_2 coated glass substrates subjected to a progressive normal load scratch of 0-50 mN with a 10 μm diamond indenter. The scratch speed is 2.5 mm/minute at a length of 2.5 mm. The sample on the left was classified as high roughness while the one on the right is classified as exhibiting low roughness.

4. CONCLUSIONS

We have attempted to illustrate the application versatility of scratch testing with the Double Cantilever Beam feature of the NST under well-adapted contact conditions. In this process, we have shown the following:

1. A diamond-like carbon layer of 100 nm deposited on a tungsten substrate loaded with a 2-μm diamond indenter at a rate of 20 mN/min. indicates a

critical failure of 4.2 mN. A penetration depth was measured at approximately 300 nm at the point of critical load and the residual depth was measured at 200 nm indicating plastic deformation in the tungsten substrate to be 100 nm.

2. Multi-pass 2.0 mN load tests conducted with a 2 μm diamond tip radius on multi-layer magnetic films of 100 nm with a 10-20 nm carbon overcoat resulted in deformation at 10 cycles and delamination at 50 cycles. Removal of the blistered area resulting from 70 loading cycles indicates that failure occurred between the substrate and first magnetic layer at a depth of 120 nm.

3. It has been shown that comparisons between glass substrates coated with 100 nm of TiO_2 of different external surface roughnesses may be characterized using a 10-μm diamond spherical indenter under normal loading conditions. The friction coefficient between the diamond tip and substrate heavily influences the delamination area, presumably due to the distribution of stress under the spherical tip, although further study of this problem is acknowledged to be necessary.

REFERENCES

1. B. Bhushan in: *Handbook of Micro/Nanotribology*, B. Bhushan (Ed.), pp. 379-380 CRC Press, Boca Raton, FL (1995).
2. S.J. Bull and D.S. Rickerby, *Surf. Coatings Technol.* **42**, 149-164 (1990).
3. P.J. Burnett and D.S. Rickerby, *Thin Solid Films* **154**, 403-416 (1987).
4. A.J. Perry, *Thin Solid Films* **78**, 77-93 (1981).
5. A.J. Perry, *Thin Solid Films* **197**, 167-180 (1983).
6. P.A. Steinmann, Y. Tardy, and H.E. Hintermann, *Thin Solid Films* **154**, 183 (1987).
7. J. Valli, *J. Vac. Sci. Technol.* A **4**, 3007-3014 (1986).
8. R. Consiglio, N.X. Randall, B. Bellaton, and J. von Stebut, *Thin Solid Films* **332**, 151-156 (1998).
9. S.V. Hainsworth, S.J. Bull and T.F. Page, *Mater. Res. Soc. Symp. Proc.* **522**, 433-438 (1998).
10. T.F. Wu, *J. Mater. Res.* **6**, 407-426 (1990).
11. S.L. Zhang, A.H. Tsou, and J.C.M. Li, *Mater. Res. Soc. Symp. Proc.* **522**, 371-376 (1998).
12. N.X. Randall, C. Julia-Schmultz, J.M. Soro, J. von Stebut, and G. Zacharie, *Thin Solid Films* **308/309**, 297 (1997).
13. H. Deng, T.W. Scharf, and J.A. Barnard, *J. App. Phys.* **81**, 5396-5398 (1997).
14. J.G. Xu, K. Kato, and T. Nishida, *Wear* **203**, 642-647 (1997).
15. G.M. Hamilton and L.E. Goodman, *J. Appl. Mech.* **88**, 371-376 (1966).
16. R. Consiglio, R. Kouitat Njiwa, and J. von Stebut, *Surface Coatings Technol.* **102**, 138-147, (1998).
17. R. Consiglio, R. Kouitat Njiwa, and J. von Stebut, *Surface Coatings Technol.* **102**, 148-153, (1998).
18. D.S. Stone, *J Mater. Res.* **13**, 3207-3213 (1998).
19. S.K. Wong and A. Kapoor, *Tribol. Int* **29**, 695-702 (1996).
20. S.K. Wong, A. Kapoor, and J.A. Williams, *Wear* **203-204**, 162-170 (1997).
21. J.A. Williams, I.N. Dyson, and A. Kapoor, *J. Mater. Res.* **14**, 1-12 (1999).
22. P. Benjamin and C. Weaver, *Proc. R. Soc. London* A **254**, 163 (1960).
23. P.J. Burnett and D.S. Rickerby, *Thin Solid Films* **157**, 195-222 (1988).
24. M.T. Laugier, *Thin Solid Films* **117**, 243 (1984).

Adhesion Measurement of Films and Coatings, Vol. 2, pp. 141–157
Ed. K.L. Mittal
© VSP 2001

Scratch adhesion testing of nanophase diamond coatings on industrial substrates

F. DAVANLOO,[1,*] C.B. COLLINS[1] and K.J. KOIVUSAARI[2]

[1] *Center for Quantum Electronics, University of Texas at Dallas, P.O. Box 830688, Richardson, TX 75083-0688, USA.*
[2] *Microelectronics and Material Physics Laboratory and Electronic Materials, Packaging and Reliability Techniques (EMPART) Research Group of Infotech Oulu, Department of Electrical Engineering, University of Oulu, PL 4500, FIN-90401 Oulu, Finland*

Abstract—Nanophase diamond coatings are deposited in vacuum onto almost any substrate by a laser ablation method which provides a highly processed plasma containing multiply-charged keV carbon ions. The high energies of condensation of these ions produce interfacial layers between the coating and substrate materials, which results in levels of adhesion that allow the protection of substrates subjected to harsh environmental conditions. In this work, a commercially available scratch tester was used to test adhesion of nanophase diamond coatings on steel and carbide substrates. A data analysis method was introduced to interpret the test measurements and assess the adhesion of coatings on these important industrial substrates.

Keywords: Scratch adhesion testing; nanophase diamond coating; laser ablation; amorphic diamond.

1. INTRODUCTION

Steel and carbide materials are widely used in a variety of practical applications where moving assemblies of tools and machine components are involved. They are often subjected to environments characterized by accelerated wear that result from the sliding and rolling of such components. Therefore, smooth protective coatings well-bonded to steel and carbide surfaces are desirable in such applications. These protective coatings should enhance mechanical performance and increase lifetime in abrasive conditions. The use of diamond coatings as protective coatings for steel materials has been suggested and attempted for practical applications in many laboratories [1-3].

Deposition of diamond coatings onto steel substrates is technologically important. However, the nucleation of a diamond phase is extremely difficult on steel materials which do not show affinity for carbon species [1]. Moreover, these im-

---

* To whom correspondence should be addressed. Phone: (972)883-2863, Fax: (972)690-1167, E-mail: fdavan@utdallas.edu

portant substrates contain carbon dissolving elements which consume carbon species needed for diamond phase nucleation [1].

Without any intermediate layers, the polycrystalline diamond coatings prepared by chemical vapor deposition (CVD) techniques have shown insufficient adhesion to steel substrates [2]. In contrast to such difficulties encountered with conventional technology, the high energy of condensation available from our laser plasma source provides for the direct bonding of nanophase diamond coatings to steel and carbide substrates. In an earlier report we detailed the adhesion and mechanical properties of nanophase diamond deposited directly on stainless steel substrates [3]. Analysis of the interface by Rutherford backscattering spectrometry (RBS) and transmission electron microscopy (TEM) indicated there was a significant intermixing of the substrate and the coating atoms [3]. The protection afforded by the nanophase diamond under harsh environmental conditions of low and high impact particle erosion was studied and it was demonstrated that mechanical performances and the lifetime against abrasion were strongly enhanced by the bonding of nanophase diamond to stainless steel surfaces [3].

The adhesion of the coatings has been the most important factor in the successes achieved in an application involving a coated sample. To quantitatively address adhesion, a diagnostic technique should measure a fundamental property that is intrinsic to the coating /substrate system and independent of the test apparatus and procedure. Many methods for adhesion measurement of thin films have been detailed [4]. However, there is no adhesion test available which can be generally used in all cases [5]. Measurements are also complicated by the complex nature of very hard materials such as nanophase diamond. Since all adhesion measurement methods including the scratch test are dependent on a wide variety of variables, they measure the practical adhesion [6-8]. The measured value for the practical adhesion is not a fundamental quantity, but it is useful for comparison purposes.

The adhesion of a thin coatings depends on the physical and chemical bonding across the film/substrate interfacial area and the intrinsic film stress. It is a macroscopic quality which is generally characterized by physical removal of the film from the substrate. Many investigations have diagnosed interfacial strengths in macroscopic experiments [9-12]. In most of the reported cases part of the interface is subjected to high stresses in an inhomogeneous deformation field. To avoid factors of uncertainty in the measurements due to the local deformation problems, the loading mechanism in the test equipment should create a uniform stress field in the vicinity of the interface to ensure the failure of the weakest link in the film/interface/substrate assembly [13].

The scratch test appears to be one of a few available practical methods to examine the adhesion of superhard coatings such as nanophase diamond. In a scratch test, a diamond stylus is drawn over the sample surface under a continuously increasing normal force until the film detaches from the substrate. The load on the indenter at which the coating is stripped cleanly from the substrate is defined as the critical load, L_c. It has been used as a qualitative measure of the practical adhesion strength of the film.

In the past, we have used several quantitative methods to assess and measure the adhesion of our films to a variety of substrates [14,15]. Parameters such as the bonding of the films has been examined by RBS and nanoindentation techniques. Resistances to wear and debonding have been estimated with a modified sand blaster, a tumbler machine and a rain erosion facility which simulate erosive environment [14]. The objective of our recent [15] and present research is to study the adhesion of nanophase diamond coating to materials such as steel and carbides which are subjected to high-impact wear as encountered in cutting and drilling. In this work we continued to study the adhesion properties of nanophase diamond coatings on these important industrial substrates by a scratch method.

2. NANOPHASE DIAMOND COATING DEPOSITION AND PROPERTIES – REVIEW

Nanophase diamond coatings have been produced by accelerating and quenching an intense laser plasma of C^{3+} and C^{4+} onto a cold substrate [16-22]. The diamond characteristics of this material have been evaluated by several analytical methods. Measurements agree in supporting sp^3 contents of higher than 75% [17-19]. The coatings are deposited without catalysts and without columnar habits of growth and the diamondlike properties of the coatings increase with the energy of condensation. Nanophase diamond seems to be a unique product of energetic condensation from C^{3+} and C^{4+} ions produced in a laser plasma. The original samples of this material were called amorphous ceramic diamond, an appellation shortened to "amorphic diamond" for convenience [15-21]. However, the subsequent improvements have succeeded in bringing the hardness of such films above 78 GPa and the term "nanophase diamond" seems more appropriate [23,24]. The importance of this nanophase diamond material has been suggested by reports of its unique mechanical properties [19,20]. It was shown that a combination of low internal stress and high bonding strength produced coatings with exceptional resistance to wear and erosion. These properties, together with the room temperature growth environment, make this material suitable for use as a protective coating in current industrial applications. The deposition of nanophase diamond by a laser plasma discharge has been realized with a Q-switched Nd:YAG laser at the University of Texas at Dallas (UTD). As described previously, the laser delivers 250-1400 mJ/ pulse to a graphite feedstock in a UHV system at a repetition rate of 10 Hz [15-21]. The beam is focused to a diameter chosen to maintain the intensity on the target near 5×10^{11} W cm^{-2} and the graphite is moved so that each ablation occurs from a new surface. A high current discharge confined to the path of the laser-ignited plasma is used to heat and process the ion flux further. Discharge current densities typically reach 10^5 - 10^6 A cm^{-2} through the area of the laser focus, but the laser power alone is sufficient to insure that the resulting plasma is fully ionized [22]. A planetary drive system for rotating the substrates within the core of the plasma where they are exposed only to ions insures the simultaneous deposition of uniform layers of nanophase diamond over several substrate disks [15-21]. Figure 1 shows a schematic representation of the deposition system. Nominal coatings with 75% sp^3 contents are produced with substrate holders 30 mm in diameter rotated in the core of the plasma at $\alpha = 80°$, where α is the deposition angle as shown schematically in Fig. 1 [15-22]. Despite the relatively high ion fluxes, bulk temperatures of the substrates monitored by a thermocouple do not exceed 35°C over deposition periods of several hours.

Modeling studies have shown that at a laser intensity around 5 x 10^{11} Wcm^{-2} the plasma is composed of multiply charged carbon ions with kinetic energies of the order of 1 keV [22]. The impact of the laser plasma upon a substrate is equivalent to an irradiation with a very high fluence ion beam. Quenching of such energetic ions yields diamond while the condensation of neutral carbon produces only graphite [17,18]. Recently, different types of laser plasma diamonds have been identified and their properties have been found to depend upon the energies of the ions condensed from the laser plasma [18,21,25]. Comparative microstructural studies of films condensed from ions passing through the core and periphery

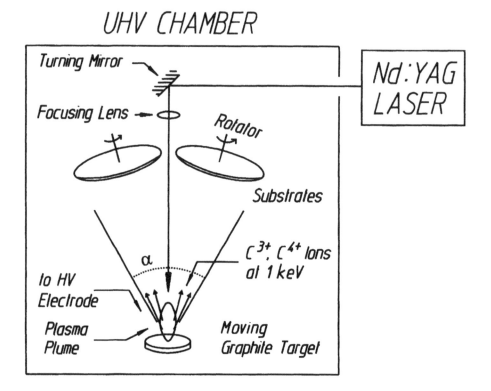

Figure 1. Schematic representation of the laser plasma discharge source used in this work to prepare nanophase diamond coatings.

of laser plasma have been performed extensively by scanning tunneling microscopy (STM). The unique nodular structure of nanophase diamond films deposited by a core of laser plasma has been also reported [18,21] from examinations of films with STM. Typically, the nodule size ranges from 10 to 50 nm in diameter and imparts the properties of diamond found in the finished films [19,20]. In contrast to the very rough surfaces developed by the CVD deposition of polycrystalline diamond films which require vigorous postprocessing, the nanophase diamond coating surfaces are smooth to the extent that the surface roughness is less than 5% of the film thickness [18,19].

3. SCRATCH ADHESION TESTING

3.1. Experimental procedure

In this work, scratch adhesion testing was performed with a Sebastian Five A tester manufactured by Quad Group of Spokane, WA, utilizing a stylometer module. In a majority of tests performed, a diamond stylus with a conical angle of 90° and a spherical tip radius of 300 μm was employed. The stylus was impressed on the coating surface with a progressively increasing normal force at a constant rate as the test sample was moved laterally also at a constant rate of travel. In all the tests performed in this work, the force rate change and the moving speed of stylus were set to 75 N/cm and 0.02 cm/sec, respectively. An acoustic transducer crystal installed on the top of the stylus mount and a force sensor mounted onto a force transducer assembly detect the acoustic noise and transverse force, respectively. The acoustic pick-up was effective over the frequency range of 4,000 to 40,000 Hz. The stylometer is a software driven device and for each test the values of normal scratch load, transverse load, effective friction and acoustic noise were recorded as a function of the scratch length.

It is commonly known that the condition of the diamond indenter can have a large effect on the scratch test results. If the stylus tip is excessively damaged or worn, the stress field created by the scratch testing may change, thus affecting the critical loads obtained [26-28]. Therefore, the scratch stylus should be inspected regularly and replaced if needed.

In this work, the indenter tip was periodically controlled by both the critical load measurements on standard samples and by the SEM examinations. Initially, we performed several scratches on a standard test sample. Results were found to be reproducible. We continued the studies with at least two scratches on each test sample examined. The standard sample was revisited for testing a few times during this work. For a total of fifty scratch tests performed, results were found reproducible within a combined uncertainty of 5%. Examination of the indenter tip by the SEM after completing the scratch tests showed no excessive wear or damage.

Test samples were prepared from coatings of 1 - 5 μm thick nanophase diamond which were bonded to stainless steel and cemented carbides. Surface preparation of stainless steel substrates involved three steps. First, a surface layer was slightly removed by a lapping and polishing with 1 μm calcined aluminum oxide powder. This was then followed by a second step comprised of ultrasonic agitation in ultrapure 1 : 1 trichlorotrifluoroethane and methanol solution for 60 min. Finally in the third step Ar plasma cleaning for 30 minutes was used. Carbide inserts were prepared only by ultrasonic agitation and Ar plasma cleaning.

3.2. Data analysis and interpretation

During each scratch test, the stylometer produces plots of the normal force applied to the stylus, the transverse drag force created, and the acoustic energy released by the scratch process. Figure 2 presents these plots for a scratch test performed on a sample of 3.12 μm thick nanophase diamond coating deposited on a 304 stainless steel substrate. A plot of the effective friction between the stylus and the coating which is simply the ratio of the transverse force to the normal load was also available but is not shown in this figure. Identifying specific loads corresponding to particular coating-substrate failure processes is the basis of the scratch-data interpretation. Depending on the coating system to be tested, these critical loads can be detected by examining the acoustic emission signal, the transverse frictional force plot and normal scratch load data.

A dramatic increase in acoustic emission represents the initiation of microcracks which can be identified by examination of the acoustic signal. In this work, the corresponding normal load was termed Crack Initiation Load (CIL) for reference and was obtained by examination of acoustic signal such as that shown in Fig. 2. However, subsequent film failure and its associated critical load cannot be easily identified from the acoustic signal as will be described later.

In a scratch test, the film failure is related to a sudden change in the transverse frictional force or normal load. However, variations are subtle and may not be readily distinguishable as was the case for the majority of samples tested in this study, especially for the coated steel samples. Therefore, numerical manipulation was performed on raw data to extract the critical load value representing the film failure.

The transverse force and normal load data such as those seen in Fig. 2 were first smoothed by a Fast Fourier Transform (FFT) filter smoothing method. This was accomplished by removing Fourier components with frequencies higher than $1/(n\Delta d)$ where n is number of data points considered at a time, and Δd is the spacing between two adjacent data points. The smoothed data were then differentiated which made small variations, not clearly seen in raw data, pronounced. The discernible abrupt changes in differentiated transverse and normal forces identified the scratch location where the film failed. The scanning electron microscopy (SEM) and optical microscopy utilized in this work confirmed the results of these analyses in a manner that any visible film failure was correlated to the changes observed in the differentiated transverse force plots.

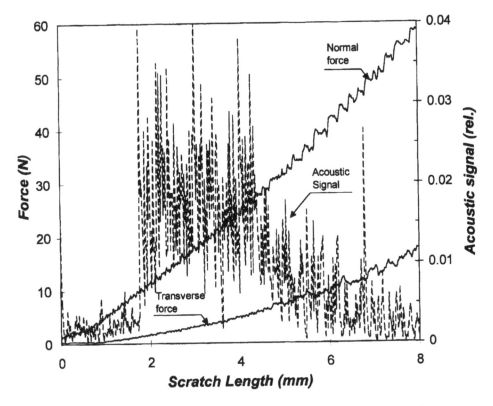

Figure 2. Plots for a scratch test performed on a sample of 3.12 μm thick nanophase diamond on a 304 stainless steel substrate. Normal scratch load, transverse force and acoustic emission plots are given as a function of scratch length.

4. RESULTS AND DISCUSSION

Generally, three major regions can be identified in a scratch test as the indenter advances the scratch length [15]. In the region I, the coating/substrate system resists the strain imparted by the stylus. The scratch mark seen in this region is due to slight plastic deformations of the film and the substrate. As the scratch track broadens the transverse force causes strain in the interfacial region which exceeds the tensile strength. At this time a microcrack is generated and emission of a burst of acoustic energy announces the beginning of the region II. The scratch process in the region II produces a substantial number of microcracks until catastrophic film failure or substrate penetration signals the start of the region III. Significant variations of transverse force and normal load signals are characteristics of the scratch process in this region which involves the substrate only.

In this work, five samples of 304 stainless steel were prepared and four coated with 1.05, 2.11, 3.12 and 5.14 μm of nanophase diamond. They were subjected to the scratch testing and data processing as described earlier. Scratch data similar to those shown in Fig. 2 were obtained and processed for these test samples. The resulting differentiated normal and transverse force plots are presented with a relative scale in Fig. 3. The dashed line separating scratch regions I and II in this figure identifies the same CIL value of 10.8 N corresponding to the scratch length of 1.7 mm. The location of this dashed line was obtained from position of the first dramatic increase in acoustic signals not presented in this figure. An examination of the differentiated transverse force plots together with optical microscopy observation of scratch tracks identified the scratch locations and critical loads for the film failure. Sudden and abrupt changes in differentiated data in Fig. 3 correspond to dashed lines separating regions II and III. The corresponding normal forces identify the critical loads for the film failure. This method is adequate to assess the practical adhesion of the coated samples. However, it should be noted that the optical microscopy and SEM examinations have also been utilized in this work to confirm the results obtained by the scratch data analyses.

The critical loads identified in Fig. 3 are presented as a function of coating thickness in Fig. 4. In contrast to the CIL value, the critical load has larger values for the samples with thicker coatings. This is in agreement with the earlier observations that the critical load increases with the coating thickness [27]. The behavior is linear for the coating thicknesses examined in this work. The dashed line in Fig. 4 was obtained by the least squares method. Its interpolation identifies a critical load of 18.4 N needed to penetrate and damage the uncoated sample with the scratch track width and depth profile similar to those of coated samples at the film failure. The corresponding scratch length location for the uncoated sample is plotted by the dashed line seen in Fig. 3. It represents the onset of transverse and normal force variations seen in this figure.

It is difficult to quantify the resistance to the scratch damage imparted by a thin film coating. However, an examination of scratch tracks at the film failure locations identified in Fig. 3 by the dashed lines separating regions II and III reveals interesting results. The scratch track widths and depth profiles measured by a surface profiler at the film failure locations were found to be approximately similar corresponding to about 180 and 12 μm, respectively. This indicated that the substrate penetration was delayed and the critical load needed to damage the 304 steel substrate to the point of film failure was almost doubled by a 5 μm coating of nanophase diamond.

The normal scratch load increases progressively in the scratch region II while the spacing between microcracks decreases until the film fails. Microscopic examinations and profilometry show a slow rate of increase for track width and depth in the scratch region II. This shows the resistance to scratch damage afforded by the nanophase diamond. It seems that eventual failure of the interface to support the stresses resulting from high substrate deformation causes the film failure at the critical load. In the region III, however, the track width and depth increase faster where most of the film has already been removed. No useful data about the adhesion of the film can be extracted in this region. However, an ex-

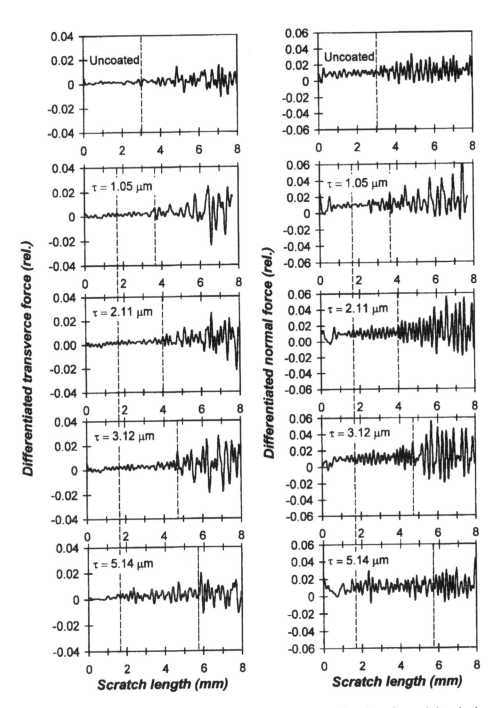

Figure 3. Plots of differentiated transverse and normal forces as a function of scratch length obtained by processing data from scratch tests performed on an uncoated sample and four nanophase diamond coated samples of 304 stainless steels.

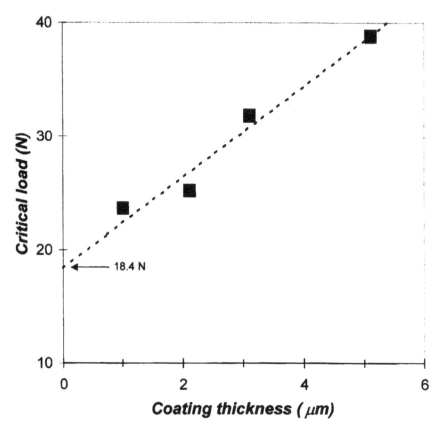

Figure 4. Critical load as a function of nanophase diamond coating thickness obtained from the plots presented in Fig. 3. The critical load needed to penetrate and damage the uncoated sample is indicated by the interpolation of the dashed line as described in the text.

amination of the scratch region III shows small pieces of film still adhering to the substrate. This indicates high adhesion quality of nanophase diamond to steel substrates.

Earlier studies have shown different film failure modes during the scratch testing of coatings on ductile substrate materials such as stainless steel [29-33]. They include spalling and buckling failures for poorly adherent films and conformal and tensile cracking for fully adherent coatings [31,32]. The conformal failure mode consists of cracking within the scratch which follows semicircular trajectories parallel to the leading edge of the indenter. They form as the stylus deforms the film and the substrate, resulting in tensile bends within the film as it is pushed under the indenter tip. Since considerable groove formation is observed for all ductile materials, the load on the diamond stylus is transferred to the front half of the indenter once sliding starts, which effectively doubles the contact stress. Because there is virtually no load on the back half of the indenter, Hertzian cracks will not be found in this region [32]. The tensile cracking mechanism is similar to the conformal cracking and the frictional stresses present behind the trailing edge of the indenter are responsible for the crack generation. Both the conformal and tensile cracks are sometimes associated with chipping and the chips stay within the scratch track.

We have found that the conformal and tensile cracks in the region II are eventually responsible for the nanophase diamond coating failure during the scratch tests performed on stainless steel substrates. We did not observe spallation or buckling failure modes in all the scratch tests performed in this work. Figure 5 presents an SEM photomicrograph of a scratch section in region II for a 1.1 µm thick nanophase diamond film on 304 stainless steel. The indenter sliding direction is to the right. The semicircular crack traces parallel to the trailing edge of the indenter are clearly seen indicating the tensile cracking. There are also a few distinguishable conformal cracks parallel to the leading edge of the indenter. It should be emphasized that we have observed similar coating failure modes when a diamond stylus with a conical angle of 90° and a spherical tip radius of 125 µm was employed in the scratch testing of our samples.

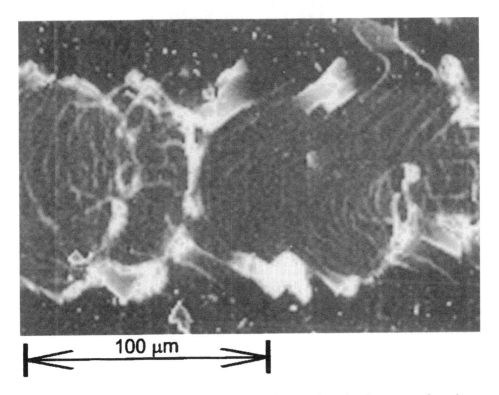

Figure 5. Photomicrograph showing an SEM image of a scratch section for a test performed on a sample of 1.1 µm thick nanophase diamond prepared on a 304 stainless steel substrate. The indenter sliding direction is to the right of this figure.

Acoustic emission signals obtained by the scratch tests on 1.05 and 5.14 μm
nanophase diamond films on 304 steel substrates are presented in Fig. 6. As mentioned earlier, a dramatic increase in acoustic emission represents the beginning
of the region II as shown by a common dashed line in this figure. However, the
film failure and critical loads cannot be readily identified from these plots.
Dashed lines corresponding to film failure were obtained from the corresponding
differentiated transverse and normal force plots together with optical microscopy
and SEM examinations.

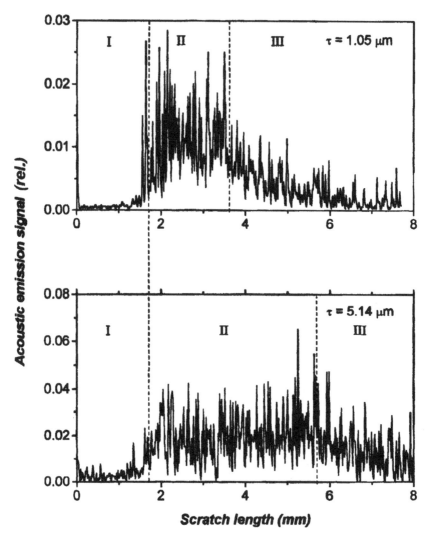

Figure 6. Plots of acoustic emission signals on a relative scale obtained from scratch tests performed on two samples of 304 stainless steel substrates with nominal diamond coatings of 1.05 and
5.14 μm thicknesses. Dashed lines separating regions II and III were obtained by examining corresponding differentiated transverse and normal force plots as described in the text.

The energy of an acoustic emission signal scales with the elastic energy re-leased in the process. For the highly adherent coatings many microcracks are gen-erated prior to and upon film failure, each emitting an acoustic energy burst. Sig-nals from these events are combined giving an overall complex acoustic emission such as that shown in Fig. 6. We attempted to numerically manipulate the acoustic data in order to obtain film failure locations. A high order polynomial was fitted to acoustic signal and then differentiated in order to find the maximum value of signal where it starts to decrease. It was expected that the corresponding scratch location was associated with the critical load for film failure. However, this pro-cedure underestimated the critical load values.

Examinations of Figs. 3 and 6 show more variations of transverse and normal forces and larger acoustic emission signals in the region II for the thicker films. This may be due to film surface roughness, which increases with the film thick-ness. Higher resistance to crack propagation afforded by the thick film may cause more pronounced variations of normal and transverse forces and release of larger elastic energy in the process as the stylus deforms the coating and the underlying substrate and pushes the film under its tip.

In our deposition system shown schematically in Fig. 1, substrates are rotated about two axes in a planetary arrangement so that their exposures to different parts of laser plasmas are averaged but with weights that depend on the geometry employed. Nominal films with 75% sp^3 contents are produced by rotating the samples in the core of the plasma at the deposition angle $\alpha = 80°$ [15-21]. In order to examine the adhesion quality of films produced from hotter core of the plasma, the planetary system was modified to coat samples at the deposition angle $\alpha = 50°$. A 304 stainless steel substrate was prepared and coated with 2.1 μm of nano-phase diamond in the modified system. Scratch tests were performed on this sam-ple and results were compared to the scratch data obtained for a 2.1 μm nominal coating deposited on a similar substrate at $\alpha = 80°$. The resulting differentiated transverse force plots for the regions II and III are presented in Fig. 7. For both samples, region II begins at the scratch length of 1.7 mm corresponding to the same CIL value of 10.8 N. However, the film failure for the sample with coating prepared from hotter core of laser plasma is delayed to the scratch length of 5 mm as shown in the top plot of Fig. 7. The corresponding critical load of 32 N is greater than the critical load value of 25 N obtained from the bottom plot in Fig. 7 corresponding to test results on the sample with the coating deposited at $\alpha = 80°$. This confirmed earlier results that the adhesion quality improves for coatings pre-pared from the inner part of laser plasmas in our deposition system [14].

The vast majority of all machining operations in the manufacturing industry are performed with cemented carbide cutting tools, usually in the form of inserts. To optimize the performance and increase the lifetime, these substrates are normally coated with single or multilayer coatings such as TiC, TiN, TiCN and Al_2O_3. Due to its unique hardness, diamond coating of cemented carbide may be an alterna-tive to these conventional coatings. In this work we studied the adhesion of nano-

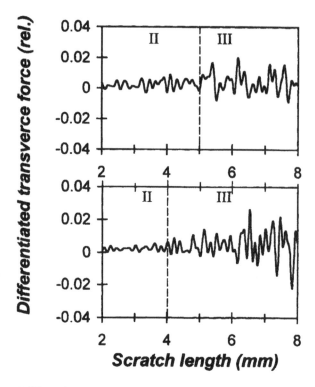

Figure 7. Plots of differentiated transverse force as a function of scratch length obtained by processing data for scratch tests performed on samples with nanophase diamond coatings of 2.1 μm on 304 stainless steel. Top plot corresponds to the sample prepared from the inner part of laser plasma at the deposition angle $\alpha = 50°$ while the bottom plot is for a sample prepared at deposition angle $\alpha = 80°$ as described in the text.

phase diamond on carbide inserts by scratch testing. A nominal film with the thickness of 1 μm was deposited at $\alpha = 80°$ on a tungsten carbide insert. Scratch test and data analyses were performed on this sample as detailed earlier. Only normal load and acoustic emission results are plotted in Fig. 8. As shown a CIL value of 23 N was obtained from examining the acoustic signal which is about twice the value obtained from the test performed on a sample of 304 stainless steel coated with a 1 μm film. The critical load for the film failure, however, was not reached for the normal scratch loads up to 50 N utilized in the scratch test. This is consistent with the earlier reports that higher values of scratch critical load are obtained for the samples with harder substrate materials [15,27]. To avoid damage to the stylus, scratch tests with larger normal force were not attempted. Optical microscopy and SEM examinations confirmed these results. It is interesting to note that there were sections in the scratch region II where the film was completely intact with no microcracks. If the local failures could have been avoided, the CIL value would have considerably increased. Similar results were also obtained for the coated TiC samples.

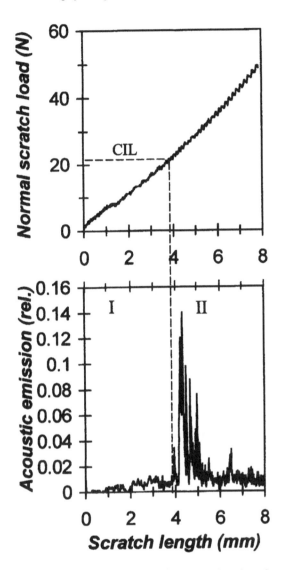

Figure 8. Plots of normal scratch load and acoustic emission as a function of scratch length obtained from a test performed on 1 μm thick nanophase diamond on a WC substrate.

The results outlined in Fig. 8 and the trend of increase for critical load seen in Fig. 4 suggest that a nominal 5 μm diamond coating on carbide substrates would result in a critical load value in excess of 90 N. Furthermore, if a 5 μm coating is deposited from the hotter parts of the laser plasma at deposition angle $\alpha = 50°$, then the critical load value is expected to approach 110 N, a value among the highest reported in the literature for a scratch test of a coated sample [34].

5. CONCLUSIONS

The scratch and indentation adhesion testing are considered among a few methods currently available for testing, hard and well adhering coatings [31]. The interpretation of test results is mainly based on determining critical loads for which the film is failed. In our earlier studies we have examined the practical adhesion of nanophase diamond coatings with a nanoindentation method [14]. In this work the adhesion property of nanophase diamond coatings deposited on stainless steel and carbide substrates was examined with a commercially available scratch tester. A method was introduced to reliably detect the critical loads from the transverse force, acoustic emission and normal scratch load plots produced by this device. The optical microscopy, surface profilometery and SEM examinations of scratch tracks confirmed the results of these analyses. It was demonstrated that substrate penetration and damage by the scratch indenter was delayed considerably by application of nanophase diamond coatings on both steel and carbide substrates. The impact of C^{3+} and C^{4+} ions carrying keV energies from the laser plasma discharge source encourages the formation of natural interfacial layers with considerable thicknesses [12] which is believed to be the cause of good practical adhesion experienced by our nanophase diamond coatings as described in this work.

Acknowledgments

This work was supported by the Texas Advanced Technology Program - 1997 under Grant No. 9741001. One of the authors, J.K. Koivusaari, would like to acknowledge the Jenny and Antti Wihuri foundation for the financial support and the encouragement provided by Prof. Seppo Leppävuori which made his visit to UTD possible.

REFERENCES

1. T. P. Ong and R. P. H. Chang, Appl. Phys. Lett., **58**, 358 (1990).
2. K. Oguri and T. Arai, J. Mater. Res., **5**, 2567 (1990).
3. F. Davanloo, H. Park, and C. B. Collins, J. Mater. Res., **8**, 2042 (1996).
4. C. C. Berndt and C. K. Lin, J. Adhesion Sci. Technol., **7**, 1235 (1993).
5. K.L. Mittal, J. Adhesion Sci. Technol., **1**, 247 (1987).
6. K.L. Mittal, in *Adhesion Measurement of Thin Films, Thick Films and Bulk Coatings*, K.L. Mittal (Ed.), STP No. 640, pp. 5-17, ASTM, Philadelphia, 1978.
7. K.L. Mittal, Electrocomponent Sci. Technol., **3**, 21(1976).
8. K.L. Mittal, in *Adhesion Measurement of Films and Coatings*, K.L. Mittal (Ed.), pp. 1-13, VSP, Utrecht, The Netherlands, 1995.
9. S. S. Chiang, D. B. Marshall, and A. G. Evans, in *Surfaces and Interfaces in Ceramic and Ceramic-Metal Systems*, J. Pask and A. G. Evans (Eds.), pp. 603-617, Plenum Press, New York, 1981.
10. M. Alam, D. E. Peebles and J. A. Ohlhausen, J. Adhesion Sci. Technol. **7**, 1309 (1993).
11. V. K. Sarin, J. Adhesion Sci. Technol. **7**, 1265 (1993).
12. S. Venkataraman, J. C. Nelson, A. H. Sieh, D. L. Kohlstedt and W. W. Gerberich, J. Adhesion Sci. Technol. **7**, 1279 (1993).

13. V. Gupta, MRS Bull., **XVI (4)**, 39 (1991).
14. F. Davanloo, T.J. Lee, H. Park, J.H. You and C.B. Collins, J. Adhesion Sci. Technol. **7**, 1323 (1993).
15. F. Davanloo, C. B. Collins, and K. J. Koivusaari, J. Mater. Res., **14**, 3474 (1999).
16. F. Davanloo, E. M. Juengerman, D. R. Jander, T. J. Lee, and C. B. Collins, J. Appl. Phys., **67**, 2081 (1990).
17. F. Davanloo, E. M. Juengerman, D. R. Jander, T. J. Lee, and C. B. Collins, J. Mater. Res., **5**, 2398 (1990).
18. C. B. Collins, F. Davanloo, D. R. Jander, T. J. Lee, H. Park and J. H. You, J. Appl. Phys., **69**, 7862 (1991).
19. F. Davanloo, T. J. Lee, D. R. Jander, H. Park, J. H. You, and C. B. Collins, J. Appl. Phys., **71**, 1446 (1992).
20. C. B. Collins, F. Davanloo, T. J. Lee, D. R. Jander, J. H. You, H. Park, and J. C. Pivin, J. Appl. Phys., **71**, 3260 (1992).
21. C. B. Collins, F. Davanloo, D. R. Jander, T. J. Lee, J. H. You, H. Park, J. C. Pivin, K. Glejbøl, and A. R. Thölén, J. Appl. Phys., **72**, 239 (1992).
22. J. Stevefelt and C. B. Collins, J. Phys. D, **24**, 2149 (1991).
23. C. B. Collins, F. Davanloo, J. H. You, and H. Park, Proc. SPIE **2097**, 129 (1994).
24. C. B. Collins, F. Davanloo, T. J. Lee, J. H. You, and H. Park, Mater. Res. Soc. Symp. Proc. **285**, 547 (1993).
25. K. J. Koivusaari, J. Levoska and S. Leppävuori, J. Appl. Phys., **85**, 2915 (1999).
26. J. Valli, J. Vac. Sci. Technol. A, **6**, 3007 (1986).
27. P. A. Steinmann, Y. Trady and H. E. Hintermann, Thin Solid Films, **154**, 333 (1987).
28. M. P. deBoer, J. C. Nelson and W. W. Gerberich, J. Mater. Res., **13**, 1002 (1998).
29. P. A. Steinmann and H. E. Hintermann, J. Vac. Sci. Technol. A, **3**, 2394 (1985).
30. J. H. Je, E. Gyarmati and A. Naoumidis, Thin Solid Films, **136**, 57 (1986).
31. P. J. Burnett and D. S. Rickerby, Thin Solid Films, **154**, 403 (1987).
32. S. J. Bull, Surface Coating Technol., **50**, 25 (1991).
33. S. J. Bull, Tribology Intl., **30**, 491 (1997).
34. A. A. Voevodin, C. Rebholz, J. M. Schneider, P. Stevenson and A. Matthews, Surface Coating Technol., **73**, 185 (1995).

Adhesion Measurement of Films and Coatings, Vol. 2, pp. 159–174
Ed. K.L. Mittal
© VSP 2001

Scratch test failure modes and performance of organic coatings for marine applications

S.J. BULL,* K. HORVATHOVA, I.P. GILBERT, D. MITCHELL,
R.I. DAVIDSON and J.R. WHITE

Department of Mechanical, Materials and Manufacturing Engineering, University of Newcastle, Newcastle-upon-Tyne, NE1 7RU, UK

Abstract—Although the scratch test is widely used for the assessment of thin hard coatings, its use for the testing of organic coatings is much less common. This is partly because these organic layers are much thicker than the hard coatings and hence interfacial loadings are much smaller and partly because the low hardness of the coatings means that they will plastically deform considerably during the test. Organic coatings for the protection of marine structures are often based on epoxy-resins or similar materials which are brittle compared to other organic resins. Water penetration to the coating/substrate interface may lead to corrosion of the underlying substrate and changes in coating adhesion. Furthermore, in service these materials undergo photo-oxidation which can increase their brittleness or introduce thin modified surface layers which have poorer properties than the as-received material. The scratch test is an ideal way of making quantitative comparisons between these materials both in the as-received state and after exposure to UV or salt water. In this study, a number of organic coatings and base resins used for marine protection have been assessed by the scratch test. Both through-thickness cracking and interfacial detachment have been observed when the coatings are less than about 200μm thick with mainly plastic deformation for thicker coatings. Interfacial detachment is usually observed as a consequence of through-thickness cracking and initiates behind the moving indenter, contrary to what is usually observed for hard coatings. The critical loads for through-thickness cracking and interfacial detachment are reduced after exposure to salt water in some coating systems. The use of the scratch test to monitor performance of these coatings will be discussed.

Keywords: Scratch test; organic coatings; critical loads.

1. INTRODUCTION

The scratch test has been used to assess the fracture resistance and practical adhesion of hard coatings on softer substrates for some time now [1-3] and was originally used for measuring the adhesion of softer metallic coatings on hard substrates [4-6]. The failure modes which occur in this test have been well documented over the years [7-9] and some attempt at quantification of the scratch

---

* To whom correspondence should be addressed. Phone: +44 191 222 7913,
Fax: +44 191 222 8563, E-mail: s.j.bull@ncl.ac.uk

adhesion test has been made [4, 5, 8, 10-15]. However, the test has not generally been applied to organic coatings on metallic substrates and the scratch test failure modes in this case are not well understood. It is important to realise that the scratch test (like all adhesion test methods) can only measure practical adhesion which is a combination of the fundamental adhesion given by chemical bonding at the interface and the effects of many other factors [16-18].

Organic coatings based on epoxy resins are widely used for corrosion protection in marine applications. Such coatings consist of a resin binder and pigment system and are designed to be used in a range of applications including in the holds of bulk carrier ships. In such locations the cargo may be loose in the hold and can do considerable damage to the coating as it moves and settles during the voyage. Scratch damage is observed including plastic deformation, through-thickness cracking and coating detachment. Thus the scratch test is an ideal method for assessing the scratch resistance of a range of coatings in the laboratory.

In this study a number of different epoxy-based coatings, similar to those used in marine applications, have been scratch tested in the as-received and aged condition in order to identify the types of failure produced in the test and determine whether the scratch test using a conventional Rockwell 'C' scratch diamond is sufficient for assessing their performance.

2. EXPERIMENTAL

2.1. Coatings investigated

All coatings investigated were two-component epoxy-amine systems, designed for application/cure under ambient conditions and used for corrosion protection in the Marine market. Commercial coatings of this sort generally contain pigments which may impart functionality to the coating (e.g. colour) or be used to reduce the amount of expensive constituents in the coating - this is known as extender pigmentation and generally consists of cheap inorganic materials like chalk or mica. Coating A is a hydrocarbon modified epoxy/ polyamide system which contains extender pigmentation and has 75% solids by volume. Coating B is an epoxy/amine adduct system, containing aluminium pigmentation as part of the pigment package and has 55% solids by volume. Coating C is a liquid epoxy/amine adduct system, containing extender pigmentation and with 85% solids by volume. Coatings D and E use the same resin system as coating C. Coating D is almost solvent free (5wt% solvent) while the Coating E was solvent thinned by the addition of 20 wt% of xylene/butanol (3/1). All coatings were deposited on one side of a 150μm thick shim steel sheet to thicknesses between 200μm and 1.2mm using a doctor blade coating technique which guaranteed good thickness uniformity.

All coatings were allowed to cure for at least three months prior to scratch testing in order to ensure that the curing process was complete. As the solvent evaporates from the coating it is put into a state of tensile stress which can be measured by changes in curvature of the thin steel shim substrate [19, 20]. Typically the stress was ≤6MPa which is insufficient to produce through-thickness cracking in any of the coatings.

2.2. Scratch testing

Coatings were scratched with a Rockwell 'C' diamond under a continuously increasing normal load in a spring-loaded automatic scratch tester. Scratches were made to three different maximum loads, 40N, 60N and 80N at a loading rate of 100N/min and a table speed of 10mm/min giving a load ramp of 10N/mm in the scratch track. The load ramp was continuously monitored and logged on the computer. Failure modes and failure loads were identified by *post facto* reflected light microscopy and scanning electron microscopy (SEM) on samples cut from the coated shims. The epoxy layers were coated with a thin layer of gold to prevent charging in the SEM. The critical load for a failure mode to be observed is defined as the lowest load at which the failure is observed to occur regularly along the scratch track. The failure load was determined by measuring the distance from the centre position of the scratch stylus at the end of the track (where the impression is always clearest) back to the low load edge of the failure and converting to a load by multiplying by the load ramp. This value is then subtracted from the maximum load to determine the failure load.

In addition, the scratch hardness, H_{scr}, of all the coatings was determined by measuring the track width, d, at 40N load using reflected light microscopy. The scratch hardness is given by [21]:-

$$H_{scr} = \frac{4kL}{\pi d^2} \tag{1}$$

where L is the normal load and k is a constant representing the fraction of the projected circular contact area which is actually in contact with the scratch diamond (Figure 1). In tests on metals and ceramics $k \approx 2$ (i.e. the load is supported on the front half of the diamond only) but due to the low elastic modulus of the coatings tested here a larger contact area is expected and $k=1$ has been used. This will overestimate the contact area and, therefore, underestimate the true scratch hardness of the system but should enable reasonable comparisons to be made.

The coefficient of friction, μ, was determined for all coatings as a function of load. The variation in observed values was small, with μ typically between 0.2 and 0.25 once the scratch was established. Given these values a simple estimate of the tensile stresses generated behind the moving indenter can be obtained by dividing the frictional force by the cross-sectional area of the scratch track (which is related to the scratch track width by the indenter geometry). Maximum tensile stresses of around 100MPa are calculated for all the samples, which are much larger than the residual stresses in the same films. It is, therefore, not surprising

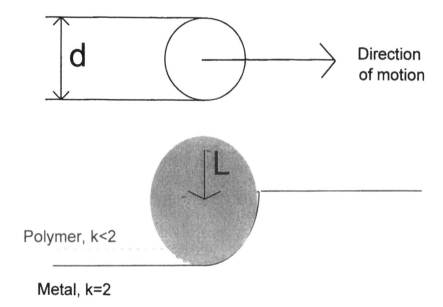

Figure 1. Definition of the parameters used in the calculation of scratch hardness. The parameter k determines the fraction of the nominal contact area which supports the normal load (see equation (1)).

that fracture occurs in the scratch test which is not observed in the as-cured samples. However, the precise values of the stresses caused by the scratch stylus are not known as there is, as yet, no good model of the scratch testing of a bulk solid which plastically deforms, let alone a layered system.

2.3. Exposure test

Coated shims were placed in a beaker of seawater immersed in a water bath held at 60°C for 24h. No attempt to stir the seawater was made during exposure. There was some evidence of corrosion of the steel and attack occurred at the epoxy-steel interface from the edges of the sample but the centre of the sample was apparently unaffected after exposure. Further scratch tests were performed in this region to see if the sea water exposure changed the type of failure and the critical load at which it first occurred.

Immersion in sea water will cause large changes in residual stress in the coating [e.g. see 20] due to swelling, change of modulus, plasticisation and relaxation. On removal from the water bath some of the changes reverse. It is hence important to control the delay between removing the samples from the bath and scratch testing. In this study the delay was fixed at one hour to allow time for the sample to cool to room temperature and dry thoroughly. Samples were stored in a vertical orientation to allow any liquid to run off in this period.

3. RESULTS

3.1. As-received coatings

3.1.1. Failure Modes. Scratch testing of the unmodified resin showed that both through-thickness cracking and adhesion failures could be induced by the scratch test if the coating was thin enough. Generally, through-thickness cracking was promoted by the presence of pigment in the coatings. This type of cracking was mostly caused by the tensile stresses behind the moving stylus but in a few cases conformal cracking was also observed. Coating detachment was only observed when the penetration of the scratch stylus was comparable to the coating thickness except in cases where obvious interfacial defects, such as bubbles in the paint, were observed. For a maximum load of 80N, as used for the coatings investigated here, this meant that adhesion failures were observed only when the coating was less than 300µm thick.

A good example of this is given by coating A deposited to a thickness of 200µm (Figure 2a). Through-thickness cracking started at a low load and formed due to the tensile stresses behind the indenter. Since coating A contains extender pigmentation the through-thickness cracking was extensive. The through-thickness cracks started towards the centre of the track and followed the curvature of the stylus. As the stylus moves forward the crack opens and extends forward giving forward-pointing chevrons of cracking. At normal loads of around 45N the stylus penetrated through the coating and it was stripped and pushed aside ahead of the moving indenter (Figures 2b and 2c).

Unpigmented coatings deposited to the same thickness, such as coating E (Figure 3a), show much less through-thickness cracking. Yet again detachment and splitting occurred at loads when the indenter penetration was comparable to the coating thickness (Figure 3b). The mechanism of detachment requires a through-thickness crack at the rear of the stylus. The coating was pushed forward as the stylus moved forward and was detached and pushed ahead and to the sides of it (Figure 4).

Increasing the thickness of the coating to >300µm prevented interfacial stripping from occurring. This is illustrated for a 500µm layer of Coating B (Figure 5a). Again this coating contains pigmentation (this time it is aluminium) so there is plenty of through-thickness cracking. However, in this case cracks are observed which follow the contours of both the front and the back of the stylus. These are conformal cracks (Figure 5b) which form as the coating bends to follow the shape of the stylus. At points where forward pointing and backward pointing through-thickness cracks intersect it is possible to detach small regions of coating but no gross spallation occurred.

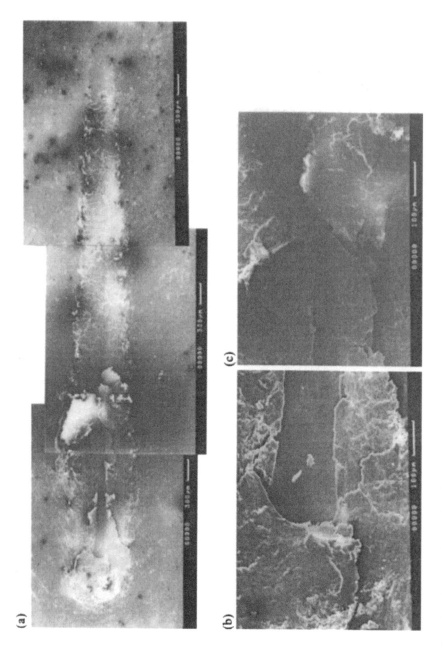

Figure 2. Scanning electron micrographs of scratches in a coating A of 200μm thickness, (a) general view, (b) end of the scratch where detached material has collected and (c) start of coating detachment from a through-thickness crack at the rear of the scratch stylus.

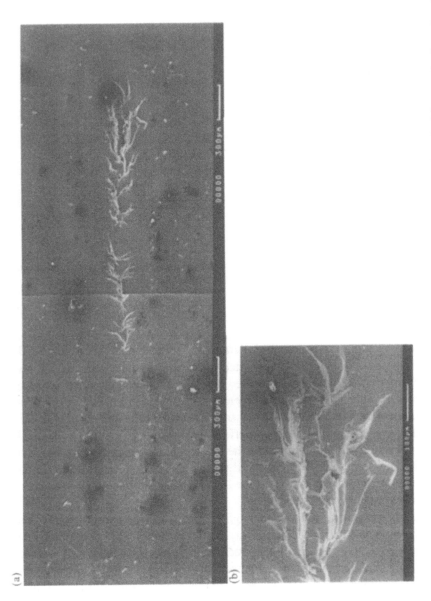

Figure 3. Scanning electron micrographs of scratches in a coating E of 200μm thickness, (a) general view, (b) end of the scratch where detachment is observed in the middle of the track.

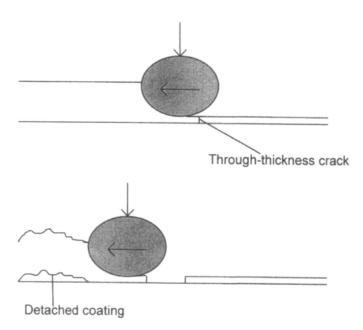

Figure 4. Schematic of the mechanism of adhesion failure. A through-thickness crack forms behind the moving stylus and is diverted along the coating/substrate interface. This is propagated forward as a shear crack as the coating is dragged forward by the stylus. Folds of detached coating are pushed forward, ahead of the moving stylus.

For thicker coatings, gross coating detachment was sometimes observed. For instance in the case of coating C deposited to 900μm thickness the scratch intersected an interfacial bubble and some localised detachment occurred in the scratch track (Figure 6a). The detachment was initiated by a through-thickness crack which formed due to the tensile stresses behind the moving stylus (Figure 6b). However, the stresses at the interface were not large enough to detach the coating so once the bubble was passed the detachment stopped (Figure 6c). Some through-thickness cracking was observed elsewhere in the scratch as this coating contains extender pigmentation but at the lowest loads used in the test the stresses are insufficient for fracture to occur.

Unpigmented coatings, such as coating D deposited to 1200μm thickness show no through-thickness cracks (Figure 7a). However, there are markings in the piled-up material at the edge of the track which may be mistaken for cracks. These are folded material and are evidence that significant plasticity has occurred (Figure 7b). At the end of the scratch track in this sample some of the coating was removed with the scratch diamond when it was lifted from the surface (Figures 7c and 7d). This is evidence that the cohesive strength of the coating is less than the adhesion strength of the diamond-coating interface.

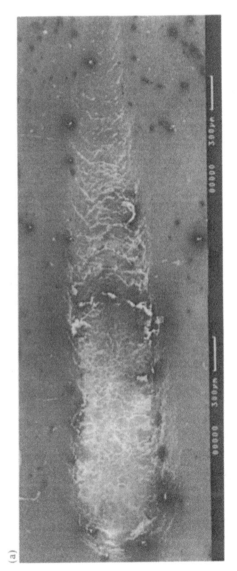

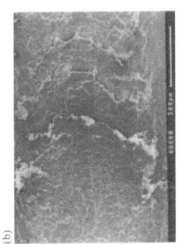

Figure 5. Scanning electron micrographs of scratches in a coating B of 500μm thickness, (a) general view, (b) tensile cracking behind the indenter. This coating is pigmented and shows plenty of through-thickness cracking which follows the contours of both the front and back of the scratch diamond (conformal and tensile cracking). Where these crack systems interact small islands of coating are produced which are relatively easy to detach.

S.J. Bull et al.

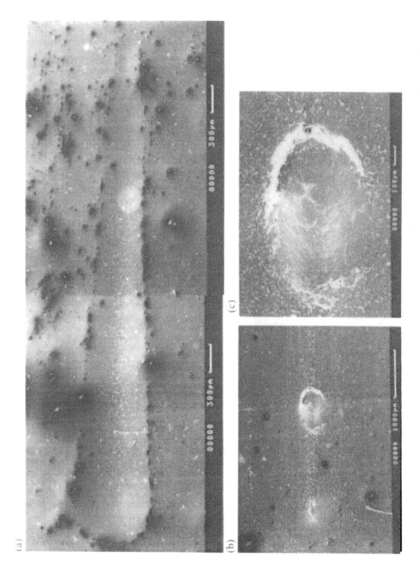

Figure 6. Scanning electron micrographs of scratches in a coating C of 900μm thickness. (a) general view of a track showing little damage, (b) localised region of damage in another scratch, (c) higher magnification of the failure in (b). Failure has started from a through-thickness crack at the rear of the stylus which occurred near an interfacial defect (bubble). The coating is pulled forward by the stylus above the defect and deformed but the stylus cannot propagate the interfacial defect and detachment is only localised.

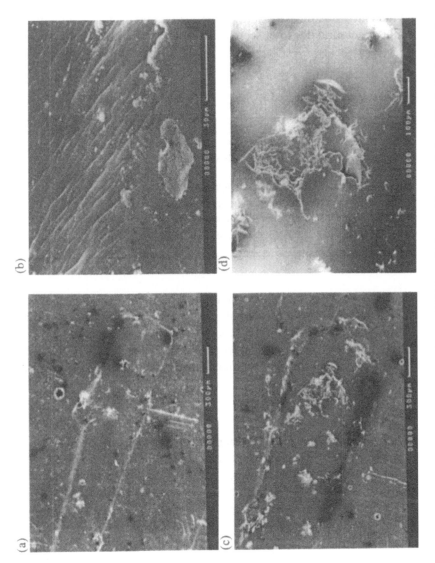

Figure 7. Scanning electron micrographs of scratches in a coating D of 1200μm thickness, (a) general view, (b) plastic deformation and the edge of the scratch track, (c) material detached at the end of the scratch track which was stuck to the stylus and removed with the stylus, (d) higher magnification image of (c).

3.1.2. Scratch Hardness. The scratch hardness of all coatings was determined at 20N normal load; at this load the scratch diamond had not penetrated through even the thinnest coatings investigated. Thus the results are not expected to be influenced by the substrate. The scratch hardness of the base resin is always greater than that of the formulated coatings (Table 1). One surprising observation is that the extender pigmented coatings are softer than their unpigmented counterparts; the aluminium pigmentation does not seem to have any major effect on hardness.

Table 1.
Scratch hardness of the coatings investigated

Coating	Description	Scratch hardness (MPa)
Base resin	Epoxy-amine	502 ± 47
A	hydrocarbon modified epoxy/polyamide with extender pigmentation	162 ± 23
B	Epoxy-amine adduct with aluminium pigmentation	268 ± 91
C	Liquid epoxy-amine with extender pigmentation	135 ± 8
D	As C but unpigmented and almost solvent free	268 ± 26
E	As C but unpigmented and solvent thinned with 20wt% xylene/butanol (3/1)	286 ± 51

It might be expected that these composite coatings would have a higher hardness than the base resin since composite materials reinforced with inorganic materials are usually harder than the base resin but this is not what is observed here. However, the extender pigmentation is a soft inorganic material (mica) and the adhesion between the pigment and the resin is poor. It is a combination of these factors which dictates the observed behaviour.

3.2. Exposed coatings

3.2.1. Failure Modes. In general through-thickness cracking is reduced by exposure to sea water at 60°C (Figure 8). This is probably because the coating has been plasticised by absorption of some water. However, interfacial detachment is made worse in most cases (Figure 8) due to corrosion at the coating-substrate interface. This occurs both by migration along the interface from exposed edges and by water absorption and diffusion through the coating. For instance, it was found that about 1% water was absorbed by the base resin after the exposure. The coatings containing pigments were much less susceptible to corrosion during ageing and showed only small changes in adhesion after the test.

300μm

(a)

(b)

Figure 8. Scanning electron micrographs showing the effect of ageing on a 200μm thick layer of coating E. As received (a) and after exposure to seawater at 60°C for 24h (b). The load at which through-thickness cracking is first observed increases after exposure but the onset of coating detachment occurs at a lower load.

3.2.2. Critical Loads for Failure. One method of characterising the extent to which changes in adhesion and through-thickness cracking occur after exposure is to determine the critical normal load at which the failure mode is first observed in the scratch test and see how this changes after exposure (Table 2). In some cases a particular failure mode is no longer observed and no comparisons can be made but in other cases a quantifiable difference is observed. For instance, the pigmented coating A shows little or no change in the load required to initiate coating detachment but the unpigmented coating of similar thickness, coating E, shows a dramatic reduction in the critical load. Similarly there is an increase in the critical load for the onset of through-thickness cracking in this sample.

Table 2.
Effect of ageing for 24h at 60°C in seawater on critical normal loads and failure modes

Coating	Thickness (μm)	As received		Aged	
		Failure mode	Critical Load (N)	Failure mode	Critical load (N)
Base resin	200 ± 15	Interfacial spall	65 ± 12	Interfacial spall	3 ± 2
A	200 ± 20	Detachment	45 ± 6	Detachment	44 ± 3
B	500 ± 32	Small spalls	25 ± 5	Small spalls	57 ± 8
C	900 ± 98	None observed	-	None observed	-
D	1200 ± 102	Adhesion to diamond	55 ± 10	None observed	-
E	200 ± 9	Detachment	81 ± 9	Detachment	9 ± 2
		Through-thickness cracks	38 ± 2	Through-thickness cracks	44 ± 3

4. DISCUSSION

The scratch test is widely used for hard coatings but it has not been applied extensively for the assessment of organic coatings because these were thought to undergo excessive plastic deformation in the scratch test, masking any adhesion effects. However, this study has shown that for relatively brittle organic coatings based on epoxy resins, both through-thickness fracture and coating detachment can be produced. Scratch tests on bulk polymers [22] have demonstrated that such fracture behaviour can be observed if the attack angle of the stylus is high enough. For the 120° cone used in this study the attack angle of 30° generates a strain, ε, in the material of about 0.11 ($\varepsilon=0.2\tan\theta$ where θ is the attack angle [21]). Under such conditions bulk polymers such as poly(methyl methactylate) (PMMA) and ultra high molecular weight polyethylene (UHMWPE) behave entirely plastically. The work presented here shows that the epoxy resin on which these coatings are based is considerably more brittle than these bulk materials. A sharper cone would be more likely to cause fracture and in addition would penetrate deeper into the coating. It would, therefore, be an advantage to optimise the stylus for a given coating system if the fracture behaviour is to be assessed most reliably.

Coating detachment was most commonly observed when the penetration of the stylus was almost equal to the coating thickness. This poses limitations for adhesion assessment of thicker coatings since this criterion cannot be achieved at sensible normal loads. The use of a sharper stylus is required to ensure deeper penetration for a given load. This further argues that a method of stylus optimisation is required if organic coatings are to be routinely assessed by the scratch test.

5. CONCLUSIONS

The scratch test using a Rockwell 'C' stylus has been applied to a number of epoxy-amine coatings developed for marine applications and it is shown that it gives information about through-thickness fracture and adhesion if the coating is sufficiently thin (i.e. the scratch stylus penetrates close to the coating/substrate interface). Extender pigmentation in the coating has a considerable effect on performance, promoting through-thickness cracking and reducing scratch hardness. However, the pigmentation reduces interfacial adhesion degradation and/or corrosion during ageing. In ageing trials the adhesion of unpigmented coatings was greatly reduced but the resistance of the same coatings to through-thickness cracking was increased. The scratch stylus needs to be optimised if the test is to be more widely used for the assessment of organic coatings.

Acknowledgements

The authors would like to thank Akzo-Nobel (International Paint) Ltd at Felling, UK for the provision of samples and useful advice and discussions with M. Buhaenko, C. Campbell and P. Jackson.

REFERENCES

1. A.J. Perry, Thin Solid Films, **107**, 167 (1983).
2. P.A. Steinmann and H.E. Hintermann, J. Vac. Sci. Technol., **43**, 2394(1985).
3. S.J. Bull, These Proceedings.
4. P. Benjamin and C. Weaver, Proc. Roy. Soc. Lond. **A254**, 177 (1960).
5. P. Benjamin and C. Weaver, Proc. Roy. Soc. Lond. **A261**, 516 (1961).
6. D.W. Butler, C.T.H. Stoddart and P.R. Stewart, J. Phys. D, **3**, 877 (1970).
7. S.J. Bull, Surf. Coat. Technol., **50**, 25 (1991).
8. P.J. Burnett and D.S. Rickerby, Thin Solid Films, **154**, 403 (1987).
9. R.D. Arnell, Surf. Coat. Technol., **43/44**, 674 (1990).
10. M.J. Laugier, Thin Solid Films, **117**, 243 (1984).
11. M.J. Laugier, J. Mater. Sci., **21**, 2269 (1986).
12. S.J. Bull, D.S. Rickerby, A. Matthews, A.R. Pace, and A. Leyland, Surf. Coat. Technol., **36**, 503 (1988).
13. S.J. Bull and D.S. Rickerby, Surf. Coat. Technol., **42**, 149 (1990).
14. S.J. Bull, Materials at High Temperatures, **13**, 169(1995).

15. S.J. Bull, Tribology Int., **7,** 491 (1997).
16. K.L. Mittal, in *Adhesion Measurement of Thin Films, Thick Films and Bulk Coatings*, K.L. Mittal (Ed.), STP No. 640, pp5-17, ASTM, Philadelphia, (1978).
17. K.L. Mittal, Electrocomponent Sci. Technol., **3**, 21 (1976).
18. K.L. Mittal, in *Adhesion Measurement of Films and Coatings*, K.L. Mittal (Ed.), pp1-13, VSP, Utrecht, The Netherlands, (1995).
19. Gu Yan and J.R. White, Polym. Eng. Sci., **39**, 1856 (1999).
20. Gu Yan and J.R. White, Polym. Eng. Sci., **39,** 1866 (1999).
21. F.P. Bowden and D. Tabor, *The Friction and Lubrication of Solids*, Clarendon Press, Oxford, Part 1 (1950); Part 2 (1964).
22. B.J. Briscoe, P.D. Evans, E. Pelillo and S.K. Sinha, Wear, **200,** 137 (1996).

Adhesion Measurement of Films and Coatings, Vol. 2, pp. 175–186
Ed. K.L. Mittal
© VSP 2001

An energetic approach for the evaluation of adhesion of sputter deposited TiC films on glass by the scratch test

A. KINBARA,* A. SATO, E. KUSANO and N. KIKUCHI

AMS R&D Center, Kanazawa Inst. Technol., Ishikawa 924-0838, Japan

Abstract—Scratch testing has been employed for the adhesion evaluation of sputter deposited Ti+C thin films of various carbon contents on glass substrates. These films showed various friction coefficients and the friction coefficient was found to affect the scratch test results. We took into account the frictional force and the internal stress in the adhesion evaluation. It was found that both of these parameters affected considerably the scratch test results and they should surely be considered in the adhesion evaluation. We have estimated adhesion energies of the Ti+C films, and these were found to range from 0.7 to 2.1 J/m^2 depending on the carbon content in the film.

Keywords: Adhesion; scratch test; TiC film.

1. INTRODUCTION

The scratch testing is one of the most useful techniques for the evaluation of adhesion of thin films to substrates and has a long history [1–5]. The critical load values for detachment obtained by this method are usually expressed in terms of N(Newton) and are valuable in a practical assessment of adhesion but are less valuable in scientific research because of the complexity of the factors influencing them. More physical quantities such as adhesion energy are required for the discussion based on the scientific research.

In the present research, we have measured the adhesion of Ti+C films sputter deposited onto non-alkali(NA) alumino-borosilicate glass substrates supplied from Nippon Sheet Glass Co. with a vibrational type scratch tester [6] and have evaluated the values obtained using an energetic approach for the physical interpretation. We focus our interest on the scratch process. In order to investigate the effect of the friction coefficient of the sample surface, we deposited Ti+C films with various compositions with varying friction coefficients. The role of friction in the scratch process has been investigated in relation to the adhesion.

---

*To whom correspondence should be addressed. Phone: 81-76-274-9250, Fax: 81-76-274-9251, E-mail: kinbara@neptune.kanazawa-it.ac.jp

Table 1.
Sputtering Conditions

DC Magnetron Sputtering	
Target Ti	99.9999%
C	99.999%
Discharge Gas Ar	<99.9999%
Ultimate Pressure	5×10^{-5} Pa
Ar Gas Flow Rate	30 sccm
Ar Gas Pressure	0.4 Pa
Discharge Current	$0.0 \sim 0.4$ A
Substrate Temperature	not controlled

2. SAMPLE PREPARATION

Ti+C films were sputter deposited in a multi-target magnetron sputtering apparatus in an Ar discharge gas atmosphere onto the NA glass substrates. The sputtering conditions are given in Table 1.

The NA glass substrates were cleaned in neutral detergent followed by cleaning in distilled water and finally dried by a nitrogen gas blower.

In the sputtering chamber, the Ti and C targets were fixed. Ti and C were alternately sputter deposited onto the substrates by rotating the substrates inside the chamber. The temperature of the substrate was not controlled during the sputtering. The schematic of the sputtering apparatus is shown in Figure 1. The film composition was controlled by adjusting the discharge current at each target. In the present work, we varied the relative sputtering atom flux ratio of C and Ti at the substrate.

The ratio of C flux to total arriving atom flux is given as follows:

$$F = \Gamma_C/(\Gamma_C + \Gamma_{Ti})$$

where Γ_C is the C flux and Γ_{Ti} is the Ti flux.

A schematic of the sample structure is shown in Figure 2. The thickness of the TiC+C layer at one deposition cycle in a multi-layered sample was about 1 nm. The deposition was repeated by rotating the substrate. The final thickness was 300 nm. The deposition temperature was ambient. However, the sample temperature will rise due to the energy of the impinging atoms. A mixing of Ti and C occurs to form TiC during deposition in the present sputtering conditions. X-ray diffraction patterns of the films are shown in Figure 3 as a function of C flux ratio, F. Diffraction peaks identified to be due to TiC were observed for a certain range of F values and beyond this range, the films seemed to be C-rich and amorphous.

Carbon was first deposited onto the substrate followed by Ti deposition. Thus the interface between the film and the substrate was possibly C-rich except for the case of $F = 0$, for which the surface of the specimen was supposed to be Ti-rich; but at present, we do not consider the inhomogeneity of the film composition. The

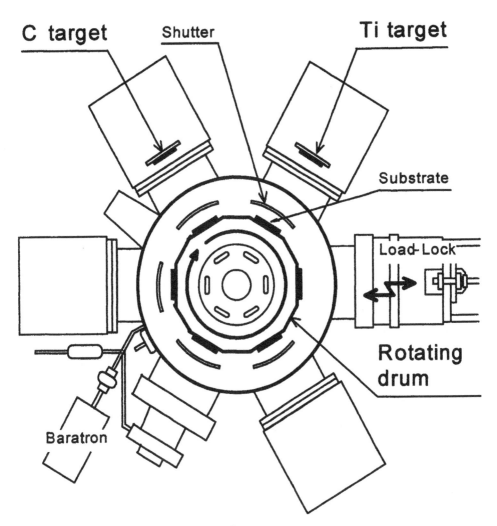

Figure 1. Schematic illustration of the sputtering apparatus.

composition of the films was examined with XPS and the concentrations of C, Ti and O are shown in Figure 4 as a function of F.

3. EXPERIMENTAL RESULTS

The critical loads to peel off the films from the substrates were measured with the scratch tester (RHESCA Co. Ltd., Tachikawa, Japan, model CSR-02). This tester is essentially the apparatus used for the frictional force measurements. A continuously increasing load is applied to the diamond stylus vibrating on the sample surface at about 30 Hz. The critical load to peel off the film is detected by the change in the vibration amplitude due to the friction force acting on the stylus [6]. Though photographs of the scratch patterns were not taken in the present experiment, the

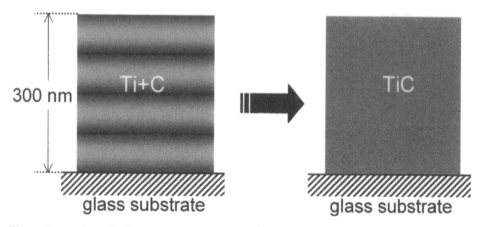

Figure 2. A schematic illustration of the sample structure. Alternate sputter deposition of Ti and C should produce a multi-layered structure shown in the left but during or after the sputtering, this structure changes to a homogeneous structure by the intermixing of Ti and C as shown in the right.

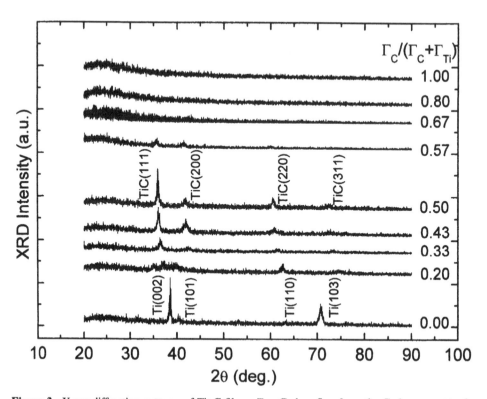

Figure 3. X-ray diffraction patterns of Ti+C films. Γ_C: Carbon flux from the Carbon target to the substrate. Γ_{Ti}: Titanium flux from the titanium target to the substrate. $\Gamma_C/(\Gamma_C + \Gamma_{Ti})$ is called C flux ratio and is taken as a parameter.

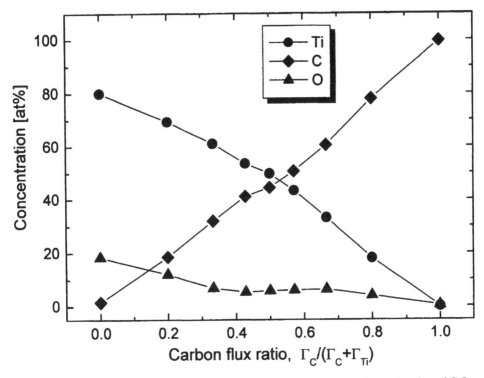

Figure 4. Composition of Ti+C film on glass substrate examined with XPS as a function of C flux ratio defined in Figure 4.

peeling of the film detected by the tester was shown to be well consistent with the microscopic observations as shown in the previous papers [7–10]. The critical load to peel off the film is shown in Figure 5 as a function of F. Pure Ti ($F = 0$) shows the smallest value and the critical load increases steeply with the increase of F. For $F \geqslant 0.4$, the adhesion strength is about a factor 4 larger than the value for the pure Ti film.

These results are rather different from the usual experience. Ti is often deposited on glass or other substrates as an intermediate layer because of its high adhesion character.

We consider that the scratch test includes rubbing of the film surface and hence, we take into account the frictional property of the film. Previously, the friction in the scratch test has been found to be important but it is more complex than the results here. The dynamic friction coefficient, μ, of the Ti, TiC and C films was measured with a wear tester (Shinto Kagaku Co., Tokyo, Japan, model HEIDON-14DR) using a steel ball with a load of 1 N. The results are shown in Figure 6. Although μ was not measured in detail as a function of F, this figure suggests a monotonous decrease of μ with an increase of F and we assume a linear change for rough approximation in the present paper.

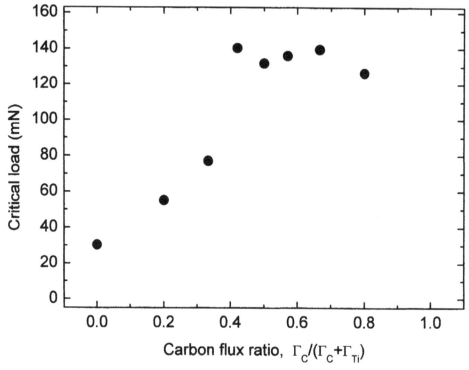

Figure 5. Critical load to peel off the Ti+C film from the glass substrates by scratching.

4. DISCUSSION

The scratch test is probably one of the most practical and useful methods for the evaluation of adhesion. But as we have mentioned before, this method sometimes gives a result that differs from most tests. In addition, the physical meaning of adhesion strength expressed in terms of critical load is not clear. In order to make clear the meaning of the scratch test, we consider the scratch process.

Perhaps, the scratch test involves several processes. One process is the application of a vertical compressive force to the film through the stylus of the tester. This process generates an elastic strain in the film and the elastic energy is stored in the film. Plastic deformation also occurs with part of the energy transferred to the film as heat during the deformation. However, its estimation is difficult and, therefore, is ignored in this report. The second process we should take into account is the process ascribed to an application of horizontal force on the film surface. If the surface is rough and the moving stylus comes in contact with protrusions on the film surface, a force directed toward the film surface acts on the film. In the macroscopic treatment, this force is regarded as a frictional force and the measurement of the friction coefficient is significant in the scratch experiment. In addition to these processes, the impact of the stylus possibly affects the scratching process. This means that momentum is transferred to the protrusions during impact. This impact

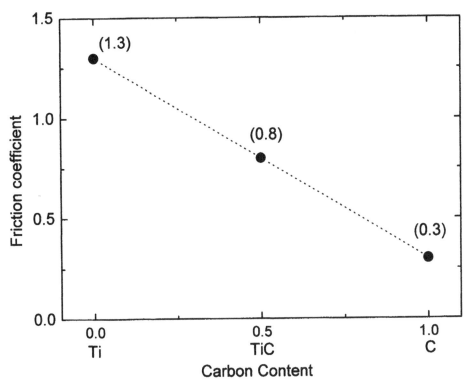

Figure 6. Friction coefficient, μ, of Ti, TiC and C films.

should have some effect on the experimental measure of adhesion. The effect is expected to depend on the velocity of the stylus but this effect will remain as a future issue in the adhesion measurement.

The scratch test is assumed to include these surface contributions and this is the reason that the results of the scratch test show a considerable scatter in the data.

When we consider the stress field in the film produced by the stylus, it should be expressed in terms of a tensor. A compressive and a shear stress components should be analyzed. The stress field depends on the shape of the stylus tip. A precise calculation of the stress requires finite element method and we need some assumptions at the interface between the film and the substrate. In order to avoid the complicated situation and to find the physical meaning of adhesion strength, we will estimate the work done by the stylus to the sample using the formula presented in a previous paper [11].

If the load W is applied to the hemisphere shaped stylus tip of radius R and Hertz contact (elastic contact) is assumed, an average vertical force p on the unit area at the center of the stylus tip and the reaction force in the film are produced. Ignoring the deformation of the diamond stylus, we obtain,

$$p = 3W/2\pi a^2 \tag{1}$$

where a is the radius of the contact surface of the stylus given by

$$a = (3WR/4E_S)^{1/3}. \tag{2}$$

E_S should be the composite Young's modulus of the substrate and the film but the effect of the film on the deformation of the sample is ignored and E_S is taken to be Young's modulus of the glass substrate, because the thickness of the film is much smaller than that of the substrate.

In the indentation process of the film, both elastic and plastic deformations occur. In the plastic deformation the work done by the stylus will be consumed as heat and will be dissipated. Thus we consider that only the elastic energy generated by the force given by eqn. (1) is stored in the film and can be used to peel off the film. When a moving stylus presses onto a rough film surface, the stylus generates not only the vertical force expressed by (1), but it also generates a horizontal/frictional force on the film surface. Hamilton and Goodman [12], and Laugier [13] have taken into account the frictional force effect in this process. We simplified the effect and expressed the frictional force, f, at the center of the stylus tip to be

$$f = \mu p$$

where μ is the friction coefficient of the film surface. Hence we assume that the total force p^* at the point under the center point of the tip is

$$p^{*2} = p^2 + f^2 = p^2 + (\mu p)^2 = (1 + \mu^2)p^2. \tag{3}$$

p^* is the force acting on the stylus tip from the film. A model of the force under the stylus is shown in Figure 7.

The stress produced by the stylus depends on its position in the film and is largest at the center of the stylus. The energetic approach in the indentation or scratching process has been performed by Laugier [15, 16] and Burnett and Rickerby [17, 18]. We also calculate the elastic energy density, u, stored right under the center of the stylus tip. The formula for the calculation is [11],

$$u = (1 - v)p^{*2}/E = (1 - v)(1 + \mu^2)p^2/E \tag{4}$$

where v and E are Poisson's ratio and Young's modulus of the film, respectively.

The energy density per unit area right under the stylus tip assuming a constant force is,

$$ud = (1 - v)(1 + \mu^2)p^2 d/E \tag{5}$$

where d is the film thickness.

Here we define the adhesion energy as the energy needed to peel off the film from the substrate. We assume that when the value of ud exceeds the adhesion energy U_0, the film peeling by scratching occurs and thus we can estimate the adhesion energy by measuring p_c and μ, where p_c is the critical force to peel off the film. This calculation is applied to the results shown in Figure 5 and, using the results

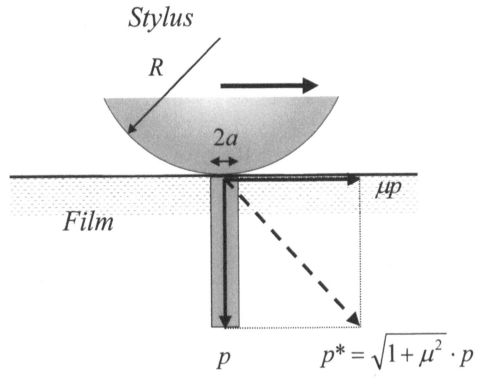

Figure 7. A model of the stress under the stylus. R: curvature radius of the stylus tip. a: radius of contact area.

shown in Figure 6, we obtain the adhesion energy shown by the closed circles in Figure 8.

In this calculation, the values of Young's moduli of the samples are needed. They could be determined by measuring the slope of the unloading curve in the nano-indentation experiment. However, the data scatter in the values of Young's modulus obtained by the experiment was considerably large and hence, we averaged the values of five samples each for Ti and TiC. They were determined to be 116 GPa and 321 GPa, respectively and for other films of $F < 0.5$, the linear dependence on F is assumed. On the other hand, the slope of the unloading curve for the films for $F > 0.5$ was unstable from sample to sample and a clear change in the loading and unloading curves could not be observed. Thus, we assumed that Young's modulus of our sample was the same and was 321 GPa to all the films for $F > 0.5$.

Figure 8 shows that the adhesion energy decreases with increasing F in an approximately monotonous manner. The highest adhesion energy by this evaluation is for Ti with a value near 1.7 J/m^2. This value is not very different from our previously obtained value [19].

For the precise evaluation of the stored elastic energy, it is necessary to evaluate the internal stress, σ, in the film. The internal stress measurement has been done and is shown in Figure 9 as a function of F. The stress varied from tensile to

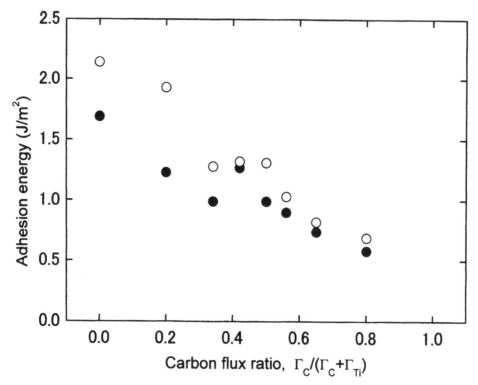

Figure 8. Calculated adhesion energy of Ti+C film as a function of Carbon flux ratio. ●: internal stress is not considered. ○: internal stress is considered.

compressive with the increase of carbon flux ratio. The maximum value was about 1 GPa. The elastic energy U due to this internal stress stored in a unit area of the substrate is roughly evaluated by the relation,

$$U = (1 - v)\sigma^2 d / E_f$$

The value of U was about 0.1 J/m². The energy due to the tensile stress is added to U_0, while the energy due to the compressive stress is reduced from U_0 because the compressive stress may be released through the indentation process by the stylus. The values of $U_0 \pm U$ are also shown by open circles in Figure 8.

In the estimation of the adhesion energy, we neglected the effect of plastic deformation and the impact of the stylus during the scratch test. In order to elucidate the whole scratch process more precisely, nano- or micro indentation experiments including the determination of the yield strength of the film and the measurement of the dependence on the stylus velocities are needed. Furthermore, in the present experiment, optical and/or SEM observations of the scratch pattern were not carried out and only the change in the friction coefficient was mainly concerned, but the failure mode during the scratching should be taken into account in a practical use of thin films.

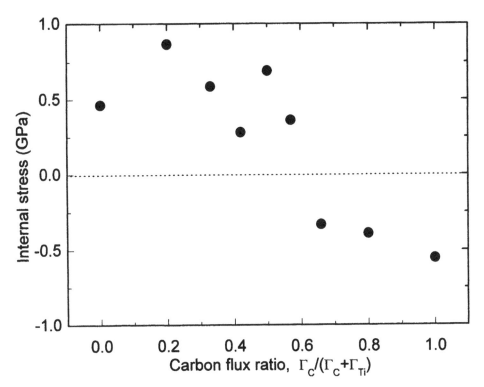

Figure 9. Internal stress in Ti+C films as a function of Carbon flux ratio.

5. CONCLUSION

The scratch test was applied to the adhesion evaluation of Ti+C films sputter deposited onto glass substrates. By controlling the C content in the film, the friction coefficient of the film surface was varied.

The critical load to peel off the Ti+C films by the scratch test seems to be small for pure Ti. But this seemed to be due to the large value of its friction coefficient. The adhesion energy is evaluated by considering the friction between the film surface and the stylus. Furthermore, the internal stress in the film was also taken into account. Introducing both effects, the adhesion energy on glass was found to decrease with the carbon content.

It seems that the scratch testing is useful for the adhesion evaluation in a practical meaning but in order to discuss the adhesion from a standpoint of physics, the critical load data are not sufficient and the friction and the internal stress should be taken into account.

Acknowledgement

The financial support from the Ministry of Education, Science and Culture is highly appreciated.

REFERENCES

1. O. S. Heavens, *J. Phys. Radium* **33**, 355 (1950).
2. K. L. Mittal, *Electrocomponent Sci. Technol.* **3**, 21–42 (1976).
3. K. L. Mittal, in *Adhesion Measurement of Thin Films, Thick Films and Bulk Coatings*, K. L. Mittal (Ed.), STP No. 640, pp. 5–17, ASTM, Philadelphia (1978).
4. J. Ahn, K. L. Mittal and R. H. MacQueen, ibid, p. 138.
5. K. L. Mittal, in *Adhesion Measurement of Films and Coatings*, K. L. Mittal (Ed.), pp. 1–13, VSP, Utrecht, The Netherlands (1995).
6. S. Baba, A. Kikuchi and A. Kinbara, *J. Vac. Sci. Technol.* **A4**, 3015 (1986), ibid. **A5**, 1860 (1987).
7. S. Baba and A. Kinbara, in *Proc. 8th Intern. Symp. Plasma Chem.*, p. 897, Tokyo, 1987.
8. A. Kinbara, S. Baba and A. Kikuchi, *J. Adhesion Sci. Technol.* **2**, 1 (1988).
9. A. Kinbara and S. Baba, *Thin Solid Films* **163**, 67 (1988).
10. F. Kimura, S. Baba, A. Kikuchi and A. Kinbara, *Thin Solid Films* **181**, 435 (1989).
11. A. Kinbara, S. Baba and A. Kikuchi, *Thin Solid Films* **171**, 93 (1989).
12. G. M. Hamilton and L. E. Goodman, *J. Appl. Mech.* **33**, 371 (1966).
13. M. Laugier, *Thin Solid Films* **76**, 289 (1981).
14. A. Somorjai, in *Mittal Festschrift on Adhesion Science and Technology*, W. J. van Ooij and H. R. Anderson (Eds.), p. 3, VSP, Utrecht, The Netherlands (1998).
15. M. T. Laugier, *Thin Solid Films* **117**, 243 (1984).
16. M. T. Laugier, *J. Mater. Sci.* **21**, 2269 (1986).
17. P. J. Burnett and D. S. Rickerby, *Thin Solid Films* **154**, 403 (1987).
18. P. J. Burnett and D. S. Rickerby, *Thin Solid Films* **157**, 233 (1988).
19. A. Kinbara and I. Kondo, *J. Adhesion Sci. Technol.* **7**, 767 (1993).

Adhesion Measurement of Films and Coatings, Vol. 2, pp. 187–204
Ed. K.L. Mittal
© VSP 2001

On the evaluation of coating-substrate adhesion by indentation experiments

B. ROTHER*

MAT GmbH Dresden, Reisstr. 3, D-01257 Dresden, Germany

Abstract—An overview of an energy-related approach for the evaluation of depth-sensing indentation measurements and their application to the evaluation of adhesion-related phenomena of thin films and coatings is given. As adhesion-related phenomena are considered internal stresses and crack formation of the coatings as well as an interface specific energy consumption during the penetration of the indenter into the coating-substrate system. The phenomena are discussed in relation to experimental effects. Experimental case studies are presented for multilayer structures of inkjet microheater systems as well as for single and multilayer hard coating systems.

Keywords: Adhesion; coating; depth-sensing indentation measurements.

1. INTRODUCTION

The technological starting point of interest in adhesion of coatings and thin films is usually the absence of adhesion or, in other words, coating delamination. Such failure would correlate to the zero point of empirical scales of adhesion and a full-scale value could be the fracture of either the coating or the substrate material. The principles of adhesion measurement (see for example [1]) can be classified into (i) destructive principles with the definition of loading conditions and finding of loading parameters leading to delamination, and (ii) non-destructive or nearly non-destructive analysis of adhesion-related responses of coating-substrate systems under defined loading conditions.

The second of the above classifications is of particular importance and it is the aim of the present paper to show the potential of depth-sensing indentation measurements as a nearly non-destructive principle for the evaluation of adhesion of physical vapor deposited (PVD) coatings.

---
* Phone: +49 (0) 351 20772-17, Fax: +49 (0) 351 20772-22, E-mail: rother@mat-dresden.de

2. EXPERIMENTAL EVALUATION OF ADHESION

PVD and CVD coatings represent a wide variety of material combinations and fields of application. The bonding principles between a coating and a substrate are generally of system specific character and involve both chemical and mechanical effects.

The response of coating-substrate systems to mechanical loading can, in principle, be described in analogy to stress-strain curves of bulk materials with the stress or fracture limits corresponding to delamination effects. Differences from the bulk analogue result from the appearance of normal and shear forces, from the superposition of coating and substrate effects as well as from the limited thickness of the coating and the interface region.

Typical loading conditions of coating-substrate systems can qualitatively be classified into elementary loading conditions as shown in Figure 1. Real mechanical loading conditions are usually combinations of such elementary loading conditions with dynamical and thermal effects as well as properties of the actual coating and substrate materials.

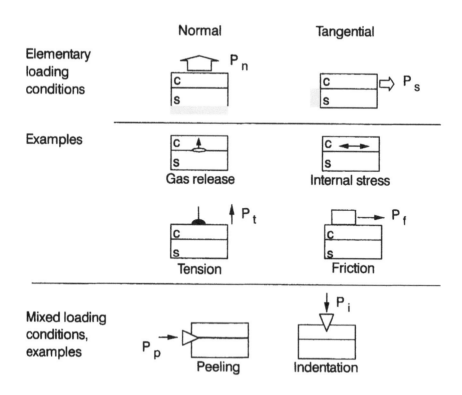

Figure 1. Loading conditions of coating-substrate systems. *c:* coating; *s:* substrate; P_n: normal force; P_s: shear force; P_t: tension force; P_f: friction force; P_p: peel force; P_i: indentation force.

From the elementary loading conditions shown in Figure 1 forces parallel to the interface with resulting shear stress in the interface region are of particular practical relevance. In this connection, important effects for coatings and thin films are: (i) internal coating stresses parallel to the interface as a result of energy deposition during film growth as well as thermally induced stresses as a result of different thermal expansions of coating and substrate materials, (ii) friction, scratch or indentation loading of the coatings with resulting force components parallel to the surface, and (iii) deformation of the substrate. These effects can be quantified by stress measurement by bending analysis [2] or x-ray diffraction [3], by scratch test with spallation pattern evaluation [4] or with acoustical emission analysis [5], by depth-sensing indentation measurements with indentation depths lower than the coating thickness [6] or higher than the coating thickness with spallation pattern characterization at the rim of the indent [7] as well as bending tests on coated samples [5].

Ideal adhesion quantification should employ non-destructive measurement methods to real samples with loading conditions close to the application of the samples. Depth-sensing indentation measurements with indentation depths smaller than the coating thickness are close to these ultimate requirements. They additionally provide an access to investigations of different classes of coating-substrate systems such as metal or ceramic coatings on metals or on plastics.

3. DEPTH SENSING INDENTATION MEASUREMENTS

Depth-sensing indentation (DSI) measurements with depth resolution in the nm-range are preferably used for the determination of hardness under load and the elastic modulus of thin films and subsurface layers (see e.g. [8]). The technique is based on classical hardness indentation measurements with additional recording of indenter displacement in dependence of the actual loading. The application of DSI measurements to the evaluation of adhesion and adhesion-related character-istics is based on an energetic interpretation of the indentation process [9]. In relation to the present discussion, a short summary of that approach will be given.

3.1. Energetic interpretation of DSI processes

3.1.1. Homogeneous Bulk Materials. The approach is based on the following assumptions.

(i)   The displacement energy of the indenter $W_d = \int\limits_0^h P\,dh_p$ with the load P, the

penetration depth h_p and the penetration depth at peak load h is converted into an increase of internal energy within a limited region of the probed material. This limited sample region is designated as virtual indenter. The concept of a virtual indenter has recently been confirmed by finite element calculations [10].

(ii) The virtual indenter is characterized by its volume V_i and its outer face area A_i as well as by the mean energy densities $\overline{w_V}$ for the volume related increase of internal energy density and $\overline{w_A}$ for the face area related increase of internal energy density. For homogeneous materials the virtual indenter is assumed to be in a most general case a segment of a spheroid with the projection face on the sample surface being determined by the diagonal(s) of the indent [11]. A schematic of such model is shown in Figure 2. During penetration of the real indenter, the virtual indenter expands against the resistance of the surrounding material and the volume of the virtual indenter is linearly related to the volume of the sample material replaced by the real indenter.

(iii) The penetration of the real indenter is considered to be quasistationary which permits to neglect the dynamic effects.

Starting from these assumptions, the balance of the displacement energy of the indenter with the increase of the internal energy of the sample region designated as virtual indenter can be written as

$$W_d = \overline{w_V}\, V_i + \overline{w_A}\, A_i \tag{1}$$

with V_i and A_i being the volume and the surface area of the virtual indenter, respectively. With respect to Figure 2 and by neglecting the real indenter, the vol-

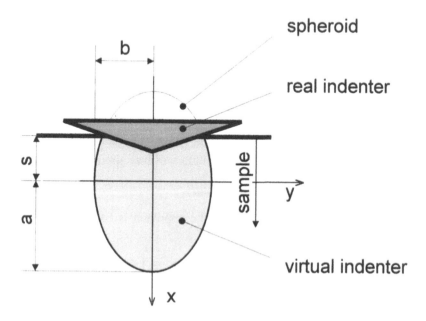

Figure 2. Schematic illustration of the indentation model with a real and a virtual indenter approximated by a cut-off of a spheroid. a, b: semi-axes of the spheroid; s: shift of the spheroid center with respect to the surface of the sample.

ume and the surface area of the virtual indenter are $V_i = \pi \int\limits_{-s}^{a} y^2 \, dx$ and

$A_i = 2\pi \int\limits_{-s}^{a} y \sqrt{1 + y'^2} \, dx$, respectively, with $y = b \sqrt{1 - \dfrac{(x-s)^2}{a^2}}$.

The integrals give for V_i the correlation

$$V_i = \frac{\pi}{3} b^2 \left(2a + 6s - 3\frac{s^2}{a} - 7\frac{s^3}{a^2} \right)$$

and for A_i a transcendental equation which can be approximated by the sum of the surface areas of one hemispheroid and the circumferential surface of the cylinder with the radius b and the height s (see Figure 2). This leads to the expression

$$A_i = \pi b \left(b + \frac{a}{\sqrt{a^2 - b^2}} \arcsin \sqrt{a^2 - b^2} \right) + 2\pi a s.$$

The two semi-axes a and b of the spheroid as well as the shift s of the spheroid have now to be correlated to the indentation process. To that aim, for pyramidal indenters and homogeneous sample materials, a linear relation between the two semi-axes according the equation $a = rb$ with the proportionality factor r is expected. Simultaneously, for the short semi-axis, a linear relation to the indentation depth h according to $b = k_i h$ with the proportionality constant k_i is assumed. The shift s of the spheroid is additionally assumed to be constant during penetration of a pyramidal indenter into homogeneous materials.

For the energy balance of Eq. 1 we thus obtain the expression

$$W_d = \overline{w_V} \, \pi \left(\frac{2}{3} r k_i^3 h^3 + 2 s k_i^2 h^2 - s^2 \frac{k_i}{r} h - \frac{7}{3 r^2} s^3 \right) + \overline{w_A} \, C(h) \tag{2}$$

with a polynomial expression for the volume term involving one cubic, one quadratic and one linear term of the indentation depth as well as with the transcendental expression $C(h)$ of the indentation depth for the surface area term. The expression is further simplified by the introduction of an indenter specific energy density e_v which is $e_v = \dfrac{2}{3} \overline{w_V} \, \pi r k_i^3$. Equation 2 can now be written as

$$W_d = e_v \left(h^3 + 3\frac{s}{r k_i} h^2 \right) - \overline{w_V} \, \pi \left(s^2 \frac{k_i}{r} h + \frac{7}{3 r^2} s^3 \right) + \overline{w_A} \, C(h) \tag{2a}$$

Eq. 2 can now be applied to the experimentally measurable quantities, load and indentation depth, by the second derivative of W_d in relation to h which is identical to the first derivative of the load in relation to h:

$$\frac{d^2 W_d}{dh^2} = \frac{dP}{dh} = 6e_V h + 6e_V \frac{s}{rk_i} + \overline{w_A} \frac{d^2}{dh^2}(C(h)) \tag{3}$$

The contributions of the three additive terms in Eq. 3 to dP/dh can be analyzed by the plot of experimental dP/dh values versus h. This approach proves, however, not to be trivial as the values of P and h are recorded, for the DSI measurements performed, at the limits of digital resolution of the available instruments. Consequently, a combination of averaging, numerical differentiation and data smoothing was developed which was designated as differential load and feed (DLF) analysis. To distinguish the results of this procedure from the pure differentiation dP/dh, the term "DLF plot" was introduced. To indicate the modification of dP/dh in the scheme of the DLF analysis, the term $\{dP/dh\}_{DLF}$ is used for the y-axis designation in the graphs.

The experimental analysis of Eq. 3 with data gained by DSI measurements with pyramidal indenters showed a linear behavior of the DLF plots for homogeneous bulk materials [9,12,13]. This experimental result suggests that the surface term of the mean internal energy density increase $\overline{w_A}$ can be neglected in relation to $\overline{w_V}$ which permits to neglect the contributions of the nonlinear term of Eq. 3 which can now, following an earlier convention [9], be simplified to

$$\frac{dP}{dh} = 6 e_V \left(h + \frac{s}{rk_i} \right). \tag{4}$$

The specific energy density e_V is closely related to the hardness under load and was, therefore, designated as hardness equivalent [9].

3.1.2. Coatings. The above considerations for homogeneous bulk materials hold, in principle, also for coating-substrate systems. This is illustrated in Figure 3. The expansion of the virtual indenter within the coating before reaching the interface region can be described according to Eq. 4 with the values of e_V, r, k_i and s being characteristics of the coating material. The situation changes after the virtual indenter has passed through the interface. Its further expansion within the coating is now limited by the coating thickness h_c whereas the expansion in the substrate follows the same rules as for bulk materials. The volume of the virtual indenter within the coating equals now the volume V_z of a zone of a spheroid defined by the semi-axes a and b as shown in Figure 2 as well as by the zone thickness which is identical to the coating thickness h_c. The volume of the virtual indenter within the coating V_{ic} is thus

$$V_{ic} = V_z = \pi b^2 (s + h_c) - \pi \frac{b^2}{a^2} \left(\frac{7}{3} s^3 + h_c s^2 - h_c^2 s + \frac{1}{3} h_c^3 \right). \tag{5}$$

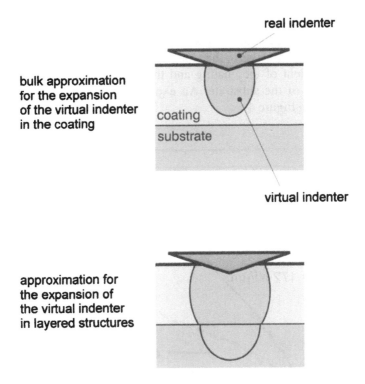

Figure 3. Schematic illustration of the penetration of a real indenter and the formation of the virtual indenter in a coating-substrate system.

The basic distinction of Eq. 5 compared to the expansion in homogeneous bulk materials as described by Eq. 2 is the disappearance of the cubic term of b and thus also of h.

An expression for the energy balance of the simultaneous expansion of the virtual indenter in both the coating and substrate, based on the discussion above and considering the relations between a, b, r, s and h as listed under 3.1.1., can now be written in analogy to Eq. 2 and the second derivative of that equation in relation to the indentation depth h gives

$$\frac{d^2 W_d}{dh^2} = \frac{d^2 W_c}{dh^2} + \frac{d^2 W_s}{dh^2} = e_{vc} \frac{3(s_c + h_c)}{r_c k_{ic}} + 6 e_{vs}\left(h + \frac{s_s}{r_s k_{is}}\right) \quad (6)$$

with the indices c and s referring to coating and substrate, respectively, and the indexed terms are analogous to the explanations in 3.1.1. Equation 6 is again a linear function of dP/dh versus h where the slope is determined only by the hardness equivalent of the substrate material. It should once more be noted that Eq. 6 stands for the simultaneous expansion of the virtual indenter in both the coating and substrate which is identical to the situation after the virtual indenter has passed the interface between coating and substrate. The expected DLF plots of coating-substrate systems should thus be determined by two successive linear ranges. Starting with h from zero, the slope of the first range should correspond to the hardness equivalent of the coating and the slope of the second range to the hardness equivalent of the substrate. An experimental confirmation for this expectation is shown in Figure 4.

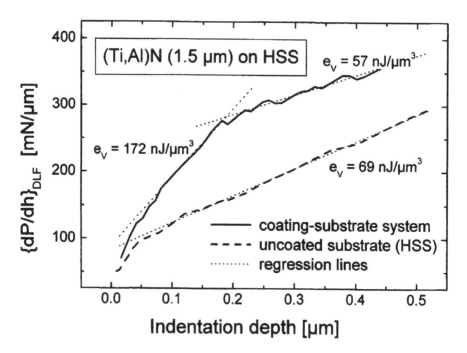

Figure 4. Experimental DLF plot for (Ti,Al)N on hardened high speed steel (HSS) together with the plot of the uncoated substrate material. e_v: hardness equivalent, see Eqs. 2 and 2a.

3.1.3. Coating-substrate Systems. The response of coating-substrate systems to the penetration of an indenter is a superposition of coating, substrate and interface effects. The proportions of these individual contributions change with increasing penetration depth. The DLF plots of coating-substrate systems should, based on the discussion above, be a combination of two adjacent linear functions as shown in Figure 4. Deviations from this estimated response can be expected to be caused by deviations from the model of two independent and homogeneous material ranges the virtual indenter passes during its penetration into coating-substrate systems. Phenomena which have to be considered in that connection are internal coating stresses, crack events and shear stress at the interface and these phenomena are again components of our complex understanding of coating adhesion.

Internal coating stresses have been shown for sputtered TiN coatings to have effects on the results of DSI measurements in that compressive internal stress causes positive vertical shift of the DLF plots [14]. This vertical shift correlates to the value of the absolute term of the linear equation which is, with respect to Eq. 4, e_V (s/rk_i). In particular, if the slope of the vertically shifted DLF plots remains constant, changes of their absolute term will, with respect to Figure 2 and the discussion in 3.1.1., correlate only to the term s/rk_i. This term describes the shape and shift of the virtual indenter independent of the indentation depth h. The above discussed changes of the term s/rk_i thus support the concept of the virtual indenter which expands against the resistance of the surrounding material with compressive stress in the surrounding material being one component of the resistance. A quantitative relation could, however, not be determined.

The crack events at the penetration of an indenter into coating-substrate systems are particularly evident with hard and brittle materials. These crack events are related to a stress relaxation which corresponds, in the picture of the virtual indenter, to a decrease of the internal energy density and, consequently, to a negative slope of the DLF plots. Such negative slopes are part of negative peaks of DLF plots. The existence of such negative peaks can be seen from the results published for example in [15] where the so-called pop-ins of load-indentation plots were correlated to crack events under the penetrating indenter. It should be mentioned, however, that such crack events do not provide information on whether the crack propagates along the interface or not. The correlation between the negative peaks in the DLF plots and the crack events in multilayer coatings as described in [16] provides, therefore, no access to the assessment of adhesion between the layers.

Besides the vertical shift of DLF plots and negative peaks, another important feature of the curves is a plateau region instead of a bending point between the coating and the substrate dominated plot ranges as shown in Figure 4. This plateau range was qualitatively discussed in terms of the resistance of the interface against the penetration of the boundary of the virtual indenter characterized by an indenter specific energy consumption [6]. The problems with this approach are, however, the definition of the plateau region as well as a comparison of the results to other adhesion assessment techniques such as scratch test [6] or Rockwell-C indent method [17].

Hence summarizing, adhesion related features of the DLF curves are vertical shifts, negative peaks and plateau ranges between the coating and substrate dominated plot ranges. A schematic illustration of these features is given in Figure 5 and the evaluation principle of the energy consumption of plateau ranges as presented in [6] is illustrated in Figure 6.

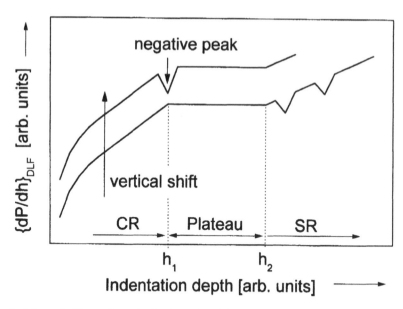

Figure 5. Schematic illustration of adhesion-related features of the DLF curves. CR: plot range where the slope is determined by the coating material; SR: plot range where the slope is dominated by the substrate material; h_1: indentation depth of the real indenter marking the end of the coating determined plot range CR; h_2: indentation depth of the real indenter marking the beginning of the substrate dominated plot range SR.

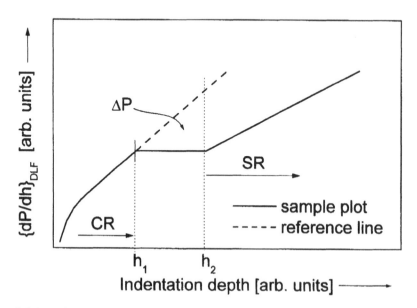

Figure 6. Schematic illustration of the evaluation principle for the energy consumption in relation to plateau ranges of the DLF plots: Calculation of indenter specific energy consumption by integration over ΔP. CR, SR, h_1, h_2: see Figure 5; ΔP: difference area (= difference force) between the linearly extrapolated coating range of an experimental DLF plot and the plateau range of the experimental plot.

3.2. Experimental case studies

The investigations in this section were performed with a microhardness tester with a load range of 0.4 to 1000 mN, a load resolution of 0.2 mN, an indenter displacement resolution of 2 nm and a stepwise quadratic increment of load over 60 steps. The waiting period between the consecutive steps was 1 s. The instrument was equipped with Vickers, Berkovich or Knoop indenters. The frame compliance of that instrument could be neglected [12] which permits to apply the above equations without modifications. All experiments were repeated at least five times and the analysis was conducted on the averaged load-indentation data.

3.2.1. Microheater Systems for Inkjet Arrays.

The main component of inkjet arrays is the microheater with multilayer systems consisting of heater, insulator and conductor materials of different thicknesses. The operation principle of such systems is a quick overheating of a very small ink volume by thermal contact and an ejection of the above liquid through a narrow nozzle. Thus the required thermal shock loads on the microheater are hard operation conditions which require high level of adhesion between the individual layers. The layers are usually sputtered in a multistage process with ion cleaning step before each layer deposition. The early detection of adhesion losses is thus of fundamental importance to the technology and the performance of the inkjet system. In the framework of a process quality program, the effects of the ion cleaning processes were determined by DSI measurements and a subsequent DLF analysis. The DSI measurements were performed with a Knoop indenter.

The following layer systems were deposited with only one section of each wafer being exposed to ion cleaning prior to deposition of the top layer:

1) Si_3N_4 700 nm / HfB_2 200 nm / thermally oxidized Si wafer,
2) Si_3N_4 700 nm /Al 600 nm / HfB_2 200 nm / thermally oxidized Si wafer,
3) Ta 600 nm / thermally oxidized Si wafer,
4) Ta 600 nm / Si_3N_4 1.5 µm / thermally oxidized Si wafer.

Despite the low layer thicknesses a high reproducibility in the DSI measurements could be achieved. The determination of linear ranges in the plots and, consequently, the calculation of a hardness equivalent failed, however, because of the low thicknesses of the individual layers. Instead, qualitative effects of ion cleaning could be shown in the DLF plots. For the systems 1, 2 and 3 the effects of ion cleaning could clearly be distinguished as a vertical shift of the DLF curves, whereas for the system 4 no changes for the ion cleaning could be determined. Typical results for the different layer arrays are shown in Figure 7. An interpretation of this result will be given in the discussion section.

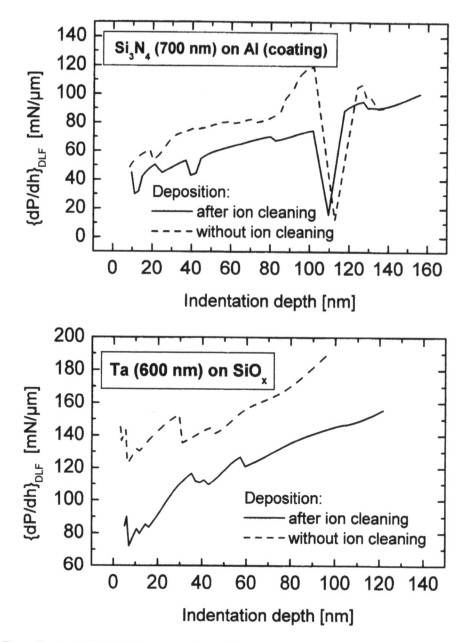

Figure 7. Typical DLF plots for the investigated layer arrangements of inkjet systems. Indenter used for DSI measurements: Knoop pyramid.

3.2.2. Hard Coatings for Wear Protection. A large field for the application of PVD and CVD hard coatings is the wear protection of tools and machinery components. Typical coating-substrate systems for this purpose consist of coating materials such as TiN, (Ti,Al)N, or, generally, transition metal nitrides or carbides and substrate materials such as high speed steel or cemented carbide. The adhesion of the coatings is of substantial importance to the performance of the systems. Process steps of the deposition with direct relation to coating adhesion are wet cleaning and ion cleaning prior to deposition as well as ion mixing between the coating and the substrate by ion bombardment during interface formation.

The efficiency of wet cleaning procedures (combination of hydrocarbon cleaning with electrochemically enhanced degreasing) was analyzed in [17] for magnetron sputtered (Ti,Zr)N coatings deposited on hardened high speed steel. Prior to deposition, the samples were exposed to three different wet cleaning procedures, and the deposition of the (Ti,Zr)N coatings was performed in one run with identical parameter sets. The different wet cleaning procedures could thus be expected to affect mainly coating adhesion with only minor influence on other properties of the coating-substrate systems. The expected effects were analyzed by DSI measurements with subsequent DLF analysis as well as by Rockwell-C indentation tests with a semiquantitative evaluation of the delamination patterns of the coating at the rim of the indent. Typical DLF plots are shown in Figure 8 where the dotted and dashed auxiliary lines shown in the bottom diagram are in relation to the scheme given in Figure 6 and indicate the area over which integration is performed to calculate the specific energy consumption for the penetration of the virtual indenter boundary through the interface. A summary of these results together with the evaluation of the Rockwell-C indents is given in Table 1.

Table 1.
Adhesion evaluation for (Ti,Zr)N coatings on high speed steel of different wet cleaning stages [16]. Calculation results of a specific penetration energy of the virtual indenter through the interface and semiqualitative evaluation of Rockwell-C indents

Sample	Wet cleaning intensity[1]	e_d in nJ	Rockwell-C adhesion class[2]
A	11 cleaning steps	0.18	HF 3
B	10 cleaning steps	0.22	HF 4
C	3 cleaning steps	1.0	HF 5

e_d: Specific penetration energy of the virtual indenter through the interface.
[1] A: 11-stage procedure of ultrasonically enhanced hydrocarbon cleaning and electrochemical degreasing. B: As for A, without neutralization between cathodic and anodic degreasing. C: 3-stage procedure of ultrasonically enhanced hydrocarbon cleaning.
[2] Classification numbers according to [7].

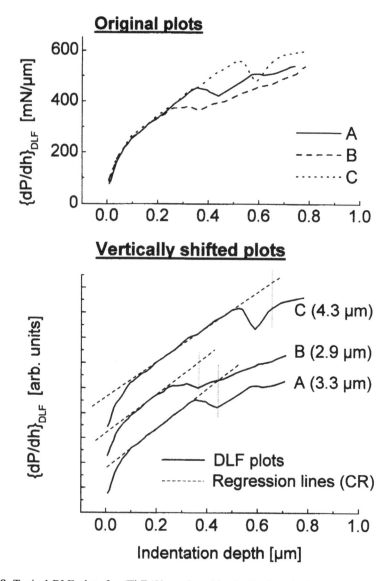

Figure 8. Typical DLF plots for (Ti,Zr)N coatings identically deposited on hardened high speed steel with different wet cleaning stages prior to deposition. Wet cleaning efficiency decreases from A to C. The numbers in parentheses are the coating thicknesses. Indenter used for DSI measurements: Vickers pyramid.

Besides wet cleaning, ion cleaning is a further process step in PVD which is directly linked to adhesion. The effects of ion cleaning time have been analyzed in connection with sputter deposited TiN on hardened high speed steel. For this purpose, the samples were prepared with identical surface finish and wet cleaning procedures. Two samples were placed in one deposition run and each of them was exposed for different durations to an Ar ion etch at -300 V substrate bias. The samples were then identically coated by magnetron sputtering [18]. The effects of the different ion cleanings on adhesion were evaluated by DSI measurements and DLF analysis. The results of these investigations are shown in Figure 9. The characteristic result for all investigated samples was the slight but reproducible vertical shift of those plots with lower cleaning time.

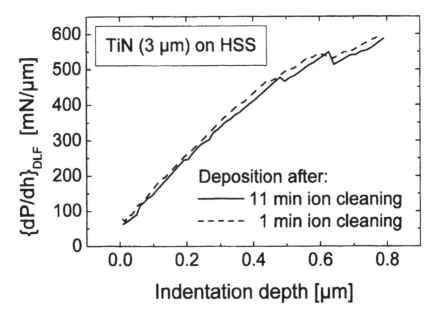

Figure 9. Typical DLF plots for TiN coatings on hardened high speed steel (HSS) deposited after different ion cleaning times. Indenter used for DSI measurements: Berkovich pyramid.

A powerful tool for the increase of adhesion between PVD coatings and metallic as well as ceramic substrates is the high energy ion bombardment during the interface formation. Ion energies in the range of some keV are thereby suitable to initiate implantation and recoil implantation of the film forming particles into the substrate material which increases generally the potential for increased adhesion [19]. This process step was again investigated with respect to its effect on DSI measurements. For this purpose, different high speed steel samples were arranged in pairs in the deposition chamber and only one of them was exposed to an Ar ion bombardment during the first minute of TiN deposition by magnetron sputtering. Before and after the high energy ion bombardment the samples were treated identically. A typical result for a pair of samples coated under identical parameters excluding the high energy ion bombardment during interface formation is shown in Figure 10. A typical result of these investigations was again a vertical shift in the DLF plot for the sample which was deposited without ion bombardment during the formation of the interface.

B. Rother

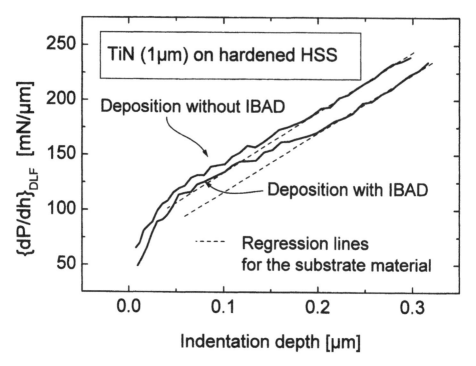

Figure 10. Typical DLF plots for TiN coatings on hardened high speed steel (HSS). Deposition with and without 10 keV Ar-ion bombardment during interface formation. Indenter used for DSI measurements: Vickers pyramid. IBAD: Ion Beam Assisted Deposition.

3.3. Discussion

The results presented here are a summary of the work on the investigation of empirical correlations between adhesion-relevant PVD process steps and features of the response of coating-substrate systems to indentation experiments. This response was analyzed by an energetic interpretation of the indentation process with the first derivative of the load in relation to the indentation depth plotted versus the indentation depth. The plot features particularly considered in relation to adhesion were (i) vertical shifts which could be expected to correlate with internal stresses, (ii) negative peaks which are known from the literature to correlate to crack events, and (iii) a plateau region between the DLF plot ranges dominated by the coating and by the substrate responses. An illustration of these features was given in Figure 5.

The dominating effect of the above features was the vertical shift of the DLF plots. The positive shifts thereby correlate with an expected reduced adhesion between the coating and underlayer. This effect is very important for hardness measurements on coatings as it is in contradiction to the generally acknowledged Bückle-rule which claims that the effect of the substrate is relevant to hardness values for indentation depths larger than 1/10 of the coating thickness [20]. The results illustrated in Figures 9 and 10 and particularly in Figure 7 show that the effects of the interface are already evident at the beginning of the penetration process. Compared to conventional hardness calculations as load per area, these positive shifts are identical to higher hardness values.

At the present state, the vertical shift of the DLF plots as an effect of adhesion can be discussed in relation to compressive stress in the coatings. This correlates to earlier results with TiN coatings deposited with different bias voltages generating different internal stresses which were determined independently from the DSI measurements [14]. The correlation between internal coating stress and adhesion follows from a qualitative interpretation of ion enhanced film growth where the ions are generally assumed to generate internal stress in the growing film [21]. This stress could be balanced by the substrate, and the transformation of the coating stress to the substrate is the more efficient the better the interface perfection is. A reduced interface perfection thus leads to accumulation of internal stress in the coating which again correlates with the origin of the vertical shift of the DLF plots as discussed above.

The consideration of the negative peaks in the DLF plots as crack events was introduced above as a further feature for an evaluation of adhesion effects. These negative peaks do not, however, distinguish between the crack propagation along or perpendicular to the interface. The negative peaks in the DLF plots are thus a more comprehensive criterion of coating-substrate systems. Therefore, in the present paper it was excluded from an extended discussion.

The plateau ranges in the DLF plots of coating-substrate systems was the feature of the plots which was first investigated for quantitative correlations with interface effects [6]. The reliability of the suggested quantification procedure depends, however, on the reliable determination of the integration boundaries and this is not fully solved yet [22]. The application of plateau criterion to the quantitative evaluation of coating adhesion is, therefore, the subject of further investigations.

4. CONCLUSIONS

The results presented illustrate the state of adhesion assessment by depth sensing indentation measurements with an energetic evaluation of the load-indentation data. It was shown that the results were particularly suitable for comparative assessment of interface effects. The sensitivity of the approach to interface effects is dependent on the nature and intensity of the effects as well as of the material combinations considered.

Effective assessments of adhesion were demonstrated for multilayer arrangements of inkjet systems. Application of the approach to hard coating evaluations was also demonstrated. At this stage, the approach can already be considered as a useful tool for adhesion assessments. Further developments of the approach require further understanding of the model of the virtual indenter combined with suitable experimental investigations.

REFERENCES

1. K.L. Mittal (Ed.), *Adhesion Measurement of Films and Coatings*, VSP, Utrecht, The Netherlands (1995).
2. C.N. Kouyumdjiev, I.V. Ivanov and R.R. Tanov, Surface Coatings Technol., **113**, 113-119 (1999).
3. R. Wiedemann, H. Oettel and M. Jerenz, Surface Coatings Technol. **97**, 313-321 (1997).
4. P.J. Burnett and D.S. Rickerby, Thin Solid Films **154**, 403 (1987).
5. H. Ollendorf and D. Schneider, Surface Coatings Technol. **113**, 86-102 (1999).
6. B. Rother and D.A. Dietrich, Thin Solid Films, **250**, 181-186 (1994).
7. Association of German Engineers, Coating (CVD, PVD) of Cold Forging Tools, VDI Guideline VDI 3198, Beuth Verlag, Berlin 1992 (in German).
8. W.C. Oliver and G.M. Pharr, J. Mater. Res., **7**, 1564-1583 (1992).
9. B. Rother and D.A. Dietrich, Phys. Stat. Sol. **142**, 389-407 (1994).
10. N. Schwarzer, F. Richter and G. Hecht, Surface Coatings Technol. **114**, 292-304 (1999).
11. J.M. Olaf, PhD Thesis, Albert-Ludwigs-Universität, Freiburg 1992 (in German).
12. B. Rother, A. Steiner, D.A. Dietrich, J. Haupt and W. Gissler, J. Mater. Res. **13**, 2071-2076 (1998).
13. B. Rother, J. Mater. Sci., **30**, 5394-5398 (1995).
14. B. Rother, Materialwiss. Werkstofftech., **26**, 477-482 (1995).
15. T.F. Page and S.V. Hainsworth, Surface Coatings Technol., **61**, 201-208 (1993).
16. M.K. Kazmanli, B. Rother, M. Ürgen and C. Mitterer, Surface Coatings Technol., **107**, 65-75 (1998).
17. B. Rother, L.A. Donohue and H. Kappl, Surface Coatings Technol. **82**, 214-217 (1996).
18. B. Rother in B. Michel and T. Winkler (Eds.): *Proceedings of the International Conference on Micromaterials*, pp. 214-217, ddp Goldenbogen, Dresden 1997 (ISBN 3-932434-05-6).
19. G.K. Wolf, P. Engel, D. Heyden, E. Vera and F. Stippich, Materialwiss. Werkstofftech. **29**, 518-524 (1998).
20. H. Bückle, *Microhardness Measurements and Their Application*, Berliner Union Verlag, Stuttgart (1965) (in German).
21. H. Oettel and R. Wiedemann, Surface Coatings Technol., **76/77**, 265-273 (1995).
22. B. Rother and M.K. Kazmanli, Surface Coatings Technol. **99**, 311-318 (1998).

Adhesion Measurement of Films and Coatings, Vol. 2, pp. 205–217
Ed. K.L. Mittal
© VSP 2001

Measurement of interfacial fracture energy in microelectronic multifilm applications

JACK C. HAY, *,† ERIC G. LINIGER and XIAO HU LIU

I.B.M. Research, P.O. Box 218, Yorktown Heights, NY 10598-0218, U.S.A

Abstract—A testing method in the literature, the modified edge lift-off test (MELT), was evaluated as a potential candidate for determining the quality of interfaces. Delamination is induced through the release of strain energy stored in an elastic superlayer which results from a large mismatch in coefficients of thermal expansion (CTE) between the film and substrate. However, several experimental issues as well as the mechanics used to describe the problem need to be considered closely. In this work a critical examination of the energy release rate was considered. Experimental data suggest that the energy release rate is independent of starting flaw size for starter flaws greater than 5% of the epoxy film thickness. This is contrary to a conventional rule of thumb that suggests the energy release rate, G, only achieves steady state conditions after the crack length exceeds 10–20 times the film thickness. Numerical models confirm this observation for epoxy films on silicon substrates. Simulations of a chromium film were also performed to (a) confirm that the models were consistent with chromium simulations in the literature and (b) to evidence that the observed behavior of the epoxy/silicon system was due to the low modulus of the epoxy film relative to the substrate. The rule of thumb does not apply when Young's modulus of the film is much smaller than the substrate Young's modulus.

Keywords: Adhesion; test; superlayer; elasticity; multifilm.

1. INTRODUCTION

It is common knowledge that the delamination of interfaces between dielectrics, caps and metals is a critical issue for the microelectronics industry. These are the building blocks for a semiconductor device. Metal lines and vias carry the signals from one device to the next and the dielectric is present as a structural material to keep the metal lines and vias separated. A cap is an additional barrier between dielectrics and metals used to prevent diffusion of the metal lines into adjacent dielectric material and the diffusion of dielectric material into the metal lines.

---

*To whom correspondence should be addressed. Phone: (865)927-3000, Fax: (865)927-3003, E-mail: jchay@ffdevices.com

†Present address: Fast Forward Devices, LLC, 11020 Solway School Road, Suite 113, Knoxville, TN 37931, U.S.A

Microelectronic devices are extremely elaborate in the variety of materials which exist in the back end of the line (BEOL) and the manner in which they interact. The effects of widely different material properties are exacerbated as newer generation low-k materials are integrated with interconnect structures.

The driving force for crack growth has several different sources and may include the residual stresses from processing, or they may be applied while conducting a variety of industry reliability tests which push, pull, shear, or thermally stress the BEOL. The challenge for the engineer is to understand the critical link between a successful adhesion characterization in the laboratory and the actual stress state in a device. At the present time this relationship is not understood, but it is assumed that the rank ordering of practical adhesion by any one test will translate to a similar ranking of interfacial performance in the device. However, it is quite another issue to predict how *much* better one interface is over another when the loading conditions change. Therefore, the question which ultimately needs to be addressed is that of the actual loading conditions in the device. It is then desirable to conduct practical adhesion tests under similar conditions with regards to mode mixity, geometry and environment.

There exist many practical adhesion tests in the literature which range in terms of mode mixity, geometry and the environment. It is clear that the total energy released during a test is dissipated in forming the new surfaces and is represented by the interfacial debond energy, Γ_i, and other dissipative processes such as plasticity and interfacial roughness. In particular, it has been shown that the critical energy release rate is a strong function of mode mixity, Ψ, when the interface is rough or when plasticity occurs with the fracture event [1, 2, 3]. Mode mixity is a measure of the shear loading to tensile loading ahead of the crack tip. Many methods exist in the literature for measuring Γ_i and can be generalized into two groups: one which provides a qualitative ranking of interfacial fracture energy, and one which provides quantitative ranking of interfaces. Examples of the former include the scratch [4–7], indentation [8–12], peel [13, 14] and blister tests [15–17]. The fundamental issue is the difficulty experienced in deconvoluting the elastic and plastic components. The peel test is one example where several groups [13, 14, 18] have quantified the elastic component of the total dissipated energy in the interface and the plastic component of the fracture energy due to plasticity in the materials peeled off the substrate. However, considerable numerical modeling is required to deconvolute the total energy into these two components. These same problems extend also to the scratch and indentation methods where considerable work may be done in plastically deforming the film or substrate, thus requiring numerical models to determine Γ_i. The condition under which the mechanics for indentation-induced delamination is valid assumes that plasticity is confined to the film [9]. This requirement makes the test difficult to use for all but the weakest interfaces, since much of the supplied energy is dissipated by plastic deformation of the film. However, Kriese et al. [12] have reported success with the technique by applying a superlayer with a high residual stress.

A preferred method for evaluating Γ_i would be one which is quantitative, allows for easy sample preparation, is described by elastic analyses, and yields unambiguous results. The tests described above meet most of these criteria, but none seems to satisfy all of them. Two popular methods in the literature which meet most of these criteria are the microstrip test [19–22] and the modified edge lift-off test (MELT) [23–27]. While the focus of the present work is to examine the MELT and the validity of the assumptions and mechanics used to describe the test, the foundations for the MELT can be found in the microstrip studies.

2. BACKGROUND MECHANICS

In both the microstrip and MELT methods the stored elastic strain energy is released when a crack initiates at an edge, and steady state conditions are attained at some crack length which depends upon geometry and the mechanical properties of the film. In its simplest form, one may consider limiting assumptions where there exists one thin film on an infinitely rigid substrate, in which case all of the strain energy in the film is released when the crack propagates. For a long narrow strip the steady state energy release rate, G_{ss}, is given by [1, 20]

$$G_{ss}E_f/\sigma_f^2 h_f = 1/2 \tag{1}$$

where E_f is Young's modulus of the film, σ_f is the tensile stress in the film, and h_f is the film thickness. When wide strips are used, the film is subject to biaxial stress and the biaxial modulus is used in Equation (1), yielding [1]

$$G_{ss}E_f/\sigma_f^2 h_f = 1 - v_f \tag{2}$$

where v_f is the Poisson ratio. When the crack length is short, edge effects must be taken into account since the energy release rate falls precipitously at small crack lengths [1, 20, 21]. In general, to avoid edge effects the energy release rate should be measured when the crack length is long relative to the film thickness, i.e., $a > h_f$.

The microstrip test was designed [19] as a means for measuring the interfacial fracture energy, assuming that all of the energy released was dissipated in the generation of the fracture surface; thus, the test can be analyzed according to Equation (1). Although the details for the test can be found elsewhere [19, 22], a brief summary of the experiment follows. A release layer, such as carbon, is deposited onto a silicon wafer and is patterned into long narrow strips. The materials for which the adhesion measurement is desired are then deposited onto the wafer in blanket form covering the carbon strips. In the methods prescribed by He et al. [21] and Bagchi et al. [19], the superlayer which provides the strain energy is chromium and is deposited directly on top of the films of interest. Chromium as deposited may have a residual tensile stress as high as 1 GPa [22]. Finally, the blanket films are patterned into long narrow strips perpendicular to the carbon release layer, and a thin slit is etched through the films along the axis of the release layer. This last step creates a starter crack beneath the test strip with an initial crack length equal to the half-width of the carbon strip. The strips delaminate from the substrate when the energy release rate is greater than the interfacial fracture energy; otherwise the crack tip is stable and will not move. Equations (1) and (2) reveal that for a given residual stress there is some critical thickness, h_f, corresponding to a critical energy release rate, thus successive changes in the chromium thickness puts bounds on the actual interfacial fracture energy.

Another test which uses a residually stressed layer to supply the energy for crack growth is the modified edge lift-off test (MELT) [23–27]. In this test one starts with a silicon wafer and the films of interest are deposited as blankets directly onto the wafer. The films of interest to the semiconductor industry are deposited by conventional techniques including CVD, PVD, PLD, electrolytic baths, or spin-on. Finally, an epoxy superlayer approximately 150 μm to 200 μm thick is coated onto the wafer using a "doctor blade", a device with a calibrated gap between itself and the wafer leaving a controlled thickness of epoxy. The epoxy is cured on a hot plate at 180°C in air and upon cooling to room temperature it develops a tensile stress of approximately 27 MPa due to the mismatch in thermal expansion coefficients between the epoxy and the silicon. Although it is beyond the scope of this paper, a study of stress relaxation at room temperature was conducted [28]. In brief the results of that study revealed an exponential-type reduction in stress yielding a 7 MPa drop in the first ten hours after curing the epoxy. Beyond 10 hours the reductions in stress were much less severe and required an additional 14 hours to obtain an additional drop of 1 MPa. The large sample is then cleaved or diced into 1–2 cm squares, placed on a cold stage and cooled at a rate of roughly 10°C/minute until delamination is observed. Typically, the samples were cured on one day and tested the next day, suggesting that the starting stress was closer to 20 MPa than the original 27 MPa. Comparing with the principles of the microstrip test, this test seeks a critical temperature, or σ_f, at which delamination occurs. It is assumed [23] in the MELT method that the crack tip is subject to plane-strain conditions. Thus, the energy release rate is given by

$$G_{ss} E_f / \sigma_f^2 h_f = (1 - v^2)/2. \tag{3}$$

The MELT method has been considered as a method for evaluating the interfacial fracture energy for materials in the BEOL. As will be discussed in the following section, the experimental observations do not support the hypothesis that the sample can be modeled as a plane strain strip since we rarely observe unstable crack growth. We have examined many aspects of the MELT method with regards to the epoxy properties and possible influences on the observed behavior, in addition to the mechanics. Modeling suggests that steady state energy release rates can be achieved at crack lengths less than the film thickness.

3. RESULTS

There are numerous issues that need to be addressed in considering the MELT method for measuring Γ_i. In this communication we explore two important aspects of the initial flaw on the test: (i) the ability of the crack to run on the desired interface when many films are present, and (ii) to determine the required crack length to achieve steady state conditions.

The crack path in a heterogeneous solid is quite complex, as discussed by Hutchinson and Suo [1]. In isotropic solids there is considerable evidence that a crack will seek a trajectory which minimizes the mode mixity at the crack tip, i.e., it will seek pure mode I opening conditions. The desired crack path in a heterogeneous solid consisting of a film and a substrate is not so simple. If the crack is trapped in a low-toughness interface then the state of stress at the crack tip will be mixed mode. The degree of asymmetry about the interface with respect to Young's moduli in addition to the loading conditions and geometry determine the severity of the mode II component experienced by the crack tip. The ability of the crack to remain in the interface depends on the relative toughnesses of the interface and of the adjacent material. Conditions that would favor kinking would include (i) a high toughness interface, (ii) low toughness materials on either side of the crack, and (iii) the presence of flaws on the interface that might allow the crack to kink at roughly 45°.

The primary reason the semiconductor industry is intent on identifying a suitable test for interfacial fracture is that the interfaces are typically low toughness. Thus, within the context of Hutchinson and Suo's work, cracks tend to get trapped in interfaces and have difficulty in kinking into other interfaces or into the substrate. In semiconductor devices and in the MELT samples there are multiple films on top of the silicon substrate, thus, there are multiple interfaces that can trap the crack. As shown schematically in Figure 1, there may be many layers deposited onto the silicon substrate and the interface of interest may be near the top film. The lower films are necessary in many cases since the materials of interest often do not adhere to silicon or to the native oxide on the silicon. Thus, liner materials, or strongly adhering films such as chromium are deposited onto the wafer first. Also shown schematically in the figure is a large chip in the silicon below the films. This is typical of what one can observe in an optical microscope when the samples are diced from the wafer.

An example of a typical delaminated sample is shown in plan view in Figure 2. The stacking sequence for the films in this example (from top down) was epoxy/interlevel dielectric (ILD)/silicon nitride/native oxide/silicon. The desired interface for fracture was the ILD/silicon nitride interface, but it is apparent from the optical micrograph that the crack propagated on many different levels. In the central portions of the sample (top of micrograph) the crack propagated along the silicon nitride/silicon interface or down into the silicon. On some initial samples a profilometer was used to determine what materials were remaining on the substrate after test was performed. Referring to pictures such as that in Figure 2 does not

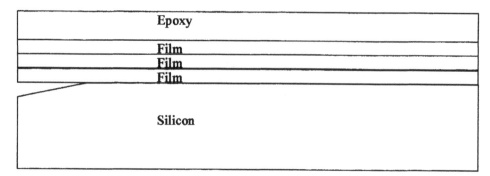

Figure 1. Schematic drawing of a diced sample illustrating the location of large flaws in the silicon due to the dicing procedure. When such flaws are present, crack growth on other interfaces is not always energetically favorable.

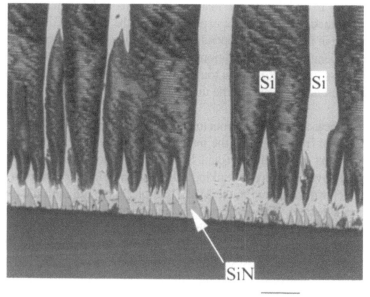

200 μm

Figure 2. Plan-view micrograph of a test sample where a delamination proceeds in many different interfaces due to initial flaws in the silicon. In some areas the crack has actually deviated into the silicon substrate.

provide out-of-plane dimensions at low magnification. However, subsequent profilometry and known film thicknesses provided the information necessary to label the picture as in Figure 2. There are other analytical tools available that can provide similar information based on the chemistry of the two sample surfaces. Those techniques can more accurately determine whether the crack ran along the interface, or through the bulk of the two adjoining materials. This information would be crucial to persons developing or characterizing adhesion promoters or pretreatments before film deposition. In this work where we are studying the edge effects on the energy release rate calculation, such chemical information is not necessary.

The process by which the crack kinks down into the silicon and propagates has been addressed by previous authors (see reference 20 for instance) for other systems and is easily explained by the large mode mixity associated with this problem. For the purposes of the MELT, however, the fracture surfaces such as that in Figure 2 were not desirable, as the energy release rate of any particular interface cannot be determined. Useful data can only be obtained from samples where the crack propagated on a single interface. From pictures at higher magnifications than that in Figure 2 one can observe large dicing flaws in the silicon, and it was our contention that these flaws were assisting the crack growth to take place on the silicon nitride/silicon interface. Summarizing, when multiple interfaces are present the crack will take the path of least resistance. Since the cracks were propagating along several different interfaces it is speculated that there was competition between the inherent flaws along the sample edges on the different interfaces.

One method we propose for dealing with the influence of the large dicing flaws in the silicon is to eliminate them by polishing the edges of the sample with 1200 grit silicon carbide grinding paper. The polishing step substantially reduces the size of flaws on these edges, but also eliminates the natural flaws in the materials at the other interfaces. This raises two questions (1) how large does the precrack on the desired interface need to be to eliminate competition from other interfaces? and (2) how large does the precrack need to be to obtain a steady state energy release rate?

We performed careful studies to determine experimentally how large a crack one needs to introduce to measure a steady state value for the energy release rate. For two systems, described next, we have polished the sample edges to remove the dicing flaws, and have followed up with a wet etch designed to remove part of the material below the interface of interest. In the case of the system described above, hydrofluoric acid (HF) removes silicon nitride from the edge of the sample. This step introduces a very sharp crack tip where the crack opening is simply the thickness of the film removed. The length of the crack is proportional to the length of etch time. Subsequent testing in the MELT results in a "clean" interfacial delamination. An example of such a sample is provided in Figure 3 where the ILD and epoxy have been removed from the sample. Profilometry can be used again to determine the thickness of the silicon nitride and compare with the known deposition thickness. The HF etch has removed silicon nitride from along the edge, providing a starter crack for which the initial crack length depends on the length of time the sample was in the etchant. In this figure the starting crack length was approximately 100 μm. In a second system consisting of epoxy/cap/copper/liner/silicon, a ferric chloride etch removes the copper from the edge producing a similar starter crack.

In Figure 4 the length of the starter crack induced by the etching (measured by optical microscopy) follows the predicted square root dependency for both of the

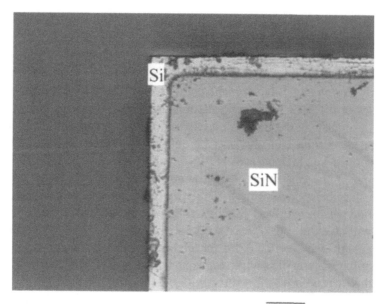

200 μm

Figure 3. Plan-view micrograph of a test sample where a delamination proceeds along the interface between SiN and the ILD, which delaminated from the SiN. In this case the sample was polished on all edges using 1200 grit SiC paper and a starter crack was introduced by etching the SiN with dilute HF.

systems. The constant of proportionality depends on the chemical activity of the etchant and the material being etched, but clearly large flaws can be induced in times less than an hour. Interestingly, when these square samples were evaluated for the adhesion, the measured energy release rate did not depend on the length of the starter crack. It was initially assumed that the energy release rate would decay rapidly as the initial flaw size was decreased, thus, requiring larger film stresses to drive the crack. Yet, the data were insensitive to the length of etching. In light of recent finite element modeling [21], this seemingly incorrect observation is supported by numerical modeling described next.

There are two possible explanations for the observation that the experimental energy release rate, G, does not depend on the initial crack length. First, the starter cracks may be too short to measure the steady state energy release rate and that the range of crack sizes investigated was too small relative to the required crack length. Second, it may be possible that all of the measured values were in the steady state regime, as the critical crack length might be a very small fraction of the film thickness. The finite element method (ABAQUS) was used to examine the dependence of energy release rate on the crack length. For the purposes of simulations we treat only one film on a silicon substrate at a time. Both epoxy and chromium are considered here to evidence the effect of Young's modulus on the required crack length for steady state conditions. Additionally, chromium has

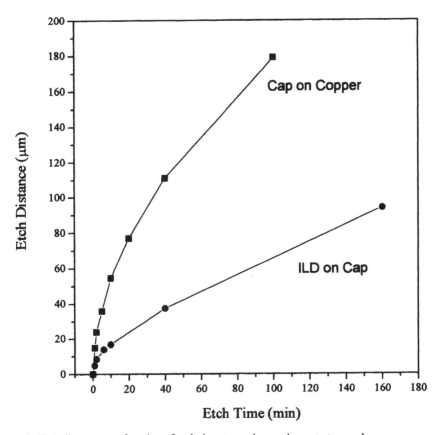

Figure 4. Etch distance as a function of etch time to produce a sharp starter crack.

the benefit of having already been modeled in the literature [21], thus providing a set of data to check consistency of the models. In this simulation the substrate is allowed to bend. It is also assumed that the constitutive behavior of both the film and substrate materials is linear-elastic, and the simulated sample is treated in plane strain as a 25 mm long strip.

In Figure 5 the normalized energy release rate, G/G_{ss}, is plotted as a function of nondimensional crack length, a/h, where G_{ss} is the steady state energy release rate of a perfectly rigid substrate with epoxy or chromium films using Equation (3). We will consider the epoxy/silicon simulation first. Similar to the numerical simulations by He et al. [21] the energy release rate is zero for zero crack length and in the limit of the converging crack. Looking first at the steady state regime for the epoxy/silicon system, $0.5 < a/h < 115$, the energy release rate never attains the value predicted by Equation (3), as there appears to be an error of approximately 4%. This discrepancy represents the error due to assuming conditions of a rigid substrate. When the modeled films are thinner or the substrate is made thicker, this error is suppressed.

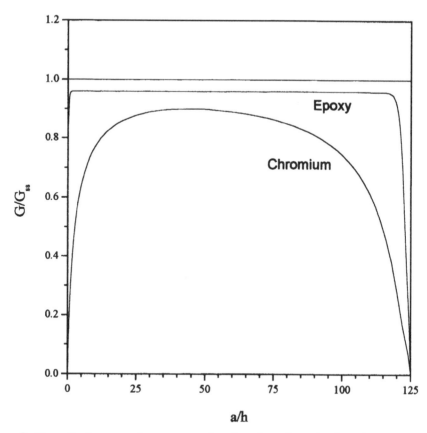

Figure 5. Normalized energy release rate as a function of nondimensional crack length, a/h, for epoxy and chromium superlayers on silicon.

In the short crack regime, $a/h < 0.5$, the energy release rate depends strongly on the crack length. The data in Figure 5 are magnified in Figure 6 in the short crack regime to reveal the details of how fast the energy release rate decreases. In the scale of Figure 6 it is noteworthy that the energy release rate does not deviate significantly from the steady state value until the crack length is less than 5% of the film thickness (10 μm) where G/G_{ss} is approximately 0.85. This observation seems to be an exception to the accepted rule-of-thumb suggesting that steady state energy release rates are achieved when the crack length is several times the film thickness. The numerical simulation was rerun using a modulus of 200 GPa, similar to chromium, to produce the second set of curves in Figures 5 and 6. Note the strong dependence of the saturation length and the converging length on Young's modulus. Although the steady state regime is not well defined for the chromium data the crack length must be at least 20–25 times the film thickness to achieve steady state conditions. The chromium data, here, agree well with the data of He et al. [21].

The implication of these data is that long cracks, 10–20 times the film thickness, are not required for the energy release rate measurements to be considered steady

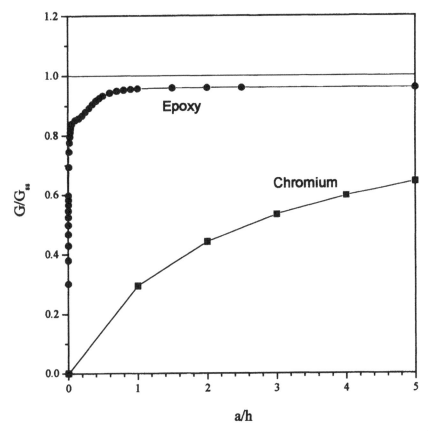

Figure 6. Magnified view of the normalized energy release rates as a function of nondimensional crack length, a/h, for epoxy and chromium superlayers on silicon at very short crack lengths.

Table 1.
Material properties used for simulations of a typical multifilm study. E is Young's modulus, h is thickness, ν is the Poisson ratio, and α is the linear thermal expansion coefficient

Material	E (GPa)	h (μm)	ν	α (ppm)
Epoxy	3	150	0.35	69
Silicon	170	700	0.25	3

state when epoxy is used as a superlayer. Referring back to the experimental data it was noted that within the control we had over the etched crack length, the energy release rate did not depend on the initial flaw size, even when the flaw sizes were much less than the film thickness. This appears to be an important attribute of the epoxy mechanical properties that can be exploited.

In Figure 7 the deformed mesh is presented for the epoxy/silicon simulation above. The assumed properties of the materials are contained in Table 1 and the temperature change was $-100°C$. In Figure 7 the displacements have been

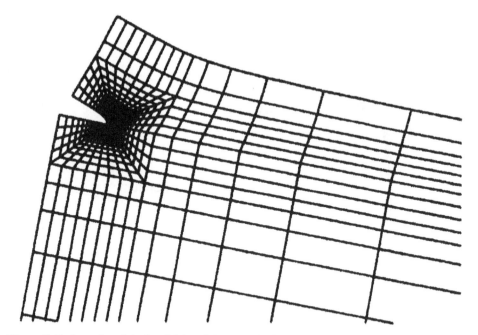

Figure 7. Deformed mesh in the vicinity of the crack tip for 200 μm epoxy on a 700 μm thick silicon substrate. Shear is present in the epoxy elements ahead of the crack tip along the interface, supporting earlier claims that this configuration has a large mode mixity.

magnified by a factor of 20 to magnify key features. First, there is bending involved in the silicon, thus storing strain energy. This energy is the source of the 4% discrepancy discussed previously. Second, elements which were initially rectangular on the interface ahead of the crack tip ended up sheared, providing evidence that there was shear on the interface ahead of the crack tip. This last point is consistent with the previous works which suggest a strong mode mixity in the problem.

4. SUMMARY

The MELT method is being considered as a test for the characterization of interfacial fracture energy. Critical to this test is the epoxy superlayer which is used to generate the strain energy which is released during crack growth, but there are instances where one needs to be aware of the mechanics, experimental assumptions and the properties of the epoxy. Specifically, the energy release rate depends on the relative Young's moduli, the effects of multiple films in the stacking sequence, the ability of the substrate to bend, and stress relaxation in the superlayer. In the current communication, we attempted to determine the precrack length required for steady state conditions. Independent of the crack lengths used in this work the energy release rate was a constant, contrary to a conventional rule of thumb. Numerical modeling revealed that there is an effect of Young's modulus on the crack length required to obtain a steady state energy release rate. In cases where $E_f/E_s \ll 1$ the required crack length for steady state conditions is very small. Considering epoxy on silicon the required crack length is 0.05 times the film thickness. In cases where $E_f/E_s \cong 1$, such as chromium on silicon, the required crack length for steady state conditions is 20–25 times the film thickness.

REFERENCES

1. J. W. Hutchinson and Z. Suo, *Adv. Apl. Mech.*, **29**, 63 (1992).
2. A. G. Evans, M. Ruhle, B. J. Dalgleish, and P. G. Charalambides, *Metall. Trans.*, **21A**, 2419 (1990).
3. Q. Ma, J. Bumgarner, H. Fujimoto, M. Lane, and R. H. Dauskardt, *Mater. Res. Soc. Symp. Proc.*, **473**, 3 (1997).
4. J. Ahn, K. L. Mittal, and R. H. MacQueen, in *Adhesion Measurement of Thin Films, Thick Films and Bulk Coatings, STP 640*, K. L. Mittal (Ed.), p. 134, ASTM, Philadelphia (1978).
5. T. W. Wu, *J. Mater. Res.*, **6**, 407 (1991).
6. S. Venkataraman, D. L. Kohlstedt, and W. W. Gerberich, *J. Mater. Res.*, **7**, 1126 (1992).
7. M. P. de Boer, J. C. Nelson, and W. W. Gerberich, *Mater. Res. Soc. Symp. Proc.*, **436**, 103 (1997).
8. C. Rossington, A. G. Evans, D. B. Marshall, and B. T. Khuri-Yakub, *J. Appl. Phys.*, **56**, 2639–2644 (1984).
9. D. B. Marshall and A. G. Evans, *J. Appl. Phys.*, **56**, 2632–2638 (1984).
10. L. G. Rosenfeld, J. E. Ritter, T. J. Lardner, and M. R. Lin, *J. Appl. Phys.*, **67**, 3291–3296 (1990).
11. A. G. Evans and J. W. Hutchinson, *Int. J. Solids Structures*, **20**, 455–466 (1984).
12. M. D. Kriese, N. R. Moody and W. W. Gerberich, *Acta mater.*, **46**, 6623–6630 (1998).
13. K. S. Kim and N. Aravas, *Int. J. Solids Structures*, **24**, 417–435 (1988).
14. K. S. Kim and J. Kim, *Trans. ASME J. Eng. Mater. Technol.*, **110**, 266 (1988).
15. B. Malyshev and R. L. Salganik, *J. Fracture Mech.*, **1**, 114–128 (1965).
16. S. J. Bennett, K. L. DeVries, and M. L. Williams, *Int. J. Fracture*, **10**, 33–43 (1974).
17. M. D. Thouless and H. M. Jensen, *J. Adhesion Sci. Technol.*, **8**, 579–586 (1994).
18. Q. D. Yang and M. D. Thouless, *Proceedings of the Society for Experimental Mechanics Annual Conference on Theoretical. Experimental and Computational Mechanics*, 616 (1999).
19. A. Bagchi, G. E. Lucas, Z. Suo, and A. G. Evans, *J. Mater. Res.*, **9**, 1734 (1994).
20. M. D. Drory, M. D. Thouless, and A. G. Evans, *Acta metall.*, **36**, 2019 (1988).
21. M. Y. He, G. Xu, and D. R. Clarke, *Mater. Res. Soc. Symp. Proc.*, **473**, 15–20 (1997).
22. G. Xu, D. D. Ragan, D. R Clarke, M. Y. He, Q. Ma, and H. Fujimoto, *Mater. Res. Soc. Symp. Proc.*, **458**, 465–470 (1997).
23. E. O. Shaffer II, M. E. Mills, D. Hawn, M. Van Gestel, A. Knorr, H. Gundlach, K. Kumar, A. E. Kaloyeros, and R. E. Geer, *Mater. Res. Soc. Symp. Proc.*, **511**, 133 (1998).
24. E. O. Shaffer II, S. A. Sikorski, and F. J. McGarry, *Mater. Res. Soc. Symp. Proc.*, **338**, 541 (1994).
25. F. J. McGarry and E. O. Shaffer II, *Mater. Res. Soc. Symp. Proc.*, **356**, 515 (1995).
26. E. O. Shaffer II, F. J. McGarry, and L. Hoang, *Polym. Eng. Sci.*, **36**, 2375 (1996).
27. E. O. Shaffer II, F. J. McGarry, and F. Trusell, *Mater. Res. Soc. Symp. Proc.*, **308**, 535 (1993).
28. J. C. Hay, E. G. Liniger, and X.-H. Liu, submitted to *J. Mater. Res.*

Adhesion Measurement of Films and Coatings, Vol. 2, pp. 219–234
Ed. K.L. Mittal
© VSP 2001

Assessment of adhesion reliability for plastic flip-chip packaging

XIANG (SAM) DAI,[1,*] MARK V. BRILLHART[2] and PAUL S. HO[3]

[1] *Hewlett-Packard Co., 1501 Page Mill Road, MS 6U-A, Palo Alto, CA 94304, U.S.A*
[2] *Cisco Systems, Inc., 170 West Tasman Drive, San Jose, CA 95134, U.S.A*
[3] *University of Texas, Lab for Interconnect and Packaging, PRC/MER2.206, MC R8650, Austin, TX 78712, U.S.A*

Abstract—In this study, we have applied both experimental and numerical techniques to assess adhesion integrity of the interfaces in a typical plastic flip-chip package. The approach is to evaluate the resistance (interface toughness) to the propagation of a pre-existing crack at the silicon passivation and underfill interface. The interface toughness curve, $G_c(\psi)$, is constructed by fitting the available experimental data to a previously proposed toughness function. The thermo-mechanical deformation of the package is examined by moiré interferometry and is used to validate a finite element analysis (FEA) model. The FEA model is then used to evaluate the driving force (energy release rate) induced by thermal excursions for a pre-existing crack. A methodology is proposed and demonstrated for providing a ranking of adhesion reliability.

Keywords: Flip-chip package; interface; toughness curve; moiré interferometry; FEA; adhesion reliability.

1. INTRODUCTION

The flip-chip area-array package, which connects the active device side of the semiconductor face-down via solder bumps on a multi-layered substrate, has been identified as an enabling technology in the National Technology Roadmap for Semiconductors [1]. The development of flip-chip area-array packages incorporates new sets of materials, processes and interconnect structures, that are not always optimized or compatible because of specific functional requirements. Although the use of underfill encapsulation significantly improves the reliability of flip-chip solder interconnections, delamination at various interfaces becomes a major reliability concern for flip-chip packages, as reviewed by Wu et al [2]. Underfill delamination from the chip and/or the board is most commonly observed and often leads to premature failure of the solder interconnects. These necessitate the assessment of interfacial adhesion reliability as part of design

---

* To whom correspondence should be addressed. Phone: (650)236-2568, Fax: (650)852-8082, E-mail: sam_dai@hp.com

and qualification of a plastic flip-chip package. However, emerging new materials and processes, together with the demands for cost reduction and fast product cycles, present serious challenges to the assessment of interfacial adhesion reliability.

To assess interfacial adhesion reliability, a natural approach is the one based on interfacial fracture mechanics. It requires knowledge of the interface toughness, the location and the length of an initial crack, and the driving force for interfacial crack growth. The preferred methodology would use the minimum number of experiments needed to determine the interface fracture envelope and provide a model that determines, under given thermal excursion conditions, the driving forces (energy release rates and mode mixities) associated with the most likely sites of delamination.

2. INTERFACE FRACTURE RESISTANCE

The resistance of an interface to fracture is characterized by the toughness curve, depicting the interface fracture energy as a function of mode-mixity. This curve can be determined by conducting interface fracture experiments to measure fracture energies at various phase angles of loading. For example, Cao and Evans [3], Liechti and Chai [4], and Liang and Liechti [5] have investigated epoxy/glass systems; Wang and Suo [6] have studied epoxy/metal and epoxy/Plexiglass systems; O'Dowd et al. [7] have examined aluminum/niobium system. Usually, it is not easy to set up and execute these experiments and it takes significant amount of time and effort to obtain meaningful data points under mixed-mode loading conditions. Hence, it is not suitable for design and reliability assessment purposes because of the increasing demands for fast product cycle and cost reduction. The toughness curve can be obtained using a so-called toughness function, which can be determined by a minimum amount of experimental data. Previously, a number of toughness functions [8, 9] have been proposed to capture the trend of G_c as a function of ψ. The toughness function proposed by Yan and Chiang [9] is adopted here because of its clear physical meaning and reasonably good representation of all the available toughness data on bimaterial interfaces in the literature.

These approaches are demonstrated here for the Silicon passivation/underfill (UF) interfaces. First, the results from double cantilever beam (DCB) and mixed-mode flexural tests are used to extract the interface fracture energy at the prescribed mixed-mode loading conditions. Then, the interface toughness curve is constructed by fitting the nearly mode-I (DCB) and mixed-mode (mixed-mode flexural) fracture energies to the toughness function.

2.1. Materials

The underfill materials, UF-3 and UF-4, were silica-filled liquid epoxies. The two underfill materials have very similar chemistry, except that the latter has twice as much organo-silane content as the former. Passivated (100) silicon (Si) wafers were used to simulate the surface conditions of silicon chips. Two types of passivation layers were studied. One was silicon nitride (SiN_x), and the other was an organic coating material, bisbenzocyclobutene (BCB). The organic substrate was epoxy-coated FR4 board (EP-PCB). Table 1 contains the material properties used in this study. The PCB used for the mixed-mode flexural test had slightly different material properties than that used in the flip-chip-on-board test package as noted in the table.

Table 1.
Material properties used in this study

Material		Modulus E (GPa)	Coefficient of Thermal Expansion CTE (ppm/°C)	Poisson's Ratio ν
Silicon		130	2.7	0.28
UF-3		3.8	22	0.36
UF-4		2.5	22	0.36
Eutectic Solder		10	21	0.40
PCB [a]	in-plane	19.1	11.0	0.20
	(out-of-plane)	12.4	38.0	0.11
PCB [b]	in-plane	15.5	14.7	0.20
	(out-of-plane)	12.4	51.2	0.11

[a] PCB used in the specimens for the mixed-mode flexural test.
[b] PCB used in the flip-chip-on-board test structure.

2.2. Results from DCB test

Delamination sandwich specimens (Figure 1) were used for the DCB test to determine mode-I interfacial fracture energy. Details of the DCB test and analysis were reported elsewhere [25]. Table 2 summarizes the results.

The results in Table 2 show that by doubling the organosilane content, the SiN_x/UF interfacial fracture energy increased by 68% from 15.6 J/m^2 to 26.2 J/m^2. It was attributed to the enhancement of intrinsic adhesion between the underfill and silicon nitride surfaces due to the increased organosilane addition [26]. In comparing SiN_x passivation and BCB coating on Si, the observed difference in underfill adhesion can be accounted for by the different surface chemistry [26].

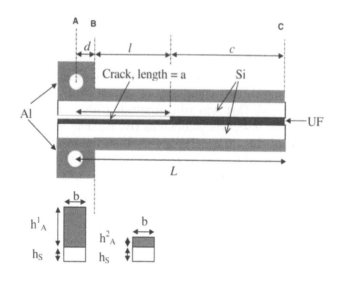

Dimension table (unit: mm)

L	d	b	h^1_A	h^2_A	h_S
70	6.0	6.0	12	2.0	0.67

Figure 1. The Si/UF/Si sandwich specimen for DCB test. The Si substrate has a passivation layer, SiN$_x$ or BCB, in contact with the underfill.

Table 2.
Mode-I interfacial fracture energies G$_c$ measured by DCB test

Interface	G_c (J/m^2)	ψ (degree)
SiN$_x$/UF-3	15.6	-0.1
SiN$_x$/UF-4	26.2	-0.1
BCB/UF-3	4.1	-0.1

2.3. Results from mixed-mode flexural test

The sandwich specimens (Figure 2) for mixed-mode flexural test were composed of a silicon piece (0.67 mm thick) and printed circuit board (1.24 mm thick) with an underfill layer (0.25 mm thick) in between. The specimens have a nominal width of 5 mm. The silicon surface had a silicon nitride layer for passivation. The PCB had an epoxy top layer. The underfill used was UF-3. The specimens were pre-cracked about 2 mm long from the two edges of the underfill along the SiN$_x$/UF interface. The two pre-cracks were created from the areas where Teflon (PTFE) mold release was applied during specimen preparation.

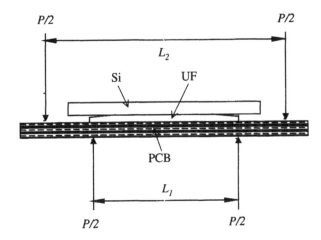

Figure 2. The Si/UF/PCB sandwich specimen with a SiN$_x$/UF interface crack for the mixed-mode flexural experiment. The Si substrate has a passivation (SiN$_x$) layer in contact with the underfill. L_1 = 30 mm and L_2 = 55 mm.

The specimens were loaded in four point bending with an inner span $L_1 = 30$ mm, an outer span $L_2=55$ mm, and a constant load-point displacement rate = 8 μm/sec. A typical load-displacement plot is shown in Figure 3. The SiN$_x$/UF interfacial crack began to propagate after a critical load was reached. But the crack did not stay at the interface. It jumped between the SiN$_x$/UF interface and underfill bulk material. This led to a saw-tooth shape in the load-displacement plot. The critical energy release rate G_c was estimated using the critical load for crack propagation ($P_c \approx 50N$).

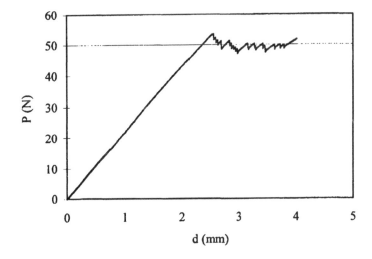

Figure 3. A typical load-displacement plot for the mixed-mode flexural experiment.

A finite element analysis was conducted to simulate the test and extract G_c from the experimentally measured critical load. It is a 2-D model with the boundary conditions and load configurations shown in Figure 4. The material properties used in the analysis were listed in Table 1. G_c is the sum of the crack separation energy G_s and the plastic dissipation energy G_p. The mode-mixity ψ was determined at the reference length $l = 35$ μm. Following the procedures described previously [25], the toughness value for the SiN$_x$/UF-3 interface was estimated to be $G_c = 285.3$ J/m^2 with the phase angle $\psi = 51.7°$.

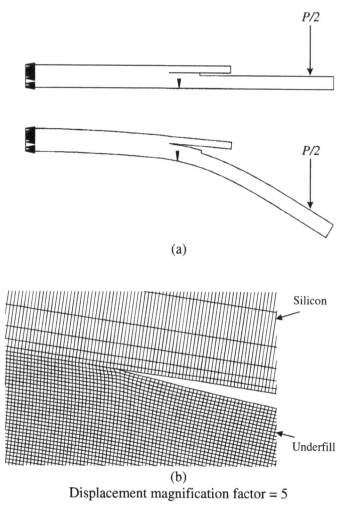

(a)

(b)

Displacement magnification factor = 5

Figure 4. A 2-D FEA model for the sandwich specimen with a SiN$_x$/UF interface crack under mixed-mode flexural experiment: (a) the configuration, load and boundary conditions for the model; (b) the deformed mesh near the crack tip.

2.4. Constructing the toughness curve

The goal is to construct the toughness curve for the entire range of the mode-mixity based on the toughness values measured at only two phase angles. The interface toughness function proposed by Yan and Chiang [9] can be used for this purpose. Although the toughness function is usually asymmetric [10], for simplicity, we assume that the toughness function is symmetrical with respect to the phase angle and the parameters have the same values for both positive and negative phase angles. Then, the toughness function takes the form of

$$G_c(\psi) = G_0\left[1 + \alpha \tan \psi\right] \tag{1}$$

with two material-specific parameters. G_0 is the toughness of an interface corresponding to mode-I loading state at the crack tip, and α is a measure of the relative strength of an interface in shear and tension. The finite element analysis conducted on the DCB SiN_x/UF-3 specimen showed that the phase angle $\psi = -0.08°$, approximately zero, at the reference length $l = 35$ μm. Therefore, the interface fracture energy obtained is assumed to be G_0. Figure 5 shows the experimental data and the predicted toughness curve for the SiN_x/UF-3 interface. The predicted curve is obtained by the toughness function with $G_0 = 15.6$ J/m^2 and $\alpha = 13.7$.

To obtain the toughness curve for the SiN_x/UF-4 interface, we assume that both SiN_x/UF-3 and SiN_x/UF-4 interfaces have the same value for α. This assumption is only a first-order approximation at this point since the difference in organosilane content may yield different α values. Based on this assumption and using $G_0 = 26.2$ J/m^2 measured by the DCB experiment, the toughness function for the silicon nitride-passivated silicon/UF-4 interface can be predicted. Similarly, the toughness function for the BCB-coated silicon/UF-3 interface is also predicted. The toughness curves for these interfaces are presented in Figure 6.

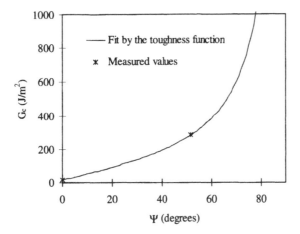

Figure 5. The experimental data and the predicted toughness curve for the SiN_x/UF-3 interface.

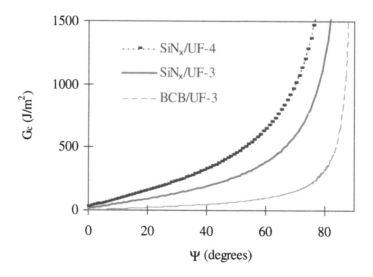

Figure 6. The toughness curves for the silicon passivation-underfill interfaces.

3. DRIVING FORCES FOR INTERFACE FRACTURE

Now that we have obtained the interface fracture resistance for the two underfill materials, the next step is to estimate the driving forces for the growth of an interfacial crack during thermal excursions.

Finite element analysis (FEA) models have been widely employed for calculating crack driving forces in electronic packages. Some [11-13] are based on the virtual crack closure method [14] while the others [15-17] are based on the contour integral method [18].

In this study, an FEA model based on the virtual crack closure method is used to calculate the total energy release rate and phase angle. To provide the validation for the FEA model, a study on the thermal deformation behavior in the flip-chip-on-board (FCOB) test structure was first conducted using moiré interferometry.

3.1. Thermal deformations in a FCOB package

3.1.1. Moiré Interferometry Results. The FCOB test package used in this study had double row peripheral solder bumps with a total of 340 I/O counts. The chip had dimensions of 12.9 mm x 12.9 mm x 0.62 mm and was mounted on a 1.5 mm thick substrate of a FR4 printed circuit board with embedded copper lines. The gap between the chip and the substrate was 75 µm, and the solder bumps, made of eutectic solder, had a 250 µm pitch. The underfill material was UF-3.

The actual specimen, shown schematically in Figures 7 and 8, was cross-sectioned for moiré studies with a grating attached at 102°C, which served as the starting temperature for tuning the null moiré displacement fields. As the temperature decreased from 102 °C to 22 °C, the fringe patterns in the U and V fields, shown in Figures 9 (a) and (b) respectively, represent the thermally induced deformations in the horizontal and vertical directions over half the package. The interference fringes represent constant displacement contours with contour interval of 0.417 μm. The fringe patterns show that differential contractions of different materials in both the horizontal and vertical directions result in a downward bending of the package, as revealed by the inclined fringes in the V displacement field. The small deformation in the silicon chip, as indicated by the low fringe density, is due to the low CTE and high modulus of the silicon. In the printed circuit board, the vertical contractions are much greater than the horizontal contractions, as shown by the higher fringe gradient in the V-field fringes, which is caused by the anisotropic CTE of the board.

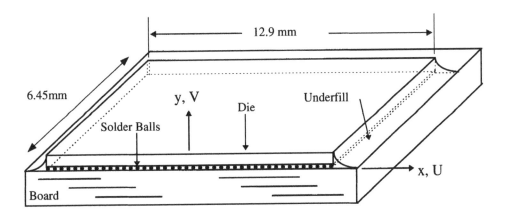

Figure 7. The schematic of the underfilled FCOB under study. *2L=12.9 mm.*

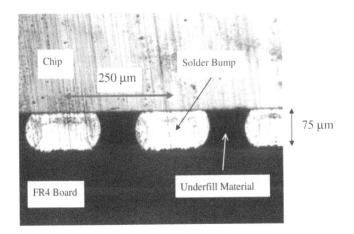

Figure 8. Cross-sectional picture of the underfilled flip-chip interconnections.

In the horizontal direction, the CTE mismatch results in a relative displacement between the chip and the board, which increases with the distance from the geometric center of the package. At the ends of the die, where the relative displacement reaches its maximum value, the highest stress and strain are to be expected in the outermost solder bumps. Detailed studies on the thermo-mechanical behaviors, shear strains over the solder bumps, and interfacial stresses can be found elsewhere [19-21].

The overall deformation can be quantitatively determined by analyzing the spatial distribution of the fringes. With a displacement sensitivity of 0.417 μm per fringe order, the sensitivity is sufficient for the displacement calculations in this study.

3.1.2. The FEA Model. A 2D finite element model was applied to analyze the deformation behavior of the test package during a uniform temperature change from 102 °C down to room temperature 22 °C. The geometry of the model was based on the actual specimen geometry in the cross section plane used for moiré interferometry study. Due to symmetry, only half of the specimen was modeled. The model, shown in Figure 10, has 4248 four-node quadrilateral elements. The stress condition is assumed to be that of plane stress (PS).

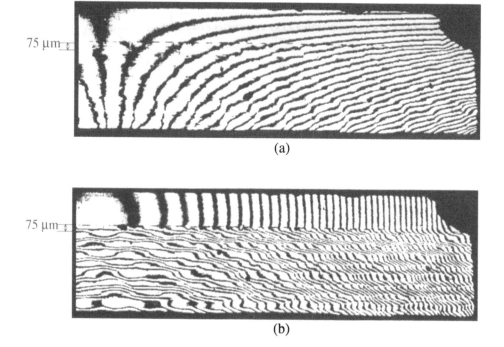

(a)

(b)

Figure 9. U and V fringe patterns under thermal loading of $\Delta T = -80$ °C over the right half of the package: (a) U; (b) V.

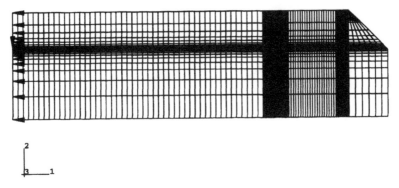

Figure 10. The configuration, mesh, and boundary conditions of the FEA model with no crack.

The glass-transition temperature was 170 ° C for the epoxy resin in the PCB and about 150-160 °C for the underfill materials. The test temperature range (22-102 °C) was below Tg's of all the materials in this study. So all the materials were assumed to behave linear elastically with temperature independent material properties within the temperature range. The silicon die and underfill were assumed to be isotropic and the FR4 board was assumed to respond as an ortho-tropic material. Preliminary analysis showed that the solder bumps contributed much less to the deformation behavior of the package than the underfill did. Thus, not all the solder bumps were modeled except the outermost two (close to the edge of the package) for study of the interfaces in the vicinity of the bumps. Per-fect adhesion was also assumed to exist for all material interfaces. This assump-tion was verified by the fringe patterns going continuously across the interfaces.

The assembly was assumed to be stress free at 102 °C and it was subsequently cooled to 22 °C. The thermo-mechanical response was computed using the finite element software ABAQUS. The material properties used for the analysis are given in Table 1.

3.1.3. Comparison of Moiré Interferometry and FEA Results. Figures 11 (a) and (b) show U and V displacements of the bottom of the die at SiN$_x$/UF interface as measured by moiré interferometry and as predicted by the FEA model. The figures show that the measured and predicted displacements are in good agree-ment. The agreement gives us confidence in our FEA model, especially in the material properties and boundary conditions used.

3.2. Delamination driving forces

The next step is to use the FEA model to calculate the driving forces for the propagation of a pre-existing interfacial crack.

There are many reasons for pre-existing interface cracks: loss of adhesion due to contamination; formation of voids, which trap moisture and, in turn, lead to va-porization; and pressure build-up and catastrophic unstable crack growth (i.e., popcorning) during high temperature thermal processing [2].

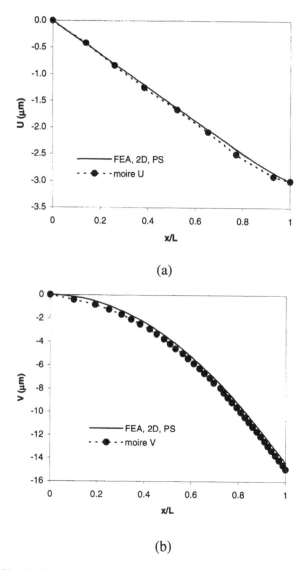

Figure 11. (a) U and (b) V displacements at SiN_x/UF interface measured by moiré interferometry and predicted by FEA (2D, plane stress (PS)). $L=6.45\ mm$.

Delaminations are flaws whose growth can ultimately lead to failure of the interconnects. A typical failure mode under deep thermal shock testing of a flip-chip-on-organic-substrate package would be delamination at the chip passivation/underfill interface, most often close to the edge of the silicon chip. This failure is often followed shortly afterwards by a fatigue failure of the solder bump near the delamination [22].

As a demonstration of the methodology, a crack at the chip passiva-tion/underfill interface, close to the edge of the silicon chip, is introduced in the FEA model. The deformed mesh for the whole model and the crack-tip region are shown in Figures 12 and 13, respectively. The energy release rates and phase an-gles were calculated for loading corresponding to a deep thermal shock from 150 °C (underfill curing temperature) to -200 °C (liquid nitrogen temperature). The results are presented in Table 3 for silicon nitride-passivated silicon to UF-3 and UF-4 interfaces, and for BCB-coated silicon to UF-3 interface. The phase angle for G was calculated at the same reference length l= 35 µm, as previously used for determining ψ in the DCB and mixed-mode flexural experiments. Because the interface toughness value is very sensitive to the phase angle and the phase angle varies with the reference length, it is very important to use the same reference length for the phase angle for consistency.

The phase angles are very high, indicating larger shear stress components. Thus, a significant portion of the fracture energy can be expected to be dissipated in underfill plasticity and interface asperity contacts. This is reflected in the high toughness values predicted at these high phase angles.

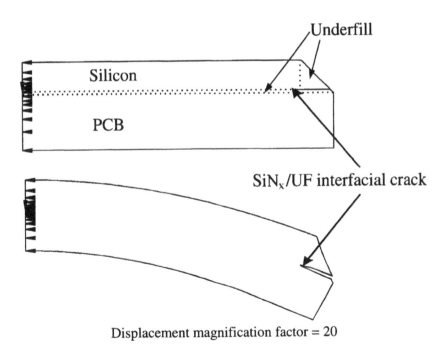

Displacement magnification factor = 20

Figure 12. The original and deformed configurations for the FEA model with a crack at the SiN$_x$/UF interface close to the edge of the package.

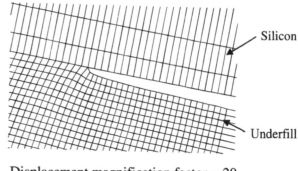

Displacement magnification factor = 20

Figure 13. The deformed mesh near the crack-tip region of the SiN$_x$/UF interface crack.

4. MODEL FOR THE ASSESSMENT OF INTERFACE INTEGRITY

In engineering fracture mechanics, safety factors are widely used for design purposes [23, 24]. To indicate the adhesion integrity of an interface against fracture, the concept of a safety factor (S) is invoked and defined here as the ratio of the delamination resistance versus driving force.

$$S = \frac{G_c(\psi)}{G(\psi)} \tag{2}$$

where $G(\psi)$ is the energy release rate at the phase angle ψ for an interfacial crack under a given thermal excursion condition, and $G_c(\psi)$ is the toughness of the interface at the same phase angle ψ. The higher the S value for an interface, the less likely the interface will fail. When $S \leq 1$, crack growth will occur. Table 3 summarizes the driving forces, resistances, and safety factors for the passivated silicon/underfill interfaces.

Once the S factors are determined for various interfaces, the interface integrity against fracture under prescribed loading conditions can then be ranked accordingly. For example, the S value for the SiN$_x$/UF-4 interface is almost twice that

Table 3.
The driving force and resistance for the delamination growth and the safety factor for adhesion reliability (ref. length l=35 µm)

Interface	G (J/m^2)	ψ (degree)	G_0 (J/m^2)	$G_c(\psi)$ (J/m^2)	S factor
SiN$_x$/UF-3	39.3	72.8	15.6	706.0	18.0
SiN$_x$/UF-4	37.1	73.3	26.2	1222.6	33.0
BCB/UF-3	39.3	72.8	4.1	185.6	4.7

of the SiN_x/UF-3 interface. Hence, the former is much more reliable in terms of interface adhesion than the latter. The BCB/UF-3 interface is much more likely to experience delamination failure than the other two interfaces because it has the lowest S value.

5. SUMMARY AND CONCLUSIONS

It is recognized that the model used here is a simplified one and needs further experimental verification using delamination detection tools such as acoustic imaging. We believe, however, that the methodology presented in assessing relative integrity of interfaces is generally valid. The methodology includes: the use of a toughness function to determine the whole toughness curve based on fracture experiments at two mode-mixities; the application of moiré interferometry to examine thermal deformations of a packaging structure and to validate the FEA model; the use of the FEA model (based on the modified crack closure method) to calculate the energy release rate of an interface crack; and defining a safety factor associated with an interface crack under prescribed thermal loading conditions to rank the relative adhesion reliability of interfaces.

The method proposed can be used to facilitate design against interface failures. Both the experimental and numerical techniques can be employed to rapidly evaluate new materials and to examine the impact of process modification on the adhesion integrity of interfaces in a wide range of electronic packaging applications.

REFERENCES

1. "The National Technology Roadmap for Semiconductors," Semiconductor Industry Association, San Jose, CA (1997).
2. T. Y. Wu, Y. Tsukada, and W. T. Chen, *Proc. 46th Electronic Comp. and Technol. Conf.* (1996).
3. H. C. Cao and A. G. Evans, *Mech. Mater.*, **7**, 295-304 (1989).
4. K. M. Liechti and Y. S. Chai, *J. Appl. Mech.*, **58**, 680-687 (1991).
5. Y.-M. Liang and K. M. Liechti, *Int. J. Solids Structures*, **32**, 957-978 (1995).
6. J.-S. Wang and Z. Suo, *Acta. Metall. Mater.*, **38**, 12279-1290 (1990).
7. N. P. O'Dowd, M. G. Stout, and C. F. Shih, *Phil. Mag.*, **A66**, 1037-1064 (1992).
8. J. W. Hutchinson and Z. Suo, *Adv. Appl. Mech.*, **29**, 63-191 (1992).
9. X. Yan and F.-P. Chiang, *Proc. ASME Symp. on Innovative Processing and Characterization of Composite Materials*, 159-173 (1995).
10. K. M. Liechti and Y. S. Chai, *J. Appl. Mech.*, **59**, 295-304 (1992).
11. G. Margaritis and F. J. McGarry, *IEEE CPMT Trans. - Part B*, **17**, 209-216 (1994).
12. S. Liu, Y. Mei, and T. Y. Wu, *IEEE CPMT Trans. - Part A*, **18**, 618-626 (1995).
13. S. Liu and Y. Mei, *IEEE CPMT Trans. - Part A*, **18**, 634-645 (1995).
14. E. F. Rybicki and M. F. Kanninen, *Eng. Fracture Mech.*, **9**, 931-938 (1977).
15. S. Liu, J. S. Zhu, M. Hu, and Y.-S. Pao, *IEEE CPMT Trans. - Part A*, **18**, 627-633 (1995).
16. R. Mahajan, E. Madenci, L. Ileri, and M. Thurston, *Mater. Res. Soc. Symp. Proc.*, **390**, 39-48 (1995).

17. H. Lee and Y. Y. Earmme, *IEEE CPMT Trans. - Part A*, **19**, 168-178 (1996).

18. J. R. Rice, *J. Appl. Mech.*, **35**, 379-386 (1968).

19. X. Dai, C. Kim, R. Willecke, and P. S. Ho, *Mater. Res. Soc. Symp. Proc.*, **445**, 167-177 (1996).

20. X. Dai and P. S. Ho, *21st IEEE/CPMT IEMT Symp. Proc.*, 326-333 (1997).

21. J.-H. Zhao, X. Dai, and P. S. Ho, *48th Electronic Comp. and Technol. Conf. Proc.*, 336-344 (1998).

22. W. T. Chen, D. Questad, D. Read, and S. Bahgat, presented at the *International Mechanical Engineering Congress and Exposition, Symposium on Application of Fracture Mechanics in Microelectronic Packaging*, Dallas, TX (1997).

23. D. Broek, *Elementary Engineering Fracture Mechanics*. Sijthoff & Noordhoff: Alphen aan den Rijn, The Netherlands (1978).

24. M. F. Kanninen and C. H. Popelar, *Advanced Fracture Mechanics*. Oxford University Press, New York (1985).

25. X. Dai, M. V. Brillhart, and P. S. Ho, *IEEE Trans. Comp. Pack. Technol.*, **23** (1), 101-116 (2000).

26. X. Dai, M. V. Brillhart, M. Roesch, and P. S. Ho, *IEEE Trans. Comp. Pack. Technol.*, **23** (1), 117-127 (2000).

Adhesion Measurement of Films and Coatings, Vol. 2, pp. 235–246
Ed. K.L. Mittal
© VSP 2001

Adhesion and abrasion of sputter-deposited ceramic thin films on glass

SUSUMU SUZUKI*

Research Center, Asahi Glass Co., Ltd., 1150 Hazawa-cho, Kanagawa-ku, Yokohama 221-0863, Japan

Abstract—This paper reviews our recent results on the adhesion and abrasion behaviors of ceramic thin films sputtered onto glass substrates. The adhesion strength at the interface of metal nitride overlayer/metal oxide underlayer: MN_x(M: Cr, Ti)/MO_y(M: Al, Sn, Ti, Zr, Ta, Zn) increases as the metal oxide is deoxidized more easily. The adhesion strength of metal nitride film: MN_x(M: Cr, Ta, Ti, Zr) to glass substrate increases as the strength of the bonding between the metal in the film and the oxide atom(M-O) increases. These results indicate the formation of chemical bonds at the interface and their importance in the adhesion of films prepared by sputtering. By using the Taber abrasion test, metal nitride thin films, MN_x(M: Cr, Ta, Ti), were abraded in the peeling mode. The peel rate increases exponentially with decreasing adhesion strength and hardness and with increasing coefficient of friction. The peel rate changes most sensitively with the hardness change.

Keywords: Adhesion strength; abrasion; sputtered thin films; glass substrate.

1. INTRODUCTION

Sputtered ceramic (metal nitride and metal oxide) thin films are widely used in coated glass products, e.g., heat reflecting glass, low-E glass and transparent conductive glass. The mechanical durability of these films is of importance since it affects production, life time and function. The goal of industrial research is to obtain thin film products with high mechanical durability. Figure 1 shows a schematic diagram for investigations of mechanical properties of thin films. The thin film products undergo mechanical degradation in various ways in actual usage and the accelerated tests for evaluating actual mechanical durability are used, such as the Taber abrasion test[1] and the sand drop test[2]. The mechanical durability is supposed to be affected by the following factors: adhesion strength, coefficient of friction and hardness of thin films, depending on the conditions in which the products are used. In turn, the basic properties of thin films, e. g., internal stress, bulk modulus, surface morphology, crystal structure, etc., affect these three factors. The important problems to be investigated in this diagram are 1)mechanism of degradation by the accelerated tests, 2)effects of adhesion

---

* Phone: 81-45-374-8761, Fax: 81-45-374-8863, E-mail: susumu@agc.co.jp

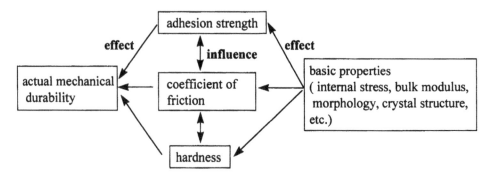

Figure 1. Scematic diagram for the investigation of mechanical properties of thin films.

strength, coefficient of friction and hardness on the actual mechanical durability, 3)mechanisms of adhesion and friction, 4)effects of basic properties on adhesion strength, coefficient of friction and hardness, and 5)mutual influences of adhesion strength, coefficient of friction and hardness, on each other. We have investigated the mechanical properties of sputtered thin films on glass from such viewpoints[3-10]. The present paper focuses on the mechanism of adhesion and the mechanism of abrasion using the accelerated test and the effects of adhesion, friction and hardness on the abrasion.

2. ADHESION

Sputtered metal nitride and metal oxide thin films show strong adhesion to glass as well as high hardness, making it difficult to measure the adhesion strength using the conventional methods such as the scratch test. We have evaluated the strength of practical adhesion[11-13] of these thin films using the direct pull-off test developed for highly adherent thin films on glass and have investigated the mechanisms of the adhesion of the films[5].

Figure 2 shows the preparation procedure and the cross section of a sample for the direct pull-off test. Ti ion-plating, Ti + Cu simultaneous evaporation and Cu evaporation are done successively on the cleaned surface of the tested film of 2mm by 2mm area. A steel wire of 0.6 mm diameter is then soldered in the area. The wire is pulled off perpendicularly to the substrate by a bond tester and the force necessary to remove the film from the substrate is measured. Usually ten measurements are done for each sample and the pull strength is obtained as an average of these measurements. It is noted that the measured value for sputtered ceramic thin films on glass is of the order of a few MPa, much smaller than expected. The reason why such a small value of the pull strength is obtained is that in our pull-off test, the pull force is not uniform over the tested area of 2mm by 2mm; rather it is concentrated in a small area on the film surface near the steel wire. Therefore, the strength obtained can only be used as a relative measure of the adhesion.

Direct pull-off test

Sample preparation   Sample cross section

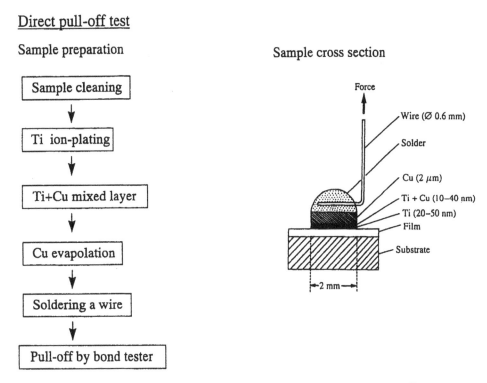

Figure 2. Sample preparation procedure and sample cross section in the direct pull-off test.

Two groups of specimens were prepared for the adhesion strength measurement. The first group was metal nitride/metal oxide/glass system, where the metal oxide underlayer was one of the following, Al_2O_3, SnO_2, Ta_2O_5, TiO_2, ZnO or ZrO_2, of 100 ± 10 nm thickness and the metal nitride toplayer was either CrN_x or TiN_x of the same thickness. The second group was metal nitride/glass system, where the metal nitride consisted of one of the following, CrN_x, TaN_x, TiN_x and ZrN_x, of 100 ± 10 nm thickness. The metal oxide films were deposited by DC reactive magnetron sputtering from an appropriate metal target in 1/1 – 2/3 Ar/O_2 gas mixtures at a pressure and an input power of 0.36 – 0.41 Pa and 1.8 – 5.5 W/cm^2, respectively, without intentional substrate heating. The metal nitride thin films were deposited using the same method in 1/1 Ar/N_2 or 100% N_2 atmosphere at a pressure and an input power of 0.35 – 0.37 Pa and 3.7 – 5.5 W/cm^2, respectively.

The separations often occurred at the metal nitride/metal oxide interfaces for the metal nitride/metal oxide/glass system. Figure 3 shows the plot of the pull strength at the metal nitride/metal oxide interface against the strength of the metal-oxygen(M-O) bond, where the bond strength is that of diatomic molecules[14]. The pull strength at the interface decreases with increasing M-O bond strength for both the CrN_x and TiN_x toplayers. In other words, the pull strength

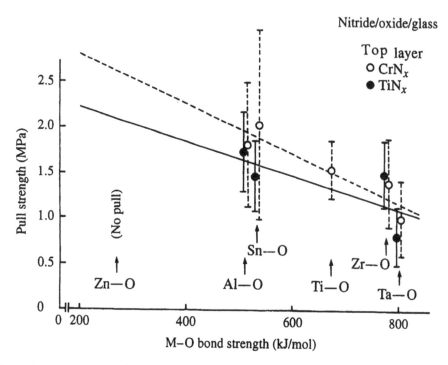

Figure 3. Pull strength at the metal nitride/metal oxide interface against the strength of metal-oxygen(M-O) bond in the metal oxide film.

becomes larger as the metal oxide underlayer is deoxidized more easily. From this finding, the mechanism of adhesion at the interface is considered, as shown in Fig. 4. At the onset of deposition of the nitride top layer, the energetically excited, sputtered target metal atoms and discharge gas nitrogen atoms striking the surface of the oxide underlayer break the metal-oxygen (M-O) bonds near the oxide surface, i.e., deoxidize the surface, as shown in (a). The oxides with weaker M-O bonds are likely to be deoxidized more easily. Then some of the target metal atoms and discharge gas nitrogen atoms are trapped by the broken M-O bonds and form new bonds at the nitride/oxide interface, as shown in (b). The number of new bonds is proportional to the number of broken M-O bonds and is larger for oxides with weaker M-O bonds. Therefore, an oxide with weaker M-O bonds shows stronger adhesion at the interface.

Figure 5 shows the dependence of the pull strength of the nitride single layer system on the strength of chemical bonding between the metal atom in the nitride and the oxygen atom(M-O). The pull-strength increases with increasing M-O bond strength, indicative of M-O bond playing as a key role for the adhesion at the interface. From this result, the mechanism of adhesion at the nitride/glass interface was suggested, as shown in Fig. 6. At the onset of deposition of the nitride film, the energetically excited sputtered target metal atoms and discharge gas ni-

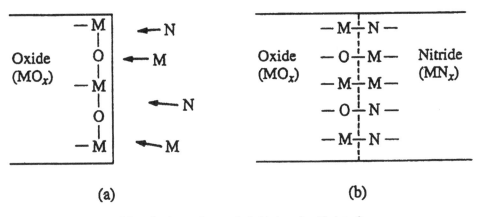

Figure 4. Mechanism of the adhesion at the metal nitride/metal oxide interface.

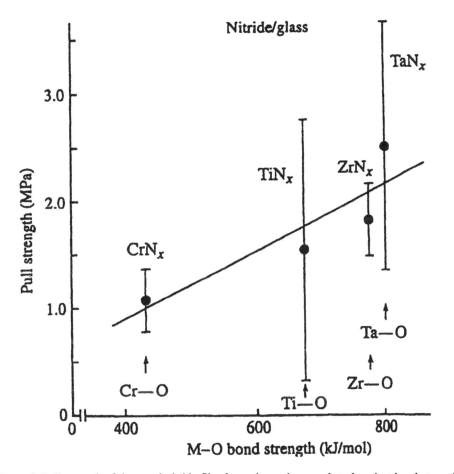

Figure 5. Pull strength of the metal nitride film from glass substrate plotted against bond strength between the metal atom in the film and the oxygen atom(M-O).

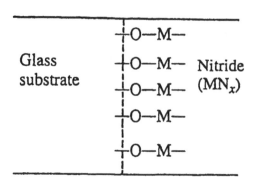

Figure 6. Mechanism of adhesion at the metal nitride/glass interface.

trogen atoms striking the glass surface produce the broken bonds of oxygen atoms, originating from glass itself or ambient water vapor. Then some of sputtered target metal atoms are trapped by the broken bonds and form M-O bonds at the nitride/glass interface. The number of M-O bonds is not very different for different nitride films and the pull strength is proportional to the strength of M-O bonds.

These results are simple but important since they provide evidence that the chemical bonds formed at the interface play an important role in the adhesion of the films deposited by sputtering.

3. ABRASION

The Taber abrasion test is popularly used as the accelerated test for actual mechanical durability of solar control glass. We have investigated the mechanism of abrasion on sputter-deposited ceramic thin films on glass and the effects of adhesion, friction and hardness on Taber abrasion[6].

Figure 7 shows the principle of the Taber test. A pair of abrading wheels of 1.27 cm in thickness and 5.08 cm in diameter are contacted with the sample surface with a suitable load. The sample rotates, and consequently, the abrading wheels rotate in an opposite direction. CS-10F type abrading wheels, made of rubber containing abrasives of several tens micrometers in diameter, are usually used with a load of 4.9 N. The thin films tested were reactively DC sputtered CrN_x, TaN_x and TiN_x films of 30, 50, 100, 150 and 200 nm thickness in similar deposition conditions as described in previous section. The degree of abrasion on the films by the Taber test was evaluated quantitatively from the increase in optical light transmission, i.e., $\Delta Tv = Tv_n - Tv_0$, where Tv_n and Tv_0 are the optical light transmissions at n Taber cycles and before the test, respectively, where a Taber cycle means total number of sample rotations. The adhesion strength, the coefficient of friction and the hardness of the films were measured using the direct pull-off test, Si_3N_4 as the opponent material, and the indentation method, respectively.

Abrading wheel

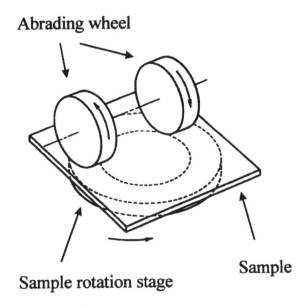

Sample rotation stage   **Sample**

Figure 7. Principle of the Taber abrasion test.

The adhesion strength and the coefficient of friction were almost constant throughout the film thickness and the hardness increased linearly with thickness(Table 1). Figure 8 shows a typical graph of ΔTv versus number of Taber cycles for the TiN$_x$ films. The slope of the graph decreases as the thickness of the film increases, but at 150 nm, ΔTv anomalously increases. The reason for this has not been clarified at present, but one possible explanation may be as follows. The Hertz contact theory predicts the occurrence of maximum shear stress in r/2 depth from the surface when two bodies are in contact in a circular area of radius r[15]. The optical micrographs of the degraded areas of the films show that there are a large number of grooves of about 0.5 μm width due to the scratching by the abrasives in the abrading wheels. Therefore, the maximum shear stress occurs around the film/substrate interface for the films of 150 nm thickness during the Taber test and the stress may cause a rapid substrate-film separation.

Table 1.
Pull strength, coefficient of friction and hardness of the TiN$_x$, CrN$_x$ and TaN$_x$ films on glass. The hardness was measured, compared with the hardness of the glass

Film	Pull strength (MPa)	Coefficient of friction		Hardness (/Hglass)	
			30nm	50nm	100nm
TiN$_x$	1.98	0.145	1.04	1.06	1.12
CrN$_x$	1.37	0.226	1.01	1.02	1.04
TaN$_x$	2.60	0.118	1.04	1.06	1.13

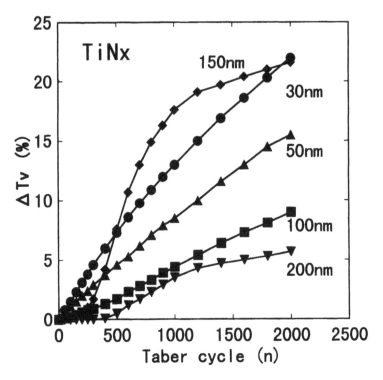

Figure 8. Increase in visible light transmission, ΔTv, as a function of number of Taber cycles for the TiN$_x$ films on glass.

It was suggested from the optical micrographs of the degraded areas of the films that the principle of the Taber test was similar to that of the scratch test[16,17]. That is, the abrasives in the abrading wheel scratch the film surface and generate a stress in the film. The stress is accumulated in the film with increasing number of Taber cycles. When the stress energy exceeds the adhesion energy of the film, the film begins to peel. However, once the film removal begins, the stress energy in the film around the peeled area is dissipated and reduced, resulting in a decrease in the possibility of film peeling.

Therefore, we have suggested the peel model as a mechanism of Taber abrasion: the film is peeled off the substrate by a peel rate p($0 \le p \le 1$) per Taber cycle in the unpeeled area without reducing the film thickness(Fig.9). The peel rate denotes the ratio of the peeled area per Taber cycle to the unpeeled area. In this model, ΔTv is expressed as,

$$\Delta Tv = [1-\exp(-pn)](Tv_s-Tv_0) \qquad (1)$$

where Tv_s is the visible light transmission of glass substrate. From this equation, the peel rate, a quantitative measure of abrasion, is obtained, if the peel rate is very much smaller than unity, by

$$p = \Delta Tv/n(Tv_s-Tv_0) \qquad (2)$$

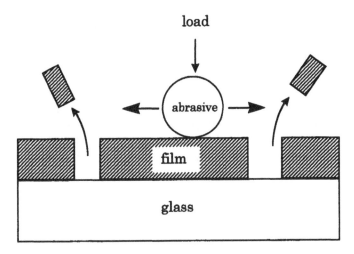

Figure 9. Peel model for the Taber abrasion test.

Table 2 shows the peel rates (/Taber cycle) for the TiN_x, CrN_x and TaN_x films.

The peel rate is considered to be affected by adhesion strength, coefficient of friction and hardness. We assume that the effects of these factors are independent of each other, and are expressed as,

$$p_A/p_B = f_W(W_A,W_B)f_F(F_A,F_B)f_H(H_A,H_B) \qquad (3)$$

where f_W, f_F and f_H are functions of adhesion, friction and hardness, respectively, and $W_{A(B)}$, $F_{A(B)}$ and $H_{A(B)}$ are the adhesion strength, the coefficient of friction and the hardness, of the film A(B), respectively. These functions should be unity when the values of these properties for the films A and B are the same.

To determine $f_F(F_A,F_B)$, we tested metal oxide(5 nm)/TiN_x(10 nm)/glass double-layer system, where metal oxide toplayer was one of the following, TiO_2, SnO_2, Ta_2O_5, $ZrSi_2Ox$ or SnSiOx. These systems showed various coefficients of friction and were assumed to have the same adhesion strength at the TiN_x/glass interface and the same hardness since the thickness of the metal oxide toplayer

Table 2.
Peel rates of the TiN_x, CrN_x and TaN_x films on glass by the Taber abrasion test

Film thickness (nm)	Peel rate (/Taber cycle)		
	TiN_x	CrN_x	TaN_x
30	2.4 E-4	4.8 E-4	1.9 E-4
50	1.4 E-4	3.1 E-4	9.7 E-5
100	3.9 E-5	9.6 E-5	2.6 E-5
150	2.8 E-4	4.7 E-3	3.3 E-6
200	5.8 E-5	6.5 E-5	1.7 E-5

was very thin. The function $f_F(F_A,F_B)$ was obtained from the dependence of the peel rate of these systems on the coefficient of friction to be $f_F(F_A,F_B) =$ $\exp[2.5(F_A-F_B)]$(Fig. 10). To elucidate $f_W(W_A,W_B)$, we tested TiN_x(40 nm)/ SnO_2(5, 10, 30 nm)/glass systems. In these systems, the pull strength of the TiN_x film to glass substrate varied from 0.6 to 1.4MPa, while the coefficients of friction and the hardnesses of the systems were measured to be the same. The reason why these systems showed different pull strengths has not been determined. The function $f_W(W_A,W_B)$ was obtained from the dependence of peel rate of these systems on the pull strength to be $f_W(W_A,W_B) = \exp[-0.3(W_A-W_B)]$. The function $f_H(H_A,H_B)$ was obtained by inserting the values of the adhesion strength, the coefficient of friction and the hardness and the peel rates of the TiN_x, CrN_x and TaN_x films shown in Tables 1 and 2 into equation (3) where $f_F(F_A,F_B)$ and $f_W(W_A,W_B)$ are replaced with $\exp[2.5(F_A-F_B)]$ and $\exp[-0.3(W_A-W_B)]$. The films more than 150 nm thick were not considered because of the anomaly in the peel rate. Finally equation (3) was obtained as

$$p_A/p_B = \exp[-0.3(W_A-W_B)+2.5(F_A-F_B)-11.3(H_A-H_B)] \qquad (4)$$

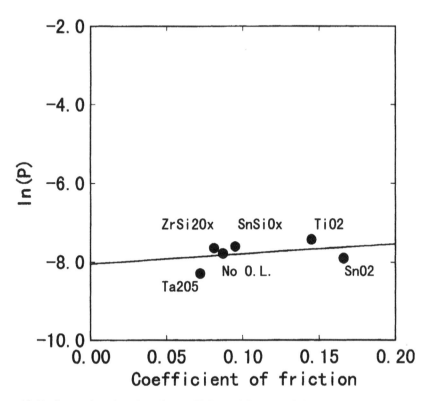

Figure 10. Peel rate plotted against the coefficient of friction of the metal oxide(5 nm)/TiN_x(10 nm)/glass double-layer systems. The materials of the metal oxide overlayers are shown. No O. L. means no overlayer.

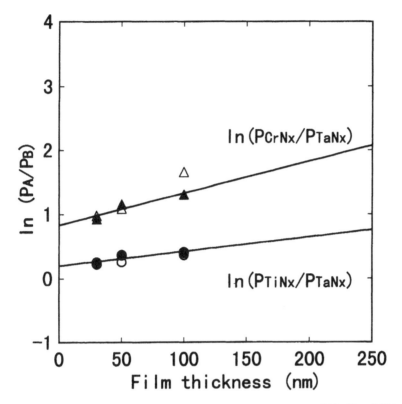

Figure 11. Logarithm of ratio of the peel rates plotted at a film thickness of 30, 50 and 100nm : (●) $\ln(P_{TiNx}/P_{TaNx})$, (▲) $\ln(P_{CrNx}/P_{TaNx})$. The ratio of the peel rates calculated by Eq. (4) : (○) $\ln(P_{TiNx}/P_{TaNx})$, (△) $\ln(P_{CrNx}/P_{TaNx})$.

Figure 11 shows the ratio of the peel rates, P_{TiNx}/P_{TaNx}, P_{CrNx}/P_{TaNx}, plotted at a film thickness of 30, 50 and 100nm, the solid and open symbols denoting experimental and calculated values, respectively. The calculated values agree well with the experimental ones. The equation shows that the peel rate increased exponentially with decreasing adhesion strength and hardness and with increasing coefficient of friction and it changes most sensitively with the change in the hardness.

4. CONCLUSIONS

1. Adhesion strength at metal oxide underlayer/metal nitride toplayer interface decreases with increasing strength of metal-oxygen(M-O) bond in the metal oxide underlayer.
2. Adhesion strength of metal nitride single layer to glass substrate increases with increasing strength of M-O bond between the metal atom in the film and oxygen atom.

3. Sputter-deposited metal nitride thin films on glass are removed by the Taber abrasion test in the peeling mode.
4. Adhesion strength, coefficient of friction and hardness all affect the peel rate in an exponential form.
5. The peel rate changes most sensitively with the hardness change.

REFERENCES

1. ASTM standard D4060, Japan Industrial Standard JIS R3221.
2. ASTM standard D968
3. S. Suzuki, N. Hashimoto, T. Oyama and K. Suzuki, *J. Adhesion Sci. Technol.*, **8**, 261-271(1994).
4. S. Suzuki, H. Ohsaki and E. Ando, *Jpn. J. Appl. Phys.*, **35**, 1862-1867 (1996).
5. S. Suzuki, Y. Hayashi, K. Suzuki and E. Ando, *J. Adhesion Sci. Technol.*, **11**, 1137-1147 (1997).
6. S. Suzuki and E. Ando, *Thin Solid Films*, **340**, 194-200 (1999).
7. S. Suzuki, *Thin Solid Films*, **351**, 194-197 (1999).
8. E. Ando, J. Ebisawa, Y. Hayashi, A. Mitsui, and S. Suzuki, *J. Non-cryst Solids*, **178**, 238-244 (1994).
9. E. Ando and S. Suzuki, *J. Non-cryst Solids*, **218**, 68-73 (1997).
10. M. Tada, Y. Hayashi, K. Suzuki and S. Suzuki, *Reports Res. Lab. Asahi Glass Co.*, **48**, 1-9 (1998).
11. K. L. Mittal, in: *Adhesion Measurement of Thin Films, Thick Films and Bulk Coatings*, K. L. Mittal (Ed.), STP No. 640, pp. 5-17. ASTM, Philadelphia (1978).
12. K. L. Mittal, *Electrocomponent Sci. Technol.*, **3**, 21-42 (1976).
13. K. L. Mittal, in: *Adhesion Measurement of Films and Coatings*, K. L. Mittal (Ed.), pp. 1-13. VSP, Utrecht, The Netherlands (1995).
14. D. R. Lide(Ed.), *Handbook of Chemistry and Physics*, 74th ed., p. 123. CRC Press, Boca Raton, FL (1993).
15. S. P. Timoshenko and J. N. Goodier, *Theory of Elasticity*, p. 413. McGraw-Hill, New York (1987).
16. M. T. Laugier, *J. Vac. Sci. Technol.*, **A5**, 67 (1987).
17. A. Kinbara, S. Baba, and A. Kikuchi, *J. Adhesion Sci. Technol.*, **2**, 1 (1988).

Adhesion Measurement of Films and Coatings, Vol. 2, pp. 247–254
Ed. K.L. Mittal

Improvement and testing of diamond film adhesion

X.C. HE,* H.S. SHEN, Z.M. ZHANG, X.J. HU and X.Q. YANG

State Key Lab of MMCM's, Shanghai Jiao Tong University, Shanghai, China, 200030

Abstract—Thin diamond films were deposited by the CVD method onto Cu/TiC layer sputter deposited on Si(111) and steel substrates. Film characterizations by X-ray, SEM and Raman spectroscopy were carried out. Tensile test as well as surface indentation were used to determine the film adhesion strength on the substrates. It is important to make sure that the fracture during the tensile test occurs only at the interface between the diamond film and the substrate to obtain quantitative results on film adhesion strength. The results indicate that the Cu/TiC sputtered intermediate layer plays an important role not only in promoting diamond nucleation but also in diamond film adhesion strength to these substrates.

Keywords: CVD diamond film; intermediate layer; adhesion.

1. INTRODUCTION

Presently, thin diamond films are under development for many applications due to their excellent hardness, strength, thermal conductivity, chemical stability and optical transmission. However, there is still some problem in growing diamond films on non-diamond substrates due to poor nucleation density and poor adhesion. It has been well recognized that a pre-treatment of the substrate or use of a transition layer before CVD diamond growth can significantly improve diamond nucleation as well as adhesion on non-diamond substrates [1]. Joffeau et al. [2] reported that during the CVD growth, there were carbide layers between the carbide-forming elements and the diamond films. This means that the transition layer may play an important role during diamond growth on these non-diamond substrates. Since the adhesion is the major obstacle in diamond film application, the intermediate layer applied prior to diamond deposition might finally prove to be one of the best possible solutions to enhance the diamond adhesion [3]. For ferrous metal substrates, it is very difficult to grow diamond films on them because first, carbon can diffuse into these metals with a relatively high diffusion rate at high temperature, and the CVD diamond nucleation density is very low. Second, ferrous metal elements, such as iron or cobalt, have a catalytic effect on the growth of graphite and nano-crystalline carbon. Third, the thermal expansion coefficients between the diamond and a ferrous metal are not

---

* To whom correspondence should be addressed. Phone: 86-21-62933163, Fax: 86-21-62822012, E-mail: xche@mail.sjtu.edu.cn

compatible and this mismatch usually causes high residual stress and poor adhesion [4]. A common way to grow a diamond film on a ferrous metal is to deposit an intermediate layer serving as a carbon diffusion barrier as well as adhesion enhancer. Spinnewyn et al. [5] used a refractory metal, such as W or Mo, as an intermediate layer. Their calculation indicated that different materials of intermediate layers needed different minimun thicknesses to restrain carbon diffusion. For example, Si and Ti required thicker layers than W, Cr and Ta. They did not compare the adhesion strength of diamond films on these different intermediate layers in their paper. The intermediate layer has to be bonded to both the diamond film and the substrate. One must consider at least these two requirements: first, diamond film must be grown on the intermediate layer without graphite formation, and second, the intermediate layer should have good adhesion strength to both the substrate and the diamond film. In some cases, the intermediate layer may relax the intrinsic stress induced by the differences in lattice parameters and thermal expansion coefficients of the diamond film and the substrate. Poor adhesion is still an impediment for the application of diamond films on ferrous substrates. Thus, how to choose and how to grow these intermediate layers is very important for diamond film growth on ferrous substrates. On the other hand, how to evaluate film adhesion strength to a substrate is also very important. The Rockwell-C indentation test [6] appears to be a simple and convenient method for evaluating the adhesion strength of diamond films; however, this method cannot give a quantitative result and is strongly affected both by the substrate as well as the applied load. The tensile test is an easy way to judge the film adhesion strength quantitatively. Because of high surface energy and complete covalent bond structure, diamond has poor adhesion to most materials [7]. It is very difficult to attain enough glue (adhesive) strength between the diamond film and the tensile bar. There was no success for diamond film adhesion measurement using a tensile test before [8]. In this paper, we sputtered Cu/TiC double layers on Si and steel substrates, and then grew diamond films on them by hot filament CVD method (HFCVD). Characterizations by X-ray, SEM and Raman spectroscopy were carried out to check film quality. The tensile test as well as indentation tests were performed to evaluate diamond film adhesion strength.

2. EXPERIMENTAL DETAILS

2.1. Sputtering of intermediate layers

The polished Si (111) and stainless steel substrates were cleaned for several minutes under 0.4 Pa pressure of Ar in the chamber before sputtering to enhance the adhesion of the sputtered film to the substrate. The intermediate layer was deposited by magnetron sputtering technique (Anelva SPC-350) on substrates using two targets made of TiC and Cu, which were sputtered one by one. The total sputtered layer thickness was 500 nm for each case. The structure of the sputtered layers was checked by grazing incident X-ray diffraction.

2.2. Thin diamond film growth

An ultrasonic treatment in acetone with 10μm diamond particles was used to create more nucleation sites on the sputtered layer surface. The diamond film was deposited by hot filament CVD method (HFCVD) in a vacuum reactor. The reactant gases used during the growth process were hydrogen and acetone vapor. In some cases, a very small amount of argon was added. The gas volume ratio of hydrogen to acetone vapor was 100:1. Total gas flow rate was 100 sccm and the gas pressure was 3-5 kPa. The operating temperature of the carbonized Ta filament was at about 2200°C. The distance between the filament and the substrate was 10-15 mm. A metal Mo slice with a Ø8 mm hole was placed on the substrate to limit the diamond film growth only in the hole area which was the same as the across area of the tensile bar. The substrate temperature was kept at about 750°C for Si(111) substrate and 600°C for steel substrate, determined by a thermocouple placed at the center of the graphite holder. The film deposition time was 4-10 hours.

2.3. Film characterization

The deposited diamond thin film was characterized by 3° grazing incident X-ray diffraction (Cu-K$_\alpha$ radiation) and Raman spectroscopy (λ=514.5nm) to obtain surface structure information. The surface morphology of the film was checked by SEM. A tensile test was carried out to evaluate diamond film adhesion strength to these substrates with sputtered intermediate layers (Fig. 1). In order to ensure that the tensile fracture occurred only at the interface between the diamond film and the substrate, a thin metal layer was deposited on diamond film surface by sputtering just after CVD diamond growth, and subsequently, a surface thermal treatment was carried out to enhance the adhesion of diamond film to the tensile bar. A high strength epoxy resin with glue strength greater than 20 MPa was chosen to bond the tensile bars to the diamond film surface as well as to the back side of the substrate. Two Ø8 mm tensile bars were laid on a special groove apparatus and were pressed by springs to keep them concentric and to have tight connection to both surfaces until the epoxy resin was solidified at 180°C. During the tensile test, the tensile bars were clamped and the load was applied at a speed of 50 N / min in an MTS machine (New810) until the fracture occurred. Finally the fracture surfaces were checked by optical microscopy to make sure that all the diamond film had been removed from the substrate. The adhesion strength **f** was determined as the critical load **F** at film fracture divided by film area **A**.

$$\mathbf{f} = \mathbf{F} / \mathbf{A} \qquad (1)$$

The Rockwell indentation test was used to obtain a rough estimate of the film adhesion. The critical applied load, at which the film cracked, is related to the film adhesion strength. As we were unable to change the applied load continuously, we only got qualitative results for comparing different films.

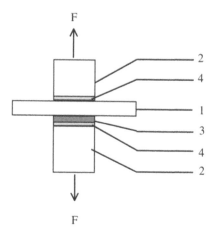

Figure 1. Schematic sketch of the tensile test: 1) substrate; 2) metal bars; 3) diamond film; 4) epoxy resin.

3. RESULTS AND DISCUSSION

3.1. Sputtered layers

No structural difference was observed from X-ray diffraction between the sputtered Cu/TiC layers with or without Ar ion beam cleaning; however, the critical applied load in the indentation test for the one with Ar ion gun treatment was about 1.5 times greater than the one without Ar treatment. This indicated that the adhesion of the Cu/TiC layer to the substrate was improved by Ar ion cleaning treatment before sputtering. Ar ion treatment played the role of both cleaning and micro-roughening the substrate surface to enhance the adhesion of sputtered layers to the substrate. Since the X-ray penetration depth in the case of 3 ° grazing incidence is only about 0.07-0.14 times that at normal incidence, the grazing X-ray diffraction spectra give only surface information on samples. In the case of 500 nm thick Cu single layer on steel, X-ray diffraction showed that there was a little amount of iron on the surface of Cu layer. This means that the 500 nm thick Cu layer was unable to prevent iron diffusion. But no Fe diffraction peak was found for the sample of Cu (200 nm) /TiC (300 nm) double layers on steel. So we can deduce that the TiC layer can effectively prevent the iron atoms from diffusing from steel substrate to the Cu surface.

3.2. CVD diamond growth on the intermediate layers

On the steel substrate, the CVD deposited product would be graphite because of the existence of Fe. Nolter et al. [7] found that there was a graphite-rich layer on Cu coated Fe substrate during CVD deposition. In our experiment, if Cu intermediate layer was 1μm or thicker, diamond film would grow on it; however, the deposited diamond film on Cu easily peels off because of the larger difference in thermal expansion coefficients. Cu/TiC double layer is a better choice as an intermediate layer on the steel substrate. The diamond nucleation density reached 10^9 / cm^2 on Cu/TiC/steel, and 10^{10} / cm^2 on Cu/TiC/Si, which was much larger than that on steel (10^{4-5}/cm^2) and higer than on Si(111) (10^{8-9}/cm^2). The quality of deposited diamond films on both Cu/TiC/steel and Cu/TiC/Si was perfect as shown by a peak at 1333cm^{-1} in the Raman spectra shown in fig.2 (a) and 2(b). Fig. 3(a) and 3(b) are the SEM pictures which reveal fine and dense diamond particles on Cu/TiC/steel and Cu/TiC/Si, respectively. This illustrated that the high quality diamond films could be grown on steel and Si substrates with Cu/TiC intermediate layers. It should be noticed that such double layers have proven to be of greater utility in diamond growth on steel substrate than a single layer of Cu or TiC. Cu and TiC intermediate layers have different functions in diamond growth: TiC layer prevents Fe diffusion into the surface area, and higher diamond nucleation density is obtained on Cu layer surface.

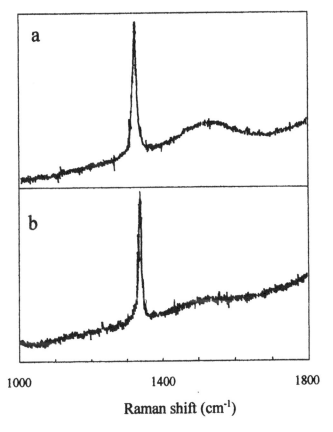

Figure 2. Raman spectra of diamond films on steel (a) and on Si(111) (b) substrates with Cu/TiC intermediate layers.

Figure 3. SEM pictures of diamond films on steel (a) and on Si(111) (b) substrates with Cu/TiC intermediate layers.

3.3. Diamond film adhesion strength

The fracture strength was determined by the tensile test on diamond films grown on both steel and Si substrates with Cu/TiC intermediate layers. We did not find any diamond fragment left on the substrate surface by optical microscopy when the film was pulled off. Diamond film adhesion strength on Cu/TiC/steel was 1-

1.5 MPa and Cu/TiC/Si was 4-5 MPa, superior to those deposited on a single intermediate layer, such as Cu or TiC, or on the substrate without any intermediate layer. For thicker diamond film, the adhesion strength tended to decrease. The tensile test is shown to be a good way to measure the diamond film adhesion strength on a brittle substrate such as Si. Since the glue strength limits the upper range of the measurement, therefore, the choice of a higher strength glue is very important. Any deviation from concentricity of the tensile bars may lead to an additional force moment and cause a decrease of film fracture load. It is pointed that the intermediate layer thickness could influence the test results. If it is too thick, the adhesion strength may decrease and the fracture during the tensile test may occur in this layer. Sputtering a proper metal layer on a fresh diamond film surface and the subsequent thermal treatment is a key to increase the glue strength between diamond film and the tensile bar, which may ensure that the fracture during tensile test occurs only at the interface between the diamond film and the substrate.

Finally, the indentation test showed that for the diamond film grown on Cu/TiC/steel, the critical applied load was about 150 kg which was 3-5 times larger than that on a single Cu or TiC layer. The critical load decreased with decreasing the thickness of the TiC intermediate layer.

4. CONCLUSIONS

Sputtering an intermediate layer is the most effective way to grow thin diamond films on non-diamond substrates, especially on steel. Cu/TiC double layer was found to be better than either single layer in our experiment. Diamond nucleation on it is much higher than on steel substrate. Fine and uniform diamond films were obtained on both steel and Si(111) substrates with Cu/TiC double intermediate layer by HFCVD growth. The adhesion of diamond film is enhanced with the existence of the intermidiate layer. For Cu/TiC double sputtered layers, the bottom sub-layer TiC may restrain the diffusion of Fe from the substrate during CVD growth. The adhesion strength of sputtered layers to substrates also affects the CVD diamond adhesion.

The tensile test can provide quantitative results on diamond film adhesion strength to the substrate. It is most important to make sure that during the tensile test the fracture occurs only between the diamond film and the substrate. In our experiment, the tensile strengths of CVD diamond on steel and on Si(111) with Cu/TiC bilayers were found to be 1-1.5 MPa and 4-5 MPa, respectively, which showed a tendency to decrease with increase in the diamond film thickness or decrease in the TiC intermediate layer.

Acknowledgements

This work was supported by the Chinese National Natural Science Foundation No. 59682001 and China High Technology 863 Foundation No. 002-0010.

REFERENCES

1. V. Buck and F. Deuerler, Diamond Related Mater., **7**, 1544 (1998).
2. P.O. Joffeau, R. Haubner and B. Lux, Int. J. Refractory Hard Mater., **7**, 186 (1988).
3. H. Chen, M.L. Nielsen, C.J. Gold, R.O. Dillon, J. DiGregorio and T. Furtak, Thin Solid Films, **212**, 169 (1992).
4. B.S. Park, Y.J. Baik, K.R. Lee, K.Y. Eun and D.H. Kim, Diamond Related Mater., **2**, 910 (1993).
5. J. Spinnewyn, N. Nesladek and C. Asinaric, Diamond Related Mater., **2**, 361 (1993).
6. K. Vandierendonck, Surface Coatings Technol., **98**, 1060 (1998).
7. S.D. Nolter, B.R. Stone and J.T. Glass, Diamond Related Mater., **3**, 188 (1994).
8. Chii Ruey Lin, Cheng Tzuo Kuo and Ruey Ming Chang, Diamond Related Mater., **7**, 1628 (1998).

Adhesion Measurement of Films and Coatings, Vol. 2, pp. 255–276
Ed. K.L. Mittal
© VSP 2001

Effect of primer curing conditions on basecoat-primer adhesion – A LIDS study

J.S. METH*

DuPont Co., Central Research & Development, P.O. Box 80328, Wilmington, DE 19880-0328, U.S.A.

Abstract—We present a new technique, laser induced decohesion spectroscopy (LIDS), which is capable of measuring the practical work of adhesion G between a transparent polymer coating and an opaque coating or substrate. In LIDS, a laser pulse directed onto the sample creates a blister at the transparent/opaque interface. The blister's internal pressure depends on the laser pulse energy, and at a critical pressure the sample fractures, creating an annular debond similar to that obtained in the standard blister test. By measuring physical variables such as the curvature of the blister, and its radius and thickness, it is possible to deduce G. Here, we have investigated the effect of primer baking conditions on the fracture of automotive panels, with the goal of understanding the physical factors that affect resistance to stone chipping. Good adhesion can be achieved by not overbaking the primer. For melamine-formaldehyde systems, this implies that the self-condensation of melamine should be avoided. Current baking temperature of 140°C is adequate. For good fracture resistance, the coating thicknesses should be substantially greater than the plastic zone sizes, which are measured here for the first time in such systems and are generally ~20-35 μm.

Keywords: Adhesion; blister test; laser; automotive coating; paint.

1. INTRODUCTION

Coatings are inherent to many technologies, from paint systems to electronics manufacturing [1-3]. One of the parameters that needs to be known is the adhesion, or resistance to fracture, of a coating to a substrate or to another coating. In general, it is difficult to quantify this adhesion. Many qualitative methods are available, such as the crosshatch-peel test and the x-hatch test for paints. Quantitative methods include instrumented peel tests, double cantilever beam methods, and blister tests. These techniques require special sample preparations, which makes them difficult to apply to paint adhesion.

We are concerned here with quantifying the adhesion between successive layers of paint, or between paint and a substrate. Currently, the adhesion is qualitatively measured using the crosshatch-peel test. It is quick and easy, and the results correlate with field tests. The disadvantage is that many modern coating systems can easily pass this test, yet there can be substantial differences in the ad-

---

* Phone: 302-695-1129, Fax: 302-695-9799, E-mail: Jeff.Meth@usa.Dupont.com

hesion, which may only become manifest when the sample is subjected to environmental stresses such as accelerated aging. As technology advances, the ability of the crosshatch-peel test to discriminate between good and excellent coating adhesion diminishes. In this sense the crosshatch-peel test has a limited dynamic range. There has long been a need for a more quantitative test method.

We have developed laser induced decohesion spectroscopy (LIDS) to quantify the adhesion in such systems. This paper presents the technique and some of the more basic results we have obtained so far. LIDS is essentially a new modification of the blister test [4-10]. In the blister test, it is necessary to generate pressure to produce a blister. This has led in the past to very ingenious ways to introduce that pressure. Unfortunately, the preparation of the samples became much more complicated. LIDS employs photothermal ablation as the mechanism for generating internal pressure. Generally in photothermal ablation, a high power, temporally short, laser pulse incident on an opaque material causes it to heat up extremely rapidly, whereupon bonds break in that material, producing gases that expel from the surface. If a transparent layer covers the opaque one, the ablation process produces an internal pressure. In our experimental configuration, Figure 1, the opaque polymer is ablated, and a pressure is generated between the opaque and transparent polymers. The close proximity of a rigid substrate prevents significant deformation of the opaque layer, but the experiment still works with a coating on a deformable substrate. This internal pressure deforms the transparent coating, producing a blister directly above the ablated area. By measuring the curvature of the blister at its top, center (see Figure 1), the internal pressure of the blister can be known. By itself, this process does not involve any debonding between the two layers outside of the ablated region. This is analogous to the small initial debond or hole that is necessary in the standard blister test. As the laser pulse energy is increased, more ablation occurs, and the internal pressure increases along with the associated strain energy stored in the system. At some critical laser pulse energy, the Griffith criterion for crack propagation [11] is exceeded, and the blister expands radially into the region of the sample that was not ablated, producing an annular debond area, similar to other blister tests. At this critical laser pulse energy, there exists a critical internal pressure that is related to the work of adhesion. While we refer to the practical work of adhesion as G in this paper, it should be understood that these values are strongly influenced by dissipative processes in the coatings, and are not a measurement of any purely interfacial property [1].

There are several advantages to the LIDS technique. Foremost, the geometry of the LIDS experiment allows testing of film systems without special sample preparation – no holes need to be drilled, no other supports need to be attached. Second, the initiating laser pulse controls the fracture rate. The rise time of the blister is ~1μs. At these rates, the response of the polymer is determined by the glassy modulus, and viscous effects are not important during the blister creation. The values measured by LIDS are applicable to the problem of chipping of paint by stones. Third, the pressures generated by the ablation process can easily be in excess of 100 MPa, allowing large values of adhesion to be measured (1-1000 J/m^2).

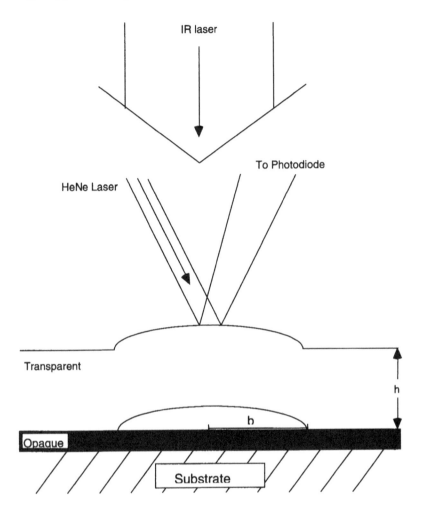

Figure 1. Schematic of LIDS experiment. An IR laser pulse creates a blister, while a HeNe beam probes the curvature of the surface.

2. EXPERIMENTAL

In LIDS (see Figure 1), the typical sample consists of a transparent polymer coating on top of an opaque polymer coating, supported by a rigid substrate (similar to the fender on an automobile). LIDS can also be performed on samples consisting of a transparent polymer covering an opaque substrate (such as primer on thermoplastic olefin, a major automotive structural material, or polyimide on copper). An IR laser pulse impinges on the sample, passing through the transparent layer and ablating a portion of the opaque layer. This produces the blister. A separate continuous wave Helium-Neon (HeNe) laser reflects from the surface of the blister. The experimental observable is the intensity of HeNe laser light that passes through a circular aperture of known diameter. This intensity is related to the diameter of the laser beam. When a laser reflects from a positively curved surface, the laser beam diverges so its diameter increases. Thus we relate the laser intensity passing through the aperture to the radius of curvature of the blister.

There is a critical laser pulse energy in the LIDS experiment, corresponding to a critical internal blister pressure. At energies below critical, the laser pulse ablates a portion of the opaque material, creating a blister in the transparent material directly above it. This is the starter crack that is necessary in all fracture mechanics experiments. At energies above critical, in addition to the starter crack, an annular debond is created around the ablation site. The sample has fractured. At the critical laser pulse energy, the pressure created by the laser pulse is just enough to begin crack propagation. The generated pressure is stored as strain energy in the blister. At the critical energy, the strain energy in the system is maximized, and any further accommodation of mechanical deformation is achieved by creating new surfaces, initiating a crack. At this critical energy, the top surface of the blister is curved, and it is this curvature that we are interested in knowing.

First we need to find this critical laser pulse energy for a given sample. This is done by exposing the sample to a laser pulse of a given energy many times (usually ten) and examining the resulting blisters under a microscope. For energies below critical, no cracks are visible under the microscope. Above the critical energy, an annular debond can be seen. So, the sample is exposed to pulses of a particular energy, and the resulting blisters are examined under the microscope. If less than half the blisters show an annular debond, then we know that the laser pulse energy is subcritical. The laser pulse energy is increased, and the sample is retested. If more than half the blisters show an annular debond, then we know that the laser pulse energy is supercritical. The laser pulse energy is then decreased, and the sample is retested. Throughout this process, the sample is translated after each laser pulse. Tens of blisters are made before the critical energy is determined. Reiterating this process eventually yields the critical energy, to $\pm 5\%$. It is noted here that for strongly adhering systems, as the laser energy continues to exceed the critical energy, the debond turns into a classic cone crack, finally resulting in the ejection of a frustum-shaped chip from the system. For weakly adhering systems, the crack stays at the interface, and the annular area continues to increase.

Once the critical laser pulse energy is obtained, we then need to measure the radius of curvature of the blister at this energy. This is done by exposing the sample to a range of laser pulse energies spanning from the threshold energy, at which a blister may just be detected, up to the critical energy, where crack propagation begins, and measuring the curvature of the blister for every exposure. In this range of energies, there is no fracture in the sample. The ablation process produces the starter crack, and we measure the radius of curvature associated with each ablation. It is important to realize that LIDS is a single shot experiment. One laser pulse impinges on one location of the sample. The sample is then moved to a fresh spot before another laser pulse hits the sample. When one plots the measured curvature versus the laser pulse energy, a linear relationship is observed, and this is fit to a line. We then extrapolate this line to the critical laser pulse energy to deduce the curvature.

It is necessary to know the thicknesses of the coatings. To do this, a small section of the plate is cut and polished in cross section. The thicknesses are measured under a microscope with a calibrated reticle, reducing the standard error to ±1%. The microscope is similarly used to measure the radii of the blisters at the critical laser pulse energy.

The ablating laser pulse is produced by a cavity-dumped regenerative oscillator running at 10 Hz, pumped by a CW, mode-locked Nd:YLF laser ($\lambda = 1.053$ µm, pulse width 50 ps). The output energy is stable to ±3%, after the beam is cleaned up by a vacuum spatial filter. A shutter system isolates a single pulse. The laser is then focused onto the sample to a $1/e^2$ intensity radius of 50 µm, thus the blisters tend to have radii of this dimension. The sample itself is mounted vertically at the focal plane of the IR laser pulse on a motorized translation stage which can be positioned reproducibly to 1 µm via computer control. A blister is formed by exposing the sample to a single IR laser pulse, and recording the reflected HeNe intensity on a digital oscilloscope (LeCroy 9374M). The scope trace is imported to a computer and analyzed to extract the curvature. The sample is then translated 1 mm before another laser pulse is incident on the sample. No position on the sample is ever exposed to more than a single laser pulse for data acquisition. The focused spot size of the HeNe laser is $1/e^2$ intensity radius of 25 µm.

Two different primer formulations, 768C50801 (referred to here as 768C) and 1K polyurethane (referred to as 1K), were examined. The 768C primer was a typical polyester resin crosslinked with methylated/butylated melamine. The 1K primer used the same polyester, but was crosslinked with hexanediisocyanate (HDI). The primers were applied to phosphated cold-rolled steel coated with M64J28 electrocoat (DuPont), and subsequently baked for 30 or 60 minutes at 130, 140, 165, 185, 205 °C. The panels were then overcoated with Oxford white tinted basecoat, baked for 17 minutes at 130 °C. This is an unpigmented solvent-borne basecoat, which allowed us to measure the adhesion between the basecoat and the primer. In all, there were twenty samples with this three-layer structure. We wanted to investigate the effect of baking conditions on basecoat-primer adhesion.

In addition, twenty more panels were prepared. These panels were similar to the first, in that they both consisted of the same two primer formulations over the same electrocoat, and baked under the same various conditions. However, these panels were then coated with solvent-borne Oxford white basecoat with the complete pigment package, then overcoated with Gen 3 clearcoat. The basecoat and clearcoat were baked together for 17 minutes at 130°C. With these samples, we were able to measure the adhesion between the clearcoat and the basecoat. Since the only variables were the primer bake and the primer formulation, we expected that the results would be the same for all samples with this four-layer structure.

3. THEORY

The theoretical analysis of the LIDS experiment consists of three links between four parameters. First, the Gaussian beam propagation theory is used to relate the measured spot size of the HeNe laser beam reflected from the blister to its radius of curvature. Next, the biharmonic equation for the mechanical deformation in a model system is solved by expanding the stress potential function in a series to the sixth order, and choosing appropriate boundary conditions to solve for the prefactors. From this analysis we are able to theoretically relate the curvature of the blister to the internal pressure. We are also able to calculate the strain energy in the system, which is similar to the far-field energy described by Andrews and Stevenson [7]. Finally, by applying Griffith's criterion for crack propagation, we are able to relate the internal pressure to the practical adhesion parameter G, which is the material property of interest.

3.1. Relating laser spot size to blister curvature

To measure the curvature of the blister, we measure the intensity of light that passes through a circular aperture placed in the beam path of the HeNe laser. The curvature is extracted by solving the complex algebraic equations for Gaussian beam propagation through the system [12]. This calculation is detailed in Appendix A. The result is an algebraic formula that is integrated into the computer program that controls data transfer from the oscilloscope. Thus, once an oscilloscope trace is transferred, the corresponding curvature is immediately calculated.

3.2. Relating blister curvature to internal blister pressure

To relate the curvature to the internal blister pressure, applicable for any laser pulse energy, we create a theoretical model that consists of a thick, circular disk, to which a uniform pressure, q, is applied at the lower surface. We use this as the model for the blister, and solve the problem using stress potential functions [13]. This solution includes stretching and bending of the disk. To account for the fact that the blister is part of a larger system, we apply various boundary conditions at the edge of the model disk. We begin with a circular disk of radius b and thickness h, diagrammed in Figure 2. The origin is taken at the center. We wish to solve the biharmonic operator for the boundary conditions of interest. The following equations represent the relationships between the stress potential function ϕ and the expressions for the normal stresses, σ, the shear stress τ, and displacements (u is the radial displacement, while w is the displacement in the z-direction), in terms of the spatial variables in cylindrical coordinates (r, θ, z), and the material properties Poisson's ratio, ν, elastic modulus, E, and the shear modulus $G = \frac{1}{2} E/(1+\nu)$.

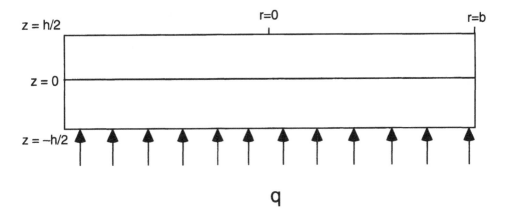

Figure 2. Diagram of theoretical model of a thick disk of radius b, thickness h, with a uniform pressure q applied to the bottom surface.

$$\nabla^2 \nabla^2 \phi = 0 \tag{1a}$$

$$\sigma_r = \frac{\partial}{\partial z}\left(v\, \nabla^2\phi - \frac{\partial^2\phi}{\partial r^2}\right) \tag{1b}$$

$$\sigma_\theta = \frac{\partial}{\partial z}\left(v\, \nabla^2\phi - \frac{1}{r}\frac{\partial\phi}{\partial r}\right) \tag{1c}$$

$$\sigma_z = \frac{\partial}{\partial z}\left((2\text{-}v)\, \nabla^2\phi - \frac{\partial^2\phi}{\partial z^2}\right) \tag{1d}$$

$$\tau_{rz} = \frac{\partial}{\partial z}\left((1\text{-}v)\, \nabla^2\phi - \frac{\partial^2\phi}{\partial z^2}\right) \tag{1e}$$

$$2\,G\,w = 2(1\text{-}v)\, \nabla^2\phi - \frac{\partial^2\phi}{\partial z^2} \tag{1f}$$

$$2\,G\,u = -\frac{\partial^2\phi}{\partial r\,\partial z} \tag{1g}$$

We expand ϕ in a series, inclusive to terms that describe quartic displacements:

$$\phi = a_3(2 z^3 - 3 r^2 z) + b_3(r^2 z + z^3)$$

$$+ a_4(8 z^4 - 24 r^2 z^2 + 3 r^4) + b_4(2 z^4 + r^2 z^2 - r^4) \qquad (2)$$

$$+ a_6(16 z^6 - 120 z^4 r^2 + 90 z^2 r^4 - 5 r^6) + b_6(8 z^6 - 16 z^4 r^2 - 21 z^2 r^4 + 3 r^6)$$

Equation 2 is substituted into Equations 1a-1g, and the appropriate boundary conditions are applied to solve for the coefficients. There are four elementary boundary conditions that are common to all the solutions: 1) $\sigma_z = 0$ for z = h/2 – there is no normal force acting at the upper, free surface; 2) $\sigma_z = $ -q for z = -h/2 – there is a uniform compressive force q acting at the lower surface; 3-4) $\tau_{rz} = 0$ for z = ±h/2 – there are no shear forces at the upper or lower surfaces. We are assuming that the ablation process produces a uniform pressure on the lower surface of the blister.

In addition to these four, two more conditions are needed. In the simply supported case, the blister is constrained to have no net bending moment along the outside edge, and to have its neutral plane in the center of the disk. In the rigidly clamped scenario, the radial displacement of the blister is zero at its upper and lower edges. These two boundary conditions represent opposite extremes of clamping, which provides limits for the possible variations in the theory. In addition to these two cases, a mixed boundary condition is examined. In this case, the radial displacement at the bottom edge of the disk is zero, while the radial stress at the top edge of the disk is zero. It turns out that in the rigidly clamped case, the radial stress is zero at the point $r/b = [(1 + v)/(3 + v)]^{1/2}$. The mixed boundary conditions case merely extends that zero point out further on the disk, to r/b = 1, allowing some radial displacement at the upper edge. This case is then intermediate to the first two. The prefactors are collected together in Table 1, normalized by the pressure q. Along with these solutions, we also examine the textbook solutions for both thin and thick disks [14]. The choice of boundary conditions does not have an extreme effect on the results. In addition to the quantities described in Equation 1, the analytical solution enables us to derive expressions for the strains, tensor invariants, and combinations thereof, most notably the hydrostatic pressure, the deformational stress, and the strain energy density.

For all cases, the relationship between pressure and curvature may be summarized by:

$$q = \frac{16 D}{R_c b^2 s(h/b)} \qquad (3a)$$

$$s(h/b) = k_1 + k_2 (h/b)^2 \qquad (3b)$$

$$D = \frac{E h^3}{12 (1 - v^2)} \qquad (3c)$$

Table 1.
Coefficients for ϕ, normalized by the pressure q

	Simply Supported	Rigidly Clamped	Mixed
a_3	$\dfrac{(5v-1)}{60(v+1)}$	$\dfrac{1}{60(v-1)}$	$\dfrac{3(v-1)(5v-7)\,b^2 - 4(v+1)\,h^2}{240(1-v^2)\,h^2}$
b_3	$\dfrac{-1}{20(v+1)}$	$\dfrac{1}{20(v-1)}$	$-\dfrac{3(v-1)\,b^2 + 2(v+1)\,h^2}{40(1-v^2)\,h^2}$
a_4	$\dfrac{5(-7v^2-13v+24)\,b^2 + (7v^2-64v-11)\,h^2}{4480(1+v)\,h^3}$	$\dfrac{-3(v-1)(7v-8)\,b^2 + (7v^2-22v+13)\,h^2}{2688(v-1)\,h^3}$	$\dfrac{3(v-1)(v+2)(7v-8)\,b^2 - (v+1)(7v^2-22v+13)\,h^2}{2688(1-v^2)\,h^3}$
b_4	$\dfrac{15(3+v)\,b^2 + 3(3-v)\,h^2}{560(1+v)\,h^3}$	$\dfrac{3(1-v)\,b^2 + (1+v)\,h^2}{112(1-v)\,h^3}$	$\dfrac{3(v+2)(1-v)\,b^2 + (1+v)^2\,h^2}{112(1-v^2)\,h^3}$
a_6	$\dfrac{(18-11v)}{10560\,h^3}$	$\dfrac{(18-11v)}{10560\,h^3}$	$\dfrac{(18-11v)}{10560\,h^3}$
b_6	$\dfrac{1}{352\,h^3}$	$\dfrac{1}{352\,h^3}$	$\dfrac{1}{352\,h^3}$

D is the flexural rigidity of the clear coat, E is the modulus, v is Poisson's ratio, h is the thickness, b is the blister radius, R_c is the measured radius of curvature, and k_1 and k_2 are constants depending on the particular boundary conditions, summarized in Table 2. From this analysis, we can relate the curvature calculated at the critical laser pulse energy to the critical internal blister pressure. This pressure is then related to the practical work of adhesion, or resistance to fracture.

3.3. Relating critical internal blister pressure to practical work of adhesion

Once the internal blister pressure is known, relating it to G is straightforward, following the literature analysis using linear elastic fracture mechanics. The total strain energy in the system is split into two terms, near-field and far-field, following Andrews and Stevenson [7]. The near-field energy is that associated with the crack tip, and is available from textbooks [15]. The far-field energy is associated with strain energy stored away from the crack tip, namely in the blister and in the opaque material that comprises the elastic foundation. From the solutions derived for the thick disk, we can calculate the strain energy in the blister. Upon equating G with dU/dA, where U is the strain energy and A is the crack area, we derive the relationship between the dimensionless parameter q^2b/EG and the aspect ratio h/b. Previous work has shown [6,7] that this relationship can be characterized by a function f(h/b). We find that f(h/b) can be expressed as a polynomial with constant coefficients:

$$\frac{q^2b}{E} = G \, f(h/b) \tag{4a}$$

$$f(h/b) = \frac{(h/b)^3}{a_1 + a_2(h/b)^2 + \dfrac{4(1-v^2)}{\pi}(h/b)^3 + a_3(h/b)^4} \tag{4b}$$

The values of a_1, a_2 and a_3 depend on the boundary conditions. Table 3 collects the various values from the different theories.

Table 2.
Summary of parameters for s(h/b)

Theory	k_1	k_2
clamped thin plate	1	0
simply supported disk	$\dfrac{(3+v)}{(1+v)}$	$\dfrac{4(2+v)}{5(1+v)}$
clamped disk	1	$\dfrac{2(2-v)}{3(1-v)}$
mixed b.c. disk	$\dfrac{(2+v)}{(1+v)}$	$\dfrac{2(2-v)}{3(1-v)}$
clamped thick plate	1	$\dfrac{2}{(1-v)}$

Table 3.
Summary of parameters for f(h/b)

Theory	a_1	a_2	a_3
thin plate	$\frac{3}{32}(1-v^2)$	0	0
simply supported	$\frac{3}{32}(7+v)(1-v)$	$\frac{3}{10}(1-v)$	$\frac{(-v^3+v^2+8v+524)}{2800}$
rigidly clamped	$\frac{3}{32}(1-v^2)$	$\frac{3}{10}(1+v)$	$\frac{(v^3-3v^2-312v+160)(1+v)}{840(1-v)}$
mixed	$\frac{3}{32}(7+v)(1-v)$	$\frac{(10+v)(1+v)}{40}$	$\frac{(v^3-3v^2-312v+160)(1+v)}{840(1-v)}$
thick plate	$\frac{3}{32}(1-v^2)$	$\frac{3}{8}(1+v)$	0

3.4. Final relationship between G and experimental observables

By combining Equations 3 and 4, a simple formula for G emerges:

$$G = \frac{16\,E}{9\,(1-v^2)^2}\frac{h^3}{R_c^2}\,g(h/b) \tag{5a}$$

$$g(h/b) = \frac{(h/b)^3}{f\,s^2} \tag{5b}$$

From this formula, we see that the theory splits into three multiplicative terms. The first term involves only E and v, material parameters. The second term involves the experimentally determined variables, h and R_c. The third term, g(h/b), is purely model dependent, accounting for the particular shape of the blister, and depends only on aspect ratio and v. Equation 5 is the primary equation of the LIDS theory.

In Figure 3, a plot of g(h/b) versus aspect ratio h/b is presented for the five sets of boundary conditions, where we have assigned v = 1/3 for convenience. Several features of the function are evident. First, aside from the thin plate theory, the theories are all close to one another, with a magnitude ~0.15. There are no discontinuities or large derivatives. As the aspect ratio of the blister varies, the theoretical factor does not vary much. Thus, the LIDS experiment is very robust with respect to the theoretical analysis. On the practical side, when the LIDS experiment is performed the aspect ratio is usually between 0.5 and 1. This is because standard coating thicknesses just happen to result in this particular range of aspect ratios. In this range of aspect ratios, the mixed boundary condition case lies between the simply supported and the rigidly clamped cases. It is for this reason, along with our belief that it best describes the actual displacements, that we choose

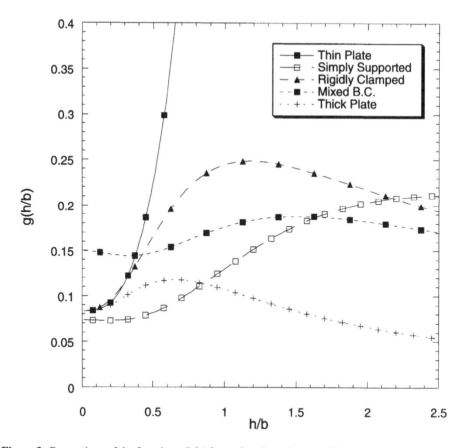

Figure 3. Comparison of the function g(h/b) for various boundary conditions.

the mixed boundary condition case for the theoretical analysis. When data are analyzed using different theories, the absolute numbers do change, but the relationships between the values are preserved. It would be desirable to see the problem solved with finite element analysis, but at this stage it is not clear how much true value would be gained from that endeavor.

It is possible to include the strain energy from the elastic foundation in the analysis. This leads to a small correction in Equation 4. The correction is small because the deformation of the opaque polymer on the rigid substrate is small. It is also worth omitting this term because any theory would need to include the thickness of the layer along with its mechanical properties, E and ν. This introduces more variables into the equation for G, and does not contribute to the essence of the experiment. When LIDS is used to examine the adhesion between a coating on a thick (several millimeter) polymer substrate, we approximate the strain energy from the substrate in the constant term describing strain at the crack tip. So we use the strain energy at the crack tip to model all substrate conditions.

4. RESULTS AND DISCUSSION

For the work presented here, we take E = 3 GPa (from low temperature DMA measurements) and ν = 1/3. The standard error is 12%. In Figure 4, we show the adhesion between unpigmented basecoat and primer as a function of baking temperature and time for the 768C primer. We see that there is no major difference between 30 minute and 60 minute baking times. However, there is a noticeable decrease in adhesion as the baking temperature increases. In Figure 5, the same results are presented for the isocyanate coating. Here we see that, for the three higher temperatures, there is no difference in adhesion for bake time or temperature. For the two lower temperatures, we see that the 60 minute bake gives G values in line with the higher temperatures, while the 30 minute bake gives very large G values that are not in accord with the rest of the data. In Figures 6 and 7, the adhesion between clearcoat and pigmented basecoat is presented. We see that the adhesion levels are different for the two primer systems. However, for each primer, the measured adhesion is independent of primer baking conditions, as expected. Note that the two 1K samples baked for 30 minutes at 130 °C and 140 °C show anomalous adhesion values for both the basecoat-primer and clearcoat-basecoat results (Figures 5 and 7).

As shown in Figure 4, the adhesion in this system is seen to be very dependent on temperature, and not so dependent on baking time. The adhesion is strong at low bake temperatures, and decreases as the baking temperature increases. The explanation for this involves the availability of unreacted functional groups remaining on the surface of the coating. At low bake temperatures, such as 130 °C, not all of the melamine and polyester functional groups have reacted. So when the unpigmented basecoat is applied, and then baked, the basecoat can form covalent bonds with the unreacted groups in the primer. As the primer bake temperature increases, there are fewer unreacted functional groups available to covalently bond to the basecoat when it is applied and baked. Thus the adhesion decreases. Another explanation can be invoked if one does not believe that covalent bonding occurs across the interface - indeed, this has not yet been proven conclusively. From that point of view, condensation of the melamine makes the melamine more covalent and less polar in nature. When melamine condenses, methanol, formaldehyde and other small, polar molecules are liberated. This reduces the polar nature of the coating, which, in turn, reduces the surface energy. Either way, the result is the same. A secondary effect occurs at the highest bake temperatures. At 210 °C the melamine begins to thermally decompose. This weakens the primer and makes the interface more susceptible to fracture. DMA and DSC data on these materials is not available. However, it is known that for melamine-polyol systems, the co-condensation reaction begins at ~110 °C. The self-condensation of melamine begins at ~150 °C. From this information, it seems reasonable to conclude that the co-condensation reaction begins at lower temperatures and proceeds to the point where further reaction in the network is limited by diffusion of reactants. At this point, there are still reactants present in the coating. When the basecoat is applied

J.S. Meth

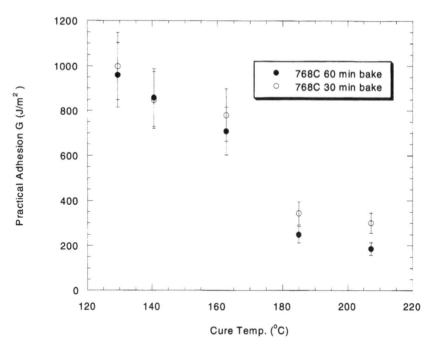

Figure 4. Practical adhesion, G, between clear basecoat and 768C primer as a function of curing.

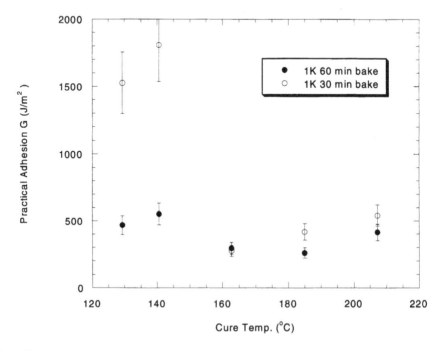

Figure 5. Practical adhesion, G, between clear basecoat and 1K primer as a function of curing.

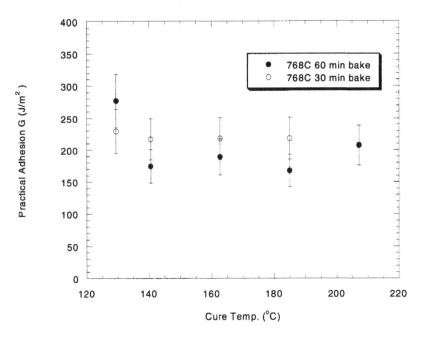

Figure 6. Practical adhesion, G, between clearcoat and basecoat with 768C primer as a function of curing.

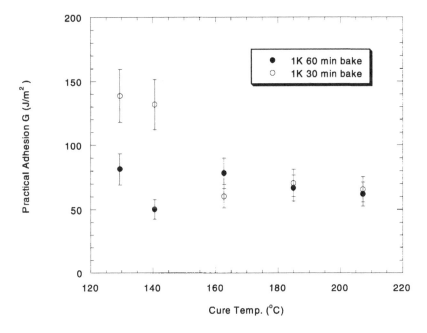

Figure 7. Practical adhesion, G, between clearcoat and basecoat with 1K primer as a function of curing.

on top of this primer, further co-condensation can occur at the interface between the two coats, because functional groups on the surface of the primer come into contact with those on the bottom surface of the basecoat. When the primer is baked at higher temperatures, so that melamine self-condensation occurs, less functionality is available to react with the functionality in the basecoat. This effect is responsible for the decrease in adhesion, seen in this experiment to begin with the samples baked at 165 °C.

As shown in Figure 5, the basecoat-primer adhesion for the 1K system varies with primer bake like the 768C, but with some curious features. The 60 minute bake at the lower temperatures has a dramatically lower adhesion than the comparable 768C panels. Or, conversely, the 30 minute bake has anomalously large adhesion. The dramatic difference between the 30 and 60 minute bakes for the lower temperatures is not entirely explainable. However, as will be discussed below, the four-layer systems analogous to these three-layer systems show the same anomaly. The absence of melamine in the 1K system precludes any thermal decomposition of the triazine ring, so the high temperature bakes have slightly higher adhesion than the 768C panels.

Excluding the 30 minutes at 130 and 140 °C data points, the adhesion is seen to vary less with baking temperature, as exemplified by the samples baked for 60 minutes. This is due to the chemical nature of the isocyanate, as opposed to the melamine. The isocyanate unblocks and begins to react with the polyester at ~120 °C. Since there is no competing reaction, there are a relatively constant number of functional groups available on the primer surface to react with the basecoat. However, the overall lower values of adhesion for the 1K system indicate that the 768C system has the capability of creating more interfacial covalent bonds.

In the four-layer system, the LIDS experiment is capable of measuring the adhesion between the clearcoat and the basecoat. Since only the primers were varied, we expected these values to be constant for all panels. This was not entirely borne out by experiment. In Figure 6, the clearcoat-basecoat adhesion is presented for the 768C primer. We see that the adhesion is independent of primer baking conditions. The average is $G = 211 \pm 10$ J/m^2. In Figure 7, the same data are displayed for the 1K primer system. Here we see that the two points that were high in the three-layer system are high in the four-layer system, also. This is very curious. There could be something anomalous about the samples as they were prepared. If these two points are excluded from the statistics, then $G = 67 \pm 4$ J/m^2. With the points included, then the average is raised to $G = 81 \pm 10$ J/m^2. In either case, there is a significant difference between the clearcoat-basecoat adhesion for the different primer systems. This is something that was not expected, and its cause is unknown.

The first hypothesis for this discrepancy was that a surfactant was included in the 1K system, which somehow migrated from the primer, through the basecoat, and to the clearcoat-basecoat interface. Unfortunately, after having examined the 1K formulation, no such material was found. The second hypothesis is that the amount of time between application of the basecoat and the clearcoat affects adhesion. If more residual solvent is in the basecoat at the time the clearcoat is applied, then the viscosity of the basecoat will be slightly lower, and more intercoat mixing could occur. It is possible that TEM analysis of the interface could discriminate the difference.

The panels that used the 1K primer, and were baked for 30 minutes at 130 and 140 °C, are anomalous. They result in larger values for the adhesion for both systems. It is possible that this is a mechanical effect that is brought out in the theory used to analyze the data. In the data analysis, we assume that the modulus of the primer is constant, independent of primer bake. This may not be true. However, no matter which model is used to analyze the data, whether we include the effect of an elastic foundation or ignore it, the results stay the same. For example, for the data point at 30 minutes, 130 °C, analysis with the elastic foundation model results in $G = 139$ J/m^2, while ignoring the elastic foundation results in $G = 150$ J/m^2. All the values for the other samples increase by the same amount proportionately.

In a previous paper, we demonstrated the effect of the thickness of the transparent coating on the measured work of adhesion [16]. This was not a problem in this study due to the control of the coating thickness, which was maintained at greater than 50 μm. In the three-layer system, the transparent coating thickness varied from 58 to 78 μm, with an average of 66 μm. In the four-layer system, the thicknesses ranged from 54 to 63 μm, with an average of 57 μm. Thus we believe that these adhesion values are mostly free from the artificial decrease in the measured work of adhesion seen with thinner (25 μm) coatings.

The plastic zone size, R_p, can be calculated for these coatings, using the expression:

$$R_p = \frac{1}{2\pi} \frac{E\,G}{\sigma_y^2}$$

where $E = 3$ GPa is assumed as the modulus, and G and the yield stress, σ_y, are determined by experiment. For the 768C primer system, $R_p = 23$ μm for the clearcoat, and 33 μm for the basecoat when the primer is baked for 30 minutes at 130 °C. The resilience of the coating system to chipping may increase if the basecoat thickness is increased. The general design criterion that becomes apparent from these studies is that the coating thickness should be substantially greater than the plastic zone size coinciding with fracture at the lower and upper interfaces.

The three-layer system experiments were performed with an unpigmented basecoat. It is not clear what the effect of pigmentation would be on the properties that we are discussing here. We would expect the modulus to increase, the yield stress to decrease, and the adhesion to remain constant. This would have the effect of increasing the plastic zone size even further. The modulus would increase because the pigmentation (TiO$_2$) has a high modulus. The modulus would be a weighted average of that of the components in a simple model. The yield stress should decrease because yielding would occur more easily at a pigment/polymer interface, where the interaction is not as strong as that of the polymer to itself. The adhesion is dependent upon binder formulation and interlayer covalent bonding, so that it would be, to first order, independent of the pigmentation.

From these experiments, it is possible to identify criteria for constructing the optimum coating system to resist chipping. First, we must realize that chipping is a matter of degree. Panels that are not damaged by stones with velocity 30 mph may be damaged by stones at 60 mph. Second, a chip that occurs at the metal/electrocoat interface is worse than a chip at any other interface because the electrocoat provides corrosion resistance to the metal. However, it would not be good to produce a system that consistently chipped at, e.g., the basecoat-clearcoat interface at very low stone velocities. So, two factors of importance in this system are: 1) to have good adhesion between individual layers; and 2) to be able to dissipate energy through plastic deformation without cracking.

To create interlayer adhesion, one wants to form covalent bonds between layers. This may be achievable by making the coatings nonstoichiometric. Using melamine, a nonstoichiometric formulation can be developed, where excess functionality is later consumed by self-condensation.

To dissipate the maximum amount of energy, one would want to create a coating with a large difference between the energy-to-break and the stored elastic energy. When a stone impacts on a coating system, the desired strategy is to dissipate as much impact energy as possible, and prevent large stresses from being produced at the metal/electrocoat interface. The stresses generated by a stone impact event will be larger than the yield stress of the polymers used in coatings. Therefore we know that the coatings will yield. The energy dissipated by the coating will be proportional to the area under the stress-strain curve from the yield strain to the final strain that the coating experiences. By having a large strain to break, the coating is less likely to fracture. By having a large yield stress, the coating can dissipate more energy. The greater the dissipation of energy by the coating system on top of the electrocoat, the less the propensity to fracture at the electrocoat/metal interface.

In addition, the coating thickness would then be 2-3x the size of the plastic zone predicted for interlayer fracture. This way, when a stone hits the car, each individual layer will hold up for as long as possible. It will eventually fail, with the crack propagating at the interface of the weakest boundary. For this reason, it would also be desirable to insure that the electrocoat/metal adhesion is not the weakest adhesion in the system. By engineering a coating system where the weakest interface is, say, the basecoat-primer, then energy dissipation can occur throughout the system, and catastrophic failure is localized at a particular interface that is away from the metal. These preliminary considerations will be refined in the future as more work is performed and our understanding of the process increases.

LIDS is not the first test to quantify paint adhesion. This has been accomplished previously using the inverted blister test [10]. In that work, Kinloch and coworkers measured the adhesion between electrocoated paint (epoxy-amine) and a steel substrate to be G = 100 J/m². This value compares favorably with the values being measured in this work. Albeit for a different system, it helps to benchmark our results.

The blister's curvature is not independent of time. The blister relaxes back to the surface of the plate over time (to within the surface noise of the plate as examined by microscopic white light interferometry) for energies below critical. Once the energy is over critical, and the sample is fractured, complete relaxation does not occur. We have performed LIDS between a thin glass plate (50 μm thick, from Schott glass) and a black epoxy substrate. These experiments were useful because the glass plate was not susceptible to plastic deformation. From these experiments, we were able to determine the time dependence of the pressure in the blister. We found that the pressure decayed exponentially with a decay time of ~1 ms. When we performed the clearcoat-basecoat experiments, we saw that the time dependence of the blister varied. At low pulse energies, where the blister is small and plastic deformation is small, we see that the curvature of the blister decays on a millisecond time scale, similar to the glass plate. As the laser pulse energy is increased, the decay time also increases and becomes nonexponential, due to viscoelastic processes in the polymer. For higher pulse energies, we have followed the blister relaxation over 10 decades of time, from 10^{-5} s to 10^{5} s, and are working on relating this time dependence to the relaxation spectrum of the polymer. In the experiments presented here, we measured the blister curvature for the first 10 μs. On this time scale it is constant, which enables us to measure the initial curvature from the ablation event.

The temperature rise in the sample is not a major concern in the LIDS experiment because the process occurs much faster than thermal diffusion. The opaque layer heats up and ablates, and the regions that do not ablate also heat up, but only by a small amount. The transparent layer does not heat up. On the time scale of the experiment, ~1 μs, heat can diffuse in a polymer only ~1 μm. Therefore the transparent coating, which is tens of micrometers, is not heated. The strain in the sample is a few percent, so the strain rate of LIDS is ~10^{5} s^{-1}. This high strain rate reduces the effect of increased temperature.

There are several parts to the LIDS experiment that can be improved. First, the HeNe laser samples a finite region of the blister, and does not measure the curvature at the exact center. This can be improved by further reduction in the HeNe spot size, or by taking into account the nonparabolic nature of the reflecting surface in the Gaussian beam propagation equations. Second, the experiment takes more effort to perform than the crosshatch-peel test, and concomitantly more time. Thus, LIDS is good for performing quantitative studies on the nature of coating adhesion, but probably would not be convenient as an everyday test method. The theoretical analysis can be improved by examining the deformation in more detail, using finite element analysis, but the current analysis is fairly robust. There is also a range of thicknesses of the transparent polymer over which LIDS will work, depending on G and the modulus of the coating. If the polymer is too thin (~10 μm for these films), then the blister shears off from the substrate before crack propagation begins. If the polymer is too thick (we have successfully performed LIDS on samples with thicknesses of 150 µm), then the curvature of the top surface will become too small to detect accurately.

5. CONCLUSIONS

There are many advantages to the LIDS experiment. Foremost, the experiment works on samples directly from the field or from automotive testing laboratories. It works on the fender of a typical automobile. Samples may be examined without drilling holes in the specimen, or adhering it to other supports. We find good correlations between the LIDS results and those from crosshatch-peel tests and gravelometer testing. We are able to examine samples that have been subjected to accelerated aging. While the results presented here are for basecoat-primer adhesion, we can use LIDS to examine other interfaces, too. We can examine the adhesion between an unpigmented primer and electrocoat, and between an unpigmented coating and a substrate such as thermoplastic olefin, copper, steel, or reactive injection molded polymers. We are able to quantify the effects of baking conditions, formulation variables, and aging on coating adhesion, which enables us to direct development to more robust coating systems.

We have demonstrated how the LIDS technique may be used to derive the practical work of adhesion for systems consisting of a transparent polymer coated onto an opaque polymer supported by a rigid substrate. LIDS is sensitive to differences in processing and formulation, while showing good reproducibility.

APPENDIX A

In this section we detail the extraction of the curvature of the blister from the intensity of the HeNe laser [12]. The HeNe laser can be successfully modeled as a Gaussian beam defined at any point by a beam diameter w and radius of curvature of the wavefront R, whose initial values are w_0 and R_0, respectively. Each optical element, be it a lens or a mirror, affects these two parameters in a known way. The variable q, not to be confused with the blister pressure, is a complex quantity characterizing the curvature and spot size of the laser by the following relation:

$$\frac{1}{q_0} = \frac{1}{R_0} + \frac{1}{i\,z_0} \qquad (A.1)$$

where $z_0 = \pi\,w_0^2\,/\,\lambda$, and wavelength $\lambda = 632.8$ nm. Since the beam is initially collimated, q_0 is a purely imaginary parameter. In the LIDS experiment, the beam then propagates a distance d_0 to a lens of focal length $f = 100.0$ mm. The beam then focuses through a distance d_2 to the sample, where it reflects from the surface possessing a radius of curvature R_c, which we wish to determine. The beam then propagates through a distance d_4 to the aperture, where its spot size is measured. These steps are summarized by the relations:

$$q_1 = q_0 + d_0 \tag{A.2}$$

$$\frac{1}{q_2} = \frac{1}{q_1} - \frac{1}{f} \tag{A.3}$$

$$q_3 = q_2 + d_2 \tag{A.4}$$

$$\frac{1}{q_4} = \frac{1}{q_3} + \frac{2}{R_c} \tag{A.5}$$

$$q_5 = q_4 + d_4 \tag{A.6}$$

By continued substitution, one arrives at an expression for q_5:

$$q_5 = \frac{d_4\left((f - d_0 - q_0) + \frac{2}{R_c}(d_2(f - d_0 - q_0) + f(d_0 + q_0))\right) + (d_2(f - d_0 - q_0) + f(d_0 + q_0))}{(f - d_0 - q_0) + \frac{2}{R_c}(d_2(f - d_0 - q_0) + f(d_0 + q_0))} \tag{A.7}$$

and the beam spot size is related to q_5 by:

$$w = \left(\frac{-\lambda}{\pi\, Im(1/q_5)}\right)^{1/2} \tag{A.8}$$

The intensity of light from a Gaussian laser beam passing through a circular aperture is given by the equation:

$$\frac{I}{I_0} = 1 - \exp\left[-2\frac{a^2}{w^2}\right] \tag{A.9}$$

where a is the radius of the aperture and I is the intensity. In the LIDS experiment, we measure this intensity ratio, invert A.9 to get w and hence q_5. Equation A.7 is inverted to solve for R_c:

$$\frac{1}{R_c} = \frac{(2\pi/\lambda)\, f\, w_0\left(w^2 k - 4\, d_4^4\, f^2\, w_0^2\right)^{1/2} - m}{2\,k} \tag{A.10}$$

where

$$m = 4\, d_4\left(f^2\left(d_2^2 + d_2\, d_4 + d_0\, d_4\right) + (f - d_2 - d_4)\left(2\, f\, d_0\, d_2 - \left(d_0^2 + z_0^2\right)(d_2 - f)\right)\right) \tag{A.11}$$

and

$$k = 4\, d_4^2\{(d_2 - f)^2\, z_0^2 + (f\,[d_0 + d_2] - d_0\, d_2)^2\} \tag{A.12}$$

These equations allow us to convert from intensity passing through a pinhole to the curvature of the blister.

Acknowledgments

The author would like to thank San Yuan for providing the samples and Dave Sanderson for technical assistance.

REFERENCES

1. K. L. Mittal (Ed.), *Adhesion Measurement of Films and Coatings*, VSP, Utrecht, 1995.
2. G. P. Anderson, S. J. Bennett and K. L. DeVries, *Analysis and Testing of Adhesive Bonds*, Academic Press, New York, 1977.
3. R. Lambourne (Ed.), *Paint and Surface Coatings: Theory and Practice*, Ellis Horwood, West Sussex, UK, 1987.
4. H. Dannenberg, J. Appl. Polym. Sci., **5**, 124 (1961).
5. M. L. Williams, J. Appl. Polym. Sci., **13**, 29 (1969).
6. S.J. Bennett, K. L. DeVries and M. L. Williams, Int. J. Fracture, **10**, 33 (1974).
7. E. H. Andrews and A. Stevenson, J. Mater. Sci., **13**, 1680 (1978).
8. M. G. Allen and S. D. Senturia, J. Adhesion, **25**, 303 (1988).
9. Y. Chang, Y. Lai and D. A. Dillard, J. Adhesion, **27**, 197 (1989).
10. M. Fernando, A. J. Kinloch, R. E. Vallerschamp and W. B. van der Linde, J. Mater. Sci. Lett., **12**, 875 (1993).
11. A. A. Griffith, Phil. Trans. Royal Soc., **221**, 163 (1921).
12. A. E. Siegman, *Lasers*, University Science Books, Mill Valley, CA 1986.
13. S. Timoshenko and J. N. Goodier, *Theory of Elasticity*, McGraw-Hill, New York, 1970.
14. S. Timoshenko and S. Woinowsky-Krieger, *Theory of Plates and Shells*, p. 74, McGraw-Hill, New York, 1959.
15. I. N. Sneddon and M. Lowengrub, *Crack Problems in the Classical Theory of Elasticity*, p. 134, Wiley, New York, 1969.
16. J. S. Meth, D. Sanderson, C. Mutchler and S. J. Bennison, J. Adhesion, **68**, 117-142 (1998).

Adhesion Measurement of Films and Coatings, Vol. 2, pp. 277–290
Ed. K.L. Mittal
© VSP 2001

Evaluation of a pulsed laser technique for the estimation of the adhesion strength of oxide coatings onto metallic substrates

GAËLLE ROSA, PANDORA PSYLLAKI and ROLAND OLTRA*

Laboratoire de Recherches sur la Réactivité des Solides, UMR 5613 CNRS, Université de Bourgogne, UFR Sciences et Techniques, B.P. 47 870, 21078 Dijon Cedex, France

Abstract—The aim of the present study was to investigate the possibility of using laser ultrasonic technique for the estimation of the adhesion of ceramic coatings, deposited onto metallic substrates by thermal spraying techniques. For this purpose, a pulsed Nd:YAG laser (λ=1064 nm, 14 ns) was used to irradiate Al_2O_3 coatings, of different thicknesses, deposited onto stainless steel substrates by atmospheric plasma spraying (APS). The so-generated acoustic waves were recorded in-situ using a laser heterodyne interferometer placed at the back surface of the metallic substrate. In order to correlate the so-obtained waveforms with the damage resulted in the bi-layered system, a physical and an analytical models have been developed. Moreover, the stress field developed at the coating/ substrate interface, due to the combined acoustic and thermo-mechanical effects, can be determined by numerical simulation, giving important information about the adhesion strength of transparent coatings deposited onto metallic substrates, by thermal spraying techniques.

Keywords: Laser ultrasonics; transparent coatings; thermoelastic interactions; interfacial failure; practical adhesion; expulsion.

1. INTRODUCTION

The determination of the adhesion strength of coatings remains a difficult task in surface engineering, although it has a great importance for the proper design of a component planned to operate under thermal and/ or mechanical loading.

According to Mittal [1], adhesion can be expressed in terms of forces or work of attachment, or in terms of forces or work of detachment. The term "fundamental adhesion" is referred to the interfacial bond strength and is the summation of all intermolecular or interatomic interactions. The term "practical adhesion" is referred to the result of an adhesion test and is a function of fundamental adhesion and a number of factors, all of which determine the work required to detach a film or a coating from a substrate.

---

* To whom correspondence should be addressed. Phone: +33-3-80396162, Fax: +33-3-80396132, E-mail: oltra@u-bourgogne.fr

In the case of thin ceramic films deposited onto metallic substrates [2], a comparative study of seven methods did not provide consistent results; it showed the practical character of them and indicated that the application of only one test method or test parameter could lead to a false evaluation of the adhesion. In the case of thick coatings deposited by thermal spraying techniques, the advantages and the inconveniences of several techniques currently used have been extensively presented, with respect to the particular structure of these coatings by Lin and Berndt [3]. However, many of the techniques applied, for both thin films and thick coatings, result in the global damage of the examined systems (destructive methods), whilst the ones inducing local damages (semi-destructive methods) give disputable results.

Ultrasonic techniques have found a wide use in non-destructive characterization of single-, bi- and multi-layered materials and in evaluating the adhesion of plasma sprayed coatings to substrates [4]. Especially, the laser ultrasonics has stimulated strong interest among researchers in the materials characterization domain, mainly due to the numerous advantages of this technique, compared to classical ultrasonics. However, some difficulties related to the choice of the appropriate laser source, as well as of the ultrasonic detector, inhibit the use of this technique on an industrial scale [5].

The use of pulsed lasers for the generation of high-frequency acoustic waves in both liquid [6] and solid media is a well-known technique. In the case of solids, up until now most of the experimental and theoretical scientific research is mainly focused on the acoustic response of single-layer materials [7, 8], while for coated substrates, the examination of the interaction between the laser and the material has been limited to the thermo-elastic regime [9].

The complexity of the propagation of acoustic waves through solids, due to the presence of longitudinal and shear, as well as of surface waves, well-known as the Rayleigh waves, has led to the development of several experimental arrangements. Until now, the great majority of research on laser ultrasonic techniques for the estimation of adhesion strength has used a piezo-electric quartz plate for the detection of the laser-generated acoustic waves [10, 11]. For example, the adhesion between a thin film and a substrate was estimated by detecting the film spallation, which takes place during the laser irradiation of the substrate [12]. The laser pulse is converted to a pressure pulse inducing the expulsion of the film. The critical stress amplitude that accomplished the removal of the coating was determined from a computer simulation of the process and it was verified by means of a piezo-electric crystal probe, capable of mapping out the profile of the stress pulse generated by the laser pulse.

Very recently, the development of laser interferometers of high accuracy turned the interest of several workers [9, 13] to the development of an experimental set-up, having the advantages of a non-contact recording of the acoustic waves. The work reported here presents the fundamental aspects for the evaluation of a semi-destructive adhesion test method, based on the laser generation of ultrasonic waves in transparent ceramic coatings deposited onto metallic substrates. The waveform changes due to different interactions between the laser beam and the examined systems were recorded in-situ at the epicenter using a laser heterodyne interferometer. A physical model, assuming that the acoustic source is placed at the absorption depth of the metallic substrate, was used to develop an analytical one, which permitted to simulate the longitudinal waves propagation, taking into account the physical properties and the thickness of both the coating and the substrate.

2. EXPERIMENTAL

A series of Al_2O_3 coatings, with different thicknesses (30, 80, 150, 230, 275 and 340 µm), were deposited onto 5 mm-thick 304L stainless steel substrates, by Atmospheric Plasma Spraying (APS), under the same deposition parameters. Additional tests were carried out on a 30 µm-thick zirconium oxide formed on pure zirconium, with a thickness of 5 mm, during high temperature oxidation, in order to verify the laser - solid interactions determined for the Al_2O_3 coatings.

The laser used was a pulsed Nd:YAG (B.M. Industrie, Lisses, France), emitting at the IR (λ=1064 nm), with a pulse duration of 34 ns. The temporal profile of the laser pulse energy was approximated by a Gaussian fit with a full width at half maximum (FWHM) of 13.5 ns, whilst the spatial distribution of the energy was also Gaussian. The laser beam was focused on the specimen surface using an optical fiber with a diameter of 1.5 mm. Experiments were carried out by applying energies from 5 to 50 mJ, which provided mean energy densities from 0.3 to 2.8 Jcm^{-2}. The acoustic signal was measured in-situ at the back surface of the specimens, using a laser heterodyne interferometer (B.M. Industrie SH130, λ=632 nm) combined with a digital oscilloscope (LeCroy 9450A). This set-up (Fig. 1a) allows detection of normal displacements at the epicenter, in the range of 0.001-10 nm with a sampling rate of 400 MHz. A typical signal recorded by this technique is presented in Fig. 1b.

Microscopic observations of the irradiated surfaces, as well as of the corresponding cross sections were carried out by means of Scanning Electron Microscopy (JEOL JSM-6400F). The analytical model was developed using Maple V software and Finite Element Modeling (FEM) was carried out using CASTEM calculation package.

3. RESULTS AND DISCUSSION

In the case that a laser beam is used to irradiate a bi-layered system, e.g. a coating on a substrate, the energy provided by the laser is partially reflected at the coating surface, partially absorbed by the first medium (coating) and partially transmitted to the second medium (substrate). The percentages of the energy reflected, absorbed and transmitted depend on the optical properties of both the coating and the substrate [14].

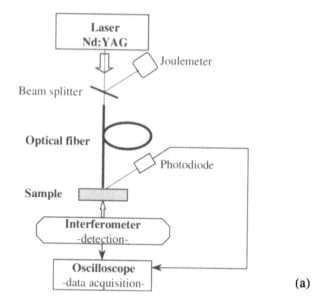

(a)

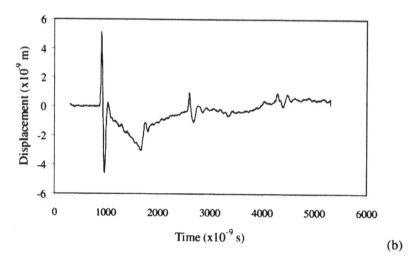

(b)

Figure 1. (a) Experimental arrangement for the laser generation of acoustic waves at the ceramic/ metal interface and their detection at the back surface of the substrate. (b) Typical acoustic waveform registered at the back surface of a ceramic coated metallic substrate.

For the wavelength of the laser beam used, all the alumina coatings, as well as the zirconium oxide, were transparent, so the laser beam energy is partially reflected and partially absorbed by the substrate. This fact allowed the assumption that the acoustic source was generated at the absorption depth ($1/\alpha$) of the metallic substrate: 21 nm below the Al_2O_3/ steel interface. For all series of specimens and in the range of energy density values applied, no ablation phenomenon occurred and, consequently, no plasma was created. In Fig. 2, the energy absorbed at the interface is plotted against the laser beam energy.

3.1. Physical model

Under the above experimental parameters, the pulsed laser irradiation created an instantaneous thermal source near the ceramic/metal interface, which led to the temperature increase, followed by the thermal expansion of the metallic substrate. This thermal expansion induced elastic vibrations, which resulted in the initiation of longitudinal (L) and shear (S) waves, in the range of ultrasonic frequencies. Their propagation through the volume of the examined systems induced the displacement of the back surface of the substrate, which was recorded in-situ at the epicenter (Fig. 3a). The first part of this signal (Fig. 3b) was indicative of the interaction between the laser radiation and the system examined [15].

In order to develop an analytical model, capable to simulate the normal displacement (z-axis) of the epicenter, a physical model based on the propagation of the longitudinal waves was used (Fig. 3c). The first displacement recorded (L^1) corresponds to the arrival of the first longitudinal wave, which depends on the longitudinal sonic velocity of the substrate and its thickness, since the acoustic wave traverses only the substrate. The second displacement (L^2) corresponds to the arrival of second longitudinal wave, which is first propagated through the coating, then is totally reflected at the surface of the coating and propagated through

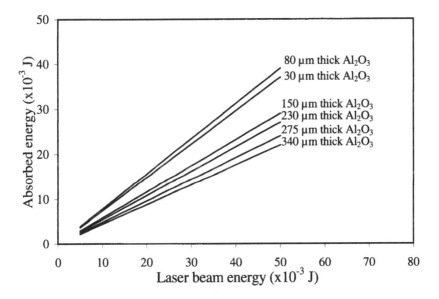

Figure 2. Absorbed energy as a function of the laser beam energy applied.

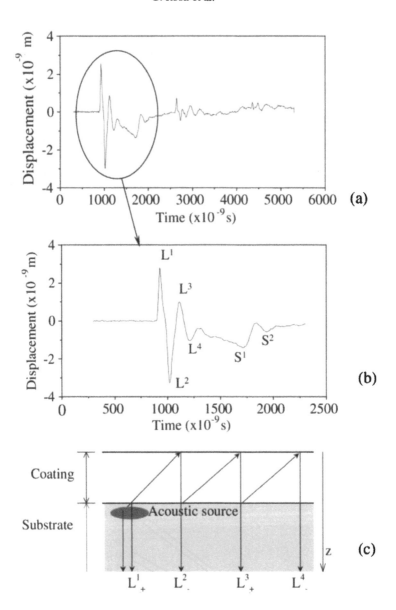

Figure 3. (a) General view of the acoustic signal recorded at the epicentre. (b) Detailed presentation of the first part of the acoustic signal. (c) Physical model explaining the displacements recorded in (b).

the substrate. On increasing the thickness of the coating, the longitudinal wave is subjected to multi-reflections at the coating/substrate interface, which appear in the waveform as successive normal displacements (L^n). The signs (+) and (-) indicate, respectively, a positive and a negative displacement. In the waveform of Fig. 3b, the shear acoustic waves (S^1 and S^2) were recorded with a delay, because of their much lower velocity compared with that of the longitudinal waves.

3.2. Analytical model

Based on the previous physical model, an analytical model describing the propagation of the longitudinal waves was developed, taking into account the acoustic properties (longitudinal velocity and acoustic impedance) and the thickness of the ceramic coating and the metallic substrate, both assumed to be isotropic. As an example, both experimental and calculated acoustic signals, when irradiating with a 15mJ laser beam energy (average energy density: 0.85 J.cm^{-2}), are illustrated in Fig. 4.

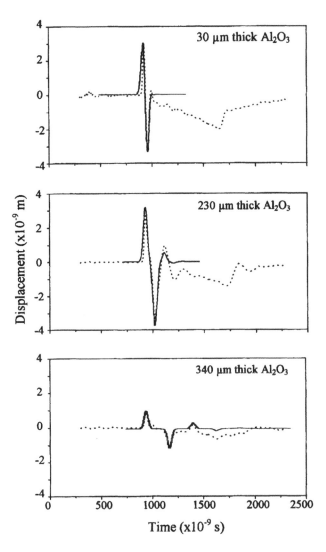

Figure 4. Experimental (dotted line) and calculated (continuous line) acoustic signals recorded at the epicentre of the back surface of the substrate, for three of the systems examined (laser beam energy: 15 mJ).

For all the samples examined, the arrival of the first longitudinal wave always took place at the same time, whilst that of the following longitudinal waves presented a delay, which increased with increasing the thickness of the coating. The excellent agreement between the experimental and the calculated waveforms validates the physical model, on which the main assumptions for the evaluation of the analytical model were based.

3.3. *Laser-induced damages in a bi-layered system*

Observations of the cross sections of the irradiated samples by means of Scanning Electron Microscopy indicated that:

(a) For laser energies lower than a critical value (E_0), the laser radiation did not induce any damage in the system examined (regime of elastic interactions).

(b) For laser energies higher than E_0 and lower than E_1, the laser radiation resulted in the local de-bonding of the ceramic coating from the underlying substrate, at the center of the laser affected interface. In the range of energies $E_0 < E < E_1$, by increasing the laser beam energy, the de-bonding area also increased (regime of interfacial failure).

(c) For laser energies higher than E_1, the laser radiation induced both the partial spallation of the ceramic coating through buckling mechanisms, as well as the crack initiation, mainly, at the borders of the laser affected area. By increasing more the laser beam energy, the spallation area also increased until the total expulsion of the coating (regime of spallation - expulsion).

The three regimes were also observed in the case of the 30 μm thick zirconium oxide. Micrographs of the de-bonded, as well as of the buckled area, together with the corresponding acoustic signals recorded, are presented in Figs. 5 and 6, respectively.

The laser beam energy threshold for the transition from one regime to the other increases with the thickness of the coating. In the case of the thicker coatings, the high laser beam energies, which are needed for the expulsion of the coating, resulted also in the damage of the metallic substrate (extensive Heat Affected Zone or even Heat Melted Zone).

The acoustic signals corresponding to the three different events are presented in Fig. 7, for the case of a 30 μm thick Al_2O_3 coating. For laser beam energies corresponding to thermoelastic interactions, the amplitude of the second wave (L^2) was important, when compared with that of the first one (L^1), Fig. 7a. For laser beam energies resulting in the failure of the interface, the former was lower than the latter (Fig. 7b) and, finally, for laser beam energies resulting in expulsion of the coating, the second wave almost disappeared (Fig. 7c).

It was found that the amplitude ratio (L^2/L^1) provided a reliable criterion for the laser effects on the integrity of the interface and, consequently, it can be used for determination of the critical laser beam energy (E_0), which leads to de-bonding. In Fig. 8, L^2/L^1 is plotted against the laser beam energy, for Al_2O_3 coating and zirconium oxide, both having a thickness of 30 μm. In both cases, the ratio remained

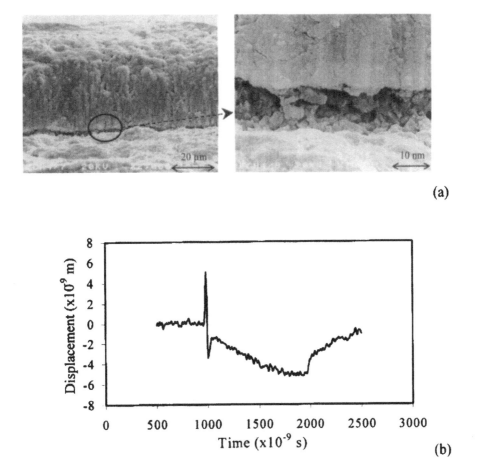

Figure 5. (a) Failure of the ceramic layer/ metal interface below the irradiated area, for a 30 μm thick zirconium oxide formed during high temperature oxidation of pure zirconium (SEM micrographs). (b) Acoustic signal corresponding to interfacial failure (laser beam energy: 20 mJ).

almost constant during the thermoelastic regime but was slightly decreased when de-bonding occurred and, finally, it rapidly reached zero, when partial and, finally, total expulsion of the coating took place.

Expressed in terms of laser beam energy, both thresholds, in general, were found to be increasing with the thickness of the Al_2O_3 coating. However, when taking into account the relationship between the applied and the absorbed energies, as presented in Fig. 2, the absorbed energy thresholds were found to be independent of the coating thickness. Fig. 9 presents the absorbed energy thresholds for the transition from thermoelastic interactions to interfacial failure (12.5 mJ), and from the latter to the expulsion of the coating (24.5 mJ), as a function of the thickness of the alumina coating. For all the examined coatings, an absorbed energy of 12.5 mJ was found to result in a 9 nm displacement of the interface.

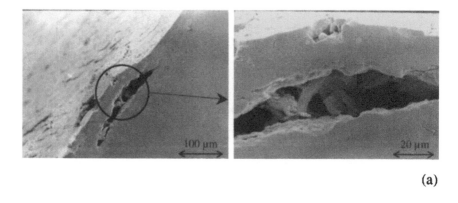

(a)

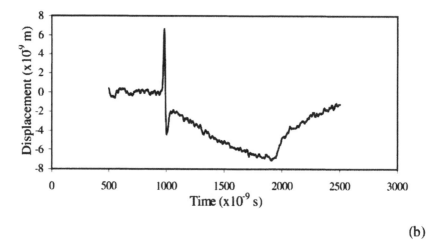

(b)

Figure 6. (a) Buckling of the ceramic layer below the irradiated area, for a 30 μm thick zirconium oxide formed during high temperature oxidation of pure zirconium (SEM micrographs). (b) Acoustic signal corresponding to buckling (laser beam energy: 25 mJ).

3.4. Thermo-mechanical Finite Element Modeling

The previously described correlation between the interfacial damage and the first two peaks of the acoustic signal recorded indicated that the former was taking place during the first few nano-seconds of the laser irradiation of the examined system. However, considering that only the acoustic effect would result in the coating de-bonding, it would lead to underestimation of the adhesion strength. The instantaneous thermal source near the ceramic/metal interface, which induced the acoustic source, had also a thermo-mechanical effect. With this in mind, our current work is focused on the determination of the thermo-mechanical field developed in the interface area during the first few nano-seconds of the laser irradiation, by Finite Element Modeling (FEM).

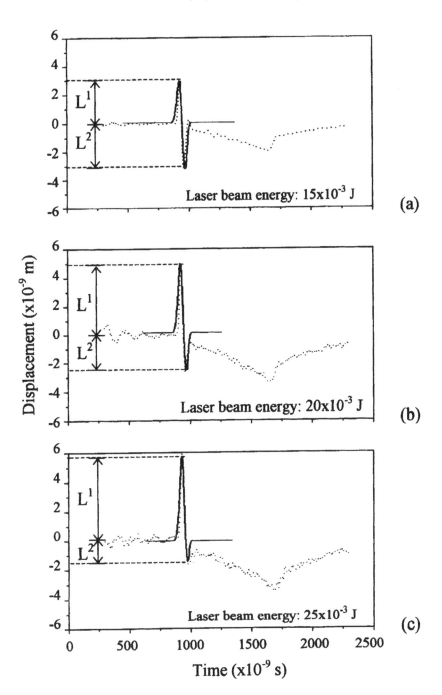

Figure 7. Experimental (dotted line) and calculated (continuous line) acoustic signals characteristic of thermoelastic regime (15 mJ), de-bonding (20 mJ) and expulsion (25 mJ) of a 30 μm thick Al_2O_3 coating.

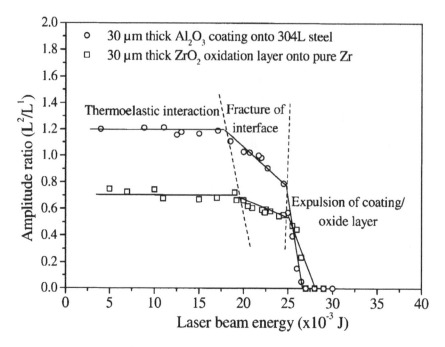

Figure 8. Amplitude ratio (L^2/L^1) as a function of the laser beam energy, for an Atmospheric Plasma-Sprayed Al_2O_3 coating and a ZrO_2 surface layer, both with a thickness of 30 µm.

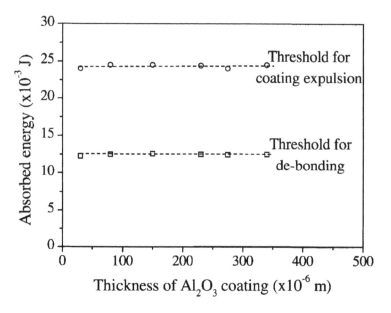

Figure 9. Absorbed energy thresholds for the transitions from thermoelastic interactions to interfacial failure and from interfacial failure to the expulsion of the coating.

The geometry of the examined systems permitted to apply a 2-D numerical solution, simulating the one-half of the specimen cross section. For the needs of this analysis, the spatial and the in-time distributions of the laser beam (thermal source) were taken into account, while the mechanical behavior of the ceramic coating was assumed to be elastic and that of the metallic substrate elasto-plastic. The thermal field is calculated for an interval of 50 ns, in steps of 1 ns, with the moment of initiation of the laser pulse as starting point (t_0=0 ns).

The thermal analysis showed that the temperature reached its maximum value after 20 ns from the pulse initiation moment (t_0) and, consequently, the maximum thermal expansion of the irradiated volume took place at the same time. This event is posterior to the interfacial damage, as determined previously. However, even during the first few nano-seconds of the laser irradiation, the rather small thermal expansion of the substrate can introduce stresses to the system, which cannot be neglected. It was found that the thermal loading did not generate any normal stress, but it was responsible for the appearance of shear stresses along the interface, as observed by other researchers, working on the effects of the laser irradiation on thin films [16].

The stress field leading to the de-bonding of the transparent coating from the underlying absorbing substrate must be considered as the combined effect of the following:

(a) the propagation of acoustic waves through the bi-layered solid, responsible for the appearance of tensile and shear stresses at the interface, and

(b) the local thermal expansion of the substrate, responsible for the appearance only of shear stresses.

4. CONCLUSIONS

In the case of transparent ceramic coatings deposited onto metallic substrates by thermal spraying techniques, the acoustic source, generated by pulsed laser irradiation, is placed at the absorption depth of the substrate, near the ceramic/metal interface. The in-situ recording of the acoustic signals at the back surface of the substrate permits to detect the transition from thermo-elastic interactions to the failure of the interface. The determination of the absorbed energy threshold, which leads to de-bonding, permits to define the stress field leading to the failure of the interface, or in other words the "practical adhesion" [1] of the coating. For this purpose, a thermo-mechanical numerical model, which takes into account the thickness, the physical properties (density, heat capacity and thermal conductivity) and the mechanical behaviour (elastic, elasto-plastic, etc) of both the transparent coating and the metallic substrate, is under development.

Compared with other "conventional" techniques, the proposed one can give a rapid, global and exact estimation of the practical adhesion strength. Among the other advantages, it must be mentioned that it is a non-contact technique, resulting only in small-sized local damages in the examined system. Moreover, a great variety of laser sources can be used, since the only limitation of the method is the transparency of the ceramic coating.

Acknowledgements

The authors would like to express their thanks to Professors C. Coddet and S. Costil (Université de Technologie Belfort-Montbéliard U.T.B.M., France), for the deposition of the plasma-sprayed alumina coatings. Dr. P. Psyllaki is grateful to the European Network for the Training and Mobility of Researchers (TMR) ERB FMRX-CT98-0188 for the financial support.

REFERENCES

1. K.L. Mittal, in: *Adhesion Measurement of Films and Coatings*, K.L. Mittal (Ed.), pp. 1-13, VSP, Utrecht, The Netherlands (1995).
2. H. Ollendorf and D. Schneider, *Surf. Coat. Technol.*, **113**, 86-102 (1999).
3. C.K. Lin and C.C. Berndt, *J. Therm. Spray Technol.*, **3**, 75-105 (1994).
4. Y. Suga, Harjanto and J. Takahashi, in: Proc. International Thermal Spray: International Advances in Coatings Technology, C.C. Berndt (Ed.), pp. 247-252, ASM International, (1992).
5. J.-P. Monchalin, in: *Review of Progress in Quantitative Nondestructive Evaluation*, D.O. Thompson and D.E. Chimenti (Eds), Vol. 12, pp. 495-506, Plenum Press, New York (1993).
6. D.A. Hutchins, in: *Physical Acoustics*, W.P. Mason and R.N. Thurston (eds), Vol. XVIII, pp. 21-123, Academic Press, New York (1988).
7. D.A. Hutchins, R.J. Dewhurst, S.B. Palmer and C.B. Scruby, *Appl. Phys. Lett.*, **38**, 677-679 (1981).
8. R.J. Dewhurst, D.A. Hutchins, S.B. Palmer and C.B. Scruby, *J. Appl. Phys.*, **53**, 4064-4071 (1982).
9. C. Grand, E. Lafond, R. Coulette, J.C. Gonthier, O. Pétillon, B. Dupont and F. Lepoutre, Proc. SPIE, **2945**, 389-401 (1996).
10. E. Lafond, R. Coulette, C. Grand, M.-H. Nadal, B. Dupont, F. Lepoutre, D. Balageas and O. Pétillon, *NDT & E Int.*, **31**, 85-92 (1998).
11. A.G. Youtsos, M. Kiriakopoulos and Th. Timke, *Theor. Appl. Fracture Mech.*, **31**, 47-59 (1999).
12. C. Sartori, R. Oltra and P. Dubief, *Surf. Coat. Technol.* **106**, 251-261 (1998).
13. V. Gupta, A.S. Argon, D.M. Parks and J.A. Cornie, *J. Mech. Phys. Solids*, **40**, 141-180, (1992).
14. C.B. Scruby and L.E. Drain, *Laser Ultrasonics: Techniques and Applications*, pp. 223-324, Adam Hilger, New York (1990).
15. G. Rosa, P. Psyllaki and R. Oltra, submitted to *Ultrasonics*.
16. M.T.A. Saif, C.Y. Hui and A.T. Zehnder, *Thin Solid Films*, **224**, 159-167 (1993).

Adhesion Measurement of Films and Coatings, Vol. 2, pp. 291–297
Ed. K.L. Mittal
© VSP 2001

The blade adhesion test applied to polyimide films onto silicon substrate

SALIM KHASAWINAH* and CHARLES G. SCHMIDT

Hewlett-Packard Company, 1000 NE Circle Blvd., Corvallis, OR 97330

Abstract—The blade adhesion test characterizes the adhesion strength by measuring the load required to remove a film from a semi-rigid substrate with a sharp blade. The blade adhesion test was used to quantify the practical adhesion strength of polyimide films spun onto Si with thickness varying from 8 μm to 30 μm. The adhesion strength was altered by applying Ta- or SiC-coatings to the silicon substrate and by subjecting the assembled specimens to environmental aging. The results indicate that the blade tangential force generally increases with increasing polyimide thickness and decreases as a result of elevated temperature exposure. The results quantify the sensitivity of the measured response to interface strength. A finite element analysis of interface separation by the blade was developed to assess the interface strength from the coating removal force data and from the geometry of the film as it is removed by the blade.

Keywords: Adhesion; blade test; scrape test.

1. INTRODUCTION

Adhesion between dissimilar materials is of utmost importance for electronic packaging applications. In many applications, adhesion between flexible polymers and rigid substrates is of particular concern. Consequently, there is a need for techniques that evaluate the quality of the polymer/rigid substrate interface before and after environmental aging (immersed in 70°C DI water). While there are many adhesion tests to choose from [1,2,3], only a few fulfill the required criteria. The appropriate adhesion test must be quantitative, repeatable, quick and straightforward requiring little or no sample preparation. The technique should be readily combined with environmental aging. Finally, the strain energy release rate of the interface should readily be derived from the practical adhesion measurements. The blade adhesion test fulfills these requirements and is particularly well suited for measuring the practical adhesion [1,2,3] of soft films on hard substrates. The blade adhesion test consists of moving a blade of a specified geometry and orientation parallel to the interface while simultaneously imposing a constant normal

---

* To whom correspondence should be addressed. Phone: (541) 715-1479, Fax: (541) 715-4199,
E-mail: salim_khasawinah@hp.com

force and measuring the tangential force exerted on the blade. The purpose of this study was to evaluate the blade adhesion test for polyimide films onto rigid substrates. The effect of polyimide thickness, substrate composition and environmental aging were considered. A finite element model was developed to derive the strain energy release rate from the measured blade removal forces.

2. EXPERIMENTAL

The blade adhesion test (see Figure 1) was executed on a computer-controlled system. The tool steel blade is approximately 5mm wide and has a rake angle (α) of 35 degrees and a relief angle of 10 degrees.

The blade was moved along the interface at 7 mm/minute and a blade normal force of 0.5 N. The 3 mm-wide test specimens consisted of polyimide films of thicknesses of 8,17 and 30 μm on tantalum or silicon carbide substrates. The samples were tested before and after environmental aging. The blade removal force was averaged over the entire length of the scan and then normalized over the sample width. Five measurements were taken per test condition.

3. MODEL OF THE BLADE ADHESION TEST

A finite element analysis was conducted to assess the strain energy release rate for separation (G_s) from the blade adhesion test results. The total energy required to separate a polymer film from a substrate can be viewed as consisting of two parts: G_s represents the energy that is required for separation at the delaminating interface, and G_p represents the energy required to plastically deform the polymer film during the delamination process. The value of G_s is expected to be constant as a function of film thickness as long as the delamination mechanism and intrinsic adhesion are unaffected. A 2-dimensional plane strain model was constructed consisting of three rows of 4-node rectangular elements. The model was cast in 2-D because the large width-to-thickness aspect ratio (100 to 1) of the polymer

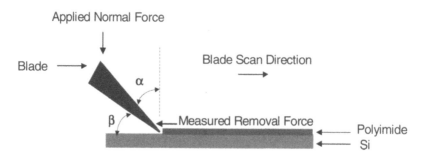

Figure 1. Schematic of the blade adhesion test.

film made edge effects small. The elements in contact with the substrate were 1.4 μm and 2.8 μm in the directions parallel and normal to the substrate, respectively. The remaining two rows were 1.4 μm parallel to the substrate and 2.6 μm, 7.1 μm, or 13.6 μm normal to the substrate for the 8-, 17-, and 30-μm-thick polyimide, respectively. An elastic-plastic constitutive model was used for the polyimide that was obtained from uniaxial tension tests on free-standing polyimide film material. Suitable viscoelastic and viscoplastic materials information was not available. Limited blade test results were obtained (not shown) at blade displacement rates of 3.5 mm/min. and 7 mm/min. that revealed no rate effect to the blade force. Although this result does not prove that rate effects are unimportant, it does suggest that the effects are small.

The FEA model was found to be mesh size independent but was crack extension step size dependent; consequently, all calculations were obtained using a single crack extension step size and mesh size. The resulting calculated G_s values are considered relative. No closed-form solution with similar boundary conditions could be found; however, the modeling approach has been validated against closed-form solutions for linear elastic blister test boundary conditions [4].

The delamination by the blade adhesion test was modeled by sequentially releasing a single node at the delamination front and moving the blade toward the delamination in successive computational steps. This procedure was repeated until a steady state blade force and G_s value were obtained (usually about 50 to 100 node releases). Decreasing or increasing the distance between the blade tip and the delamination front in separate computational runs produced higher or lower G_s values.

The G_s was computed from the change in the force and displacement of the node at the delamination front immediately before and after node release as follows.

$$G_S = \frac{f \bullet \delta}{2 \bullet \Delta A} \tag{1}$$

where f is the nodal force before separation

   δ is the nodal displacement after separation

   ΔA is the additional crack face area produced by delamination front advance.

Opening and shear mode G_s values were computed and summed to produce the reported G_s values. The dominant mode was opening.

4. RESULTS AND DISCUSSION

The blade removal energy, which is the total blade energy less the energy expended to overcome the blade-to-substrate friction, was plotted as a function of polyimide thickness (see Figure 2). The blade-to-substrate friction was measured in two ways. First, after removing the polyimide from the substrate, the test was repeated over the previously tested region. Second, substrates that were never coated were used. Both techniques indicated that the energy required to over-come the blade-to-substrate friction was approximately 200 J/m^2. There were no significant differences in the friction energy values for silicon carbide or tanta-lum.

The polyimide removal energy was found to decrease after aging for both the tantalum and silicon carbide substrates which suggests that the blade adhesion test is an effective means of assessing the aging environment effects on delamination.

The blade removal energy consistently increased with increasing thickness only for the tests on Ta in the initial condition. All other test conditions exhibited a decline in blade energy for the 17-μm- and 30-μm-thicknesses. To find an expla-nation for these trends, the morphologies of the separated interfaces were ana-lyzed using optical microscopy at a magnification of 500X. The substrates (tan-talum and silicon carbide) at the interface for all the test conditions were found to be free of polyimide and free of any mechanical damage that could be associated with the blade. However, the polyimide morphology at the interface was found to vary with removal energy and polyimide thickness.

All of the 8-μm-, 17-μm-thick samples, and the 30-μm-thick SiC-initial sam-ples shown in Figure 2 were seen to have ridges and grooves on the interface side of the polyimide. The polyimide interfaces from the remaining samples were featureless (i.e., all 30-μm-thick samples except the SiC-initial condition). The difference in the appearance of the polymer surfaces indicates that there are two different de-adhesion mechanisms: one that is more or less continuous, the other that indicates an interrupted propagation of the delamination front.

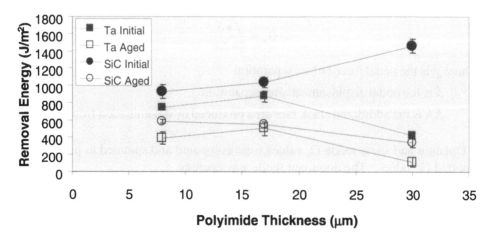

Figure 2. Blade Removal Energy vs. Polyimide Thickness for Tantalum and Silicon Carbide Sub-strates.

The extremes in the range of differences in the macroscopic deformation of the polyimide films are shown schematically in Figure 3. The polymer was removed as a scroll at high blade forces. For a given polyimide thickness, the radius of curvature decreased for increasing removal energy. At extremely low adhesion levels, the polyimide was removed as a sheet. In general, the blade tip-to-delamination front distance was found to decrease as the removal force increased. This occurred for all substrate surfaces and polyimide thicknesses tested. The trends in curl diameter as a function of the polyimide removal energy for different polyimide thicknesses are shown in Figure 4.

Figure 4 shows that for a given polymer thickness, the curl diameter increases for decreasing blade removal energy. In addition, for a given blade removal energy, the curl diameter of the removed polyimide increases as the polyimide thickness increases.

The values of G_s, obtained from the finite element modeling, are compared to the measured values of the blade removal energy in Figure 5. Arrows are used (see Figure 5) to demarcate samples of identical polyimide thickness. Dashed ovals (see Figure 5) are used to group samples that have the same G_s. Since G_s is expected to be independent of thickness, each dashed oval should encompass three data points (representing the three different thicknesses). The model shows that the value of G_s is nearly independent of thickness for the 8-μm- and 17-μm-thick specimens for all test conditions and for the 30-μm-thick SiC-initial specimen. The 30-μm-thick specimens in conditions other than SiC-initial exhibit sub-

Figure 3. Polyimide removal as a function of blade removal energy at a fixed polyimide thickness.

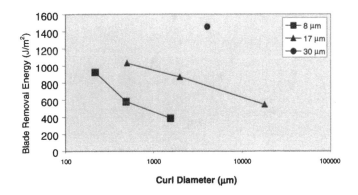

Figure 4. Blade removal energy as a function of polyimide curl diameter and polyimide thickness.

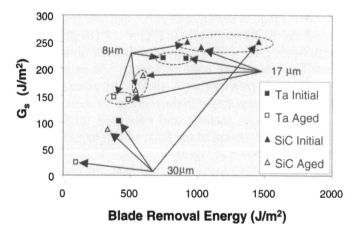

Figure 5. Model output (G$_s$) as a function of the experimentally measured blade removal energy.

stantially lower G$_s$ values. The tests that exhibited low G$_s$ values were the only tests to exhibit featureless polyimide delamination surfaces. This suggests that a unique delamination mechanism might be the reason for the low G$_s$ values observed in some 30-μm-thick specimens. Modeling adjustments that might predict the transition in delamination mechanism are under investigation. Nevertheless, the independence of G$_s$ on thickness for all tests with similar fracture surfaces suggests that G$_s$ is an effective way of quantifying the adhesion strength that is independent of geometry.

Figure 6 compares the blade removal energy and curl diameter obtained from the simulations. A qualitative agreement with the experimental data was observed (cf, Figure 6 and Figure 4).

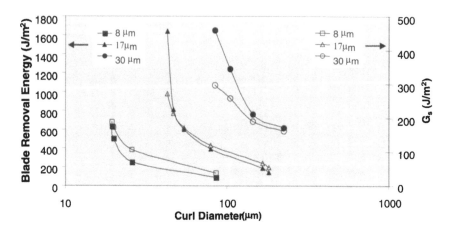

Figure 6. Trends in blade removal energy and G$_s$ as a function of curl diameter from the finite element model.

5. SUMMARY

The blade adhesion test was examined as a technique to characterize the effects of substrate type, polyimide thickness, and aging on the adhesion of polyimide films onto rigid substrates. In general, it was found that the measured adhesion decreased after aging and was lower for the Ta substrate than for the SiC substrate. Two different polyimide surface morphologies were observed after separation: a flat featureless surface and a surface with ridges and valleys.

A finite element analysis model was developed to derive the strain energy release rate for separation (G_s) from the experimental data. The G_s values were found to be independent of polyimide thickness in only one out of four test cases. The macroscopic deformation observed empirically as a function of removal energy and polyimide thickness was reproduced qualitatively by the model.

REFERENCES

1. K.L. Mittal, in *Adhesion Measurement of Thin Films, Thick Films and Bulk Coatings,* K.L. Mittal (Ed.), STP 640, pp. 5-17, ASTM, Philadelphia (1978).
2. K.L. Mittal, *Electrocomponent Sci. Technol.,* **3**, 21-42 (1976).
3. K.L. Mittal, in *Adhesion Measurement of Films and Coatings,* K.L. Mittal (Ed.), pp. 1-13, VSP, Utrecht, The Netherlands, (1995).
4. Audrey Finot, private communication.

Adhesion Measurement of Films and Coatings, Vol. 2, pp. 299–328
Ed. K.L. Mittal
© VSP 2001

Mechanics of the JKR (Johnson-Kendall-Roberts) adhesion test

C.Y. HUI,* J.M. BANEY and Y.Y. LIN

Thurston Hall, Theoretical and Applied Mechanics, Cornell University, Ithaca, NY 14853, U.S.A.

Abstract—The mechanics of the Johnson, Kendall and Roberts (JKR) adhesion test is reviewed in this article. We first re-derive the basic JKR theory using a fracture mechanics approach. This is followed by a detailed discussion of the assumptions underlying the JKR theory and its applicability. These discussions include finite specimen effects, modeling of surface forces using cohesive zone descriptions, nonlinear elasticity, large deformation effects, hysteresis due to surface effects, and bulk viscoelasticity. Some new results on viscoelastic contact and their applicability to adhesion test are discussed.

Keywords: JKR; contact mechanics; adhesion.

1. INTRODUCTION

The JKR theory of contact between solid elastic lenses has been used extensively to characterize the surface energy and tack of materials (for examples, see [1-15]). The JKR theory is an extension of the Hertzian theory of contact [16], in that it accounts for adhesion between the lenses by including the effect of attractive surface forces that act when the bodies are in contact. These surface forces are modeled by adding a tensile stress distribution to the usual compressive Hertzian pressure that exists in the contact region. Because it is only within the contact region that the stresses are modified, the JKR theory neglects any surface forces that act outside the contact zone, and, therefore, may not be applicable in all situations (e.g. to hard materials that produce small contact regions). However, the JKR theory has proven to be extremely useful in modeling the adhesion contact of polymeric materials (typically elastomers), where the effect of the surface forces can be measured fairly easily to obtain the work of adhesion, W, of the materials. The work of adhesion is often used to quantify the adhesion properties of the system.

A schematic diagram of a JKR experiment is illustrated in Figure 1, which shows two smooth spherical lenses of radii R_1 and R_2 and thicknesses d_1 and d_2 being brought into contact under an applied force P. The contact zone is a circle

---

*To whom correspondence should be addressed. Phone: (607) 255 3718, Fax: (607) 255 2011, E-mail: ch45@cornell.edu

of radius a, and the displacements at the points where materials 1 and 2 are loaded are denoted by δ_1 and δ_2, respectively. The normal approach δ of the two lenses is the sum of δ_1 and δ_2. In a typical JKR experiment the contact radius first increases as the lenses are brought into contact and then decreases as they are pulled apart. We define the phase of the experiment in which the contact radius is an increasing function of time as the bonding phase, and that in which the contact radius decreases with time as the debonding phase. Throughout both phases of the experiment, the contact radius a and either the force P or the normal approach δ are monitored. Typical experimental setups for doing this are described in [7, 13-17]. Since P and a are measured experimentally in a load controlled test, and if the moduli and radii of the samples are known, the work of adhesion, W, can be determined using the JKR equation

$$a^3 = \frac{3R}{4E^*}\left[P + 3\pi RW \pm \sqrt{9\pi^2 R^2 W^2 + 6\pi PRW}\right] \tag{1}$$

The negative sign of the square root in (1) describes the unstable branch of the JKR solution. In a load-controlled test, the hemispheres will jump into contact and this branch of the solution cannot be observed. Similarly, in a displacement controlled test δ is measured and W can be found using

$$\delta = \frac{a^2}{R} - \sqrt{\frac{2\pi Wa}{E^*}} \tag{2}$$

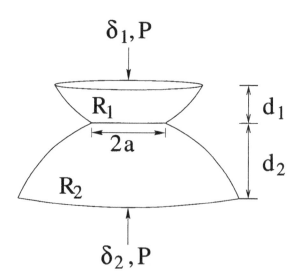

Figure 1. A schematic diagram of the JKR test, in which hemispherical lenses are compressed into contact under an applied load P or applied displacements δ_1 and δ_2. The resulting contact area is a circle of radius a.

In (1) and (2) E^* and R are the effective modulus and radius of curvature, respectively, i.e.,

$$1/E^* = (1-v_1^2)/E_1 + (1-v_2^2)/E_2 \tag{3}$$

$$1/R = 1/R_1 + 1/R_2 \tag{4}$$

where E and v denote Young's modulus and Poisson's ratio, respectively.

The standard approach to the JKR theory is based on global energy balance, i.e. the work of breaking (or making) contact is equated to the change in strain energy of the elastic bodies due to the decreasing (or increasing) contact zone. Reviews of this approach can be found in [18-20]. In this work the basic equations of the JKR theory will be derived using a fracture mechanics approach similar to that of [21] and [22]. The role of local stresses near the edge of the contact zone is emphasized in this approach, although these two approaches are equivalent when there is no hysteresis. During this review we will discuss the underlying assumptions involved in the theory, and comment on their validity. Additionally, we will review some recent attempts to extend the JKR theory to dissipative systems, where the energy balance approach can become difficult but the fracture mechanics based approach still seems tractable.

2. ASSUMPTIONS OF THE JKR THEORY

The basic kinematic assumptions of the JKR theory (with the exception of assumption D below) are based on the Hertzian theory of contact [16]. They are:

A. The deformation of the lenses is assumed to be small so that a linearized continuum theory can be used. In particular, small strains are required so that there is no distinction between the deformed and undeformed configurations of the bodies so far as force equilibrium is concerned.

B. The materials are assumed to be linearly elastic, isotropic and homogeneous. The bonding and debonding processes are assumed to be reversible.

C. The contact surfaces are assumed to be frictionless so that the shear tractions acting on the contact surfaces are identically zero.

D. The contact radius is large compared with the size of the region at the edge of the contact zone where the surface forces due to intermolecular attraction act, since the JKR theory neglects any surface forces that act outside the contact zone. Furthermore, the energy available for breaking (or forming) contact can be calculated based on a continuum solution that contains no reference to the details of the local failure (or adhesion) process.

E. The contact radius is small in comparison with the dimensions of the lenses, i.e., $a/R \ll 1$ and $a/d \ll 1$.

The nature of the surface forces in assumption D needs to be clarified. Although all surface forces are assumed to act inside the contact zone in the JKR theory, in reality some attractive forces will act across the air gap outside the edge

of contact. The region where these forces, which come from short-range molecular interactions, act must be much smaller than the contact radius for the JKR approximation to be a good one. Additionally, it is implicitly assumed in the JKR
theory that the balance of potential and surface energy is independent of the detailed distribution of these forces. In other words, the amount of elastic strain energy available to make or break a contact can be computed without detailed
knowledge of the bonding or failure process. Later, we shall show that this concept cannot be applied to materials that exhibit inelastic behavior.

Assumption E allows the local deformation of the lenses to be calculated by re-
garding them as linear elastic half spaces loaded by the contact pressure. Consistent with this assumption, the displacements at the remote points of load application (i.e. at infinity) due to the contact pressure are identically zero (see figure 2).
Thus δ_1 and δ_2, the relative displacements between these points and the points on
each body at the center of contact, are simply the displacements of the contact
center points themselves. Hence the interpretation of δ that is consistent with assumption E is that it represents the sum of the displacements of the points at the
center of contact on each surface.

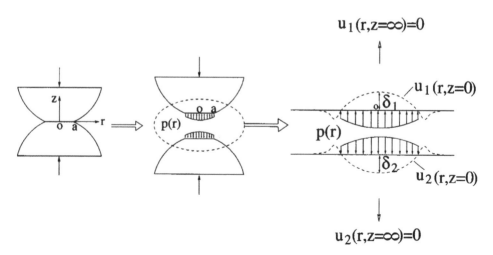

Figure 2. A schematic illustration of the normal approach $\delta = \delta_1 + \delta_2$. The lenses are modeled as
half-spaces and δ_1 and δ_2 are caused by the pressure distribution in the contact zone.

3. DERIVATION OF THE BASIC EQUATIONS

3.1. Three-dimensional case

Consistent with assumptions A and E, the spherical surfaces before contact are
approximated by paraboloids. Let u_1, u_2 denote the displacements of the contact
surfaces from its initial configuration (see Fig. 2). Using the definition of the
normal approach defined above and $\delta = \delta_1 + \delta_2$, the contact condition is

$$(\delta_1 - u_1) + (\delta_2 - u_2) \equiv \delta - (u_1 + u_2) = \frac{r^2}{2}(R_1^{-1} + R_2^{-1}) \equiv \frac{r^2}{2R} \qquad r < a \qquad (5)$$

where r is the in-plane radial distance from the center of the contact circle. The problem has thus been reduced to one of finding the pressure distribution $p(r)$ which produces surface displacements that are consistent with (5). This problem is solved using assumptions C and E. According to the theory of linear elasticity, the surface displacement of an elastic half space due to a pressure distribution

$$p(r) = \begin{cases} p_o(1 - r^2/a^2)^{1/2} + p_o'(1 - r^2/a^2)^{-1/2} & r < a \\ 0 & r > a \end{cases} \qquad (6)$$

must have the form

$$u = \frac{(1-v^2)\pi a}{E}[p_o' + p_o] - \frac{(1-v^2)\pi p_o}{4Ea}r^2 \qquad (7)$$

In (6) and (7) p_o' must be negative to avoid surface displacements that interfere with each other, so the second term in the pressure distribution (6) represents tensile stresses. Comparing (5) and (7) gives

$$\frac{\pi a}{E^*}\left[p_o' + \frac{p_o}{2}\right] = \delta \qquad (8)$$

and

$$\frac{\pi p_o}{2E^* a} = \frac{1}{R} \qquad (9)$$

Using (6) and the condition of force balance, i.e., $P = 2\pi \int_0^a p(r)r dr$, we have

$$2\pi a^2 [p_o' + \frac{p_o}{3}] = P \qquad (10)$$

Equations (6), (9) and (10) show that the contact pressure is completely determined in a force controlled test (i.e., no knowledge of δ is required). Likewise, in a displacement controlled test, the contact pressure is completely determined by (6), (8) and (10). From a theoretical viewpoint, one would expect to be able to determine δ, a, p_o and p_o' if the applied load P were known. However, (8)-(10) provide only three equations for the four unknowns δ, a, p_o and p_o'. Therefore, an extra equation is required. In Hertzian contact, the extra equation comes from the assumption that no tensile stress can be sustained by the interface, or $p_o' = 0$. In the JKR theory (where adhesion is considered), the contact pressure has an inverse square root singularity at the edge of contact zone, i.e.,

$$p(r) \to \frac{-K_I}{\sqrt{2\pi(a-r)}} \qquad \text{as } r \to a \qquad (11)$$

with

$$K_I = -p'_o\sqrt{\pi a} \qquad (12)$$

The negative sign in (12) is due to the contact mechanics convention that tensile stress is considered negative. Thus, the second part of the pressure distribution in (6) can be regarded as the stresses due to the loading of an external crack, which occupies the planar region $z = 0$ and $r > a$ with the stress intensity factor given by $K_I = -p'_o\sqrt{\pi a}$. The constant p'_o is related to the work of adhesion W by the condition that breaking a unit area of contact requires an amount of work equal to the work of adhesion of the two surfaces i.e., $G = W$. Here assumption D is used, since the effect of the cohesive forces at the contact edge is characterized only by W, and the detailed distribution of these forces is not considered. The connection between the amount of elastic strain energy released (or gained) G during decreasing (or increasing) contact and the local stress fields (i.e., K_I) is provided by the path independent J integral. The concept of the J integral is illustrated in Figure 3, which shows a mathematically sharp crack moving in the x_1 direction with velocity $\dot{a}$. For an elastic material, the time rate of energy flow to the crack tip, $G\dot{a}$, is the difference between the rate of traction work on the contour Γ and the rate of stress work on the fixed set of material points which coincide, instantaneously, with the time dependent region denoted by Ω. For an elastic material homogeneous in the x_1 direction, Rice [23] showed that

$$G = J_\Gamma \equiv \int_\Gamma (wn_1 - \sigma_{ij}n_ju_{i,1})ds \qquad (13)$$

where w is the strain energy density, n_i is the ith component of the unit outward normal vector to the curve Γ, u_i is the ith component of the displacement field and $u_{i,1} \equiv \partial u_i / \partial x_1$. Furthermore, Rice showed that J_Γ was independent of path [23]. By choosing a path close to the crack tip so that the local field given by (7) is dominant (see Fig. 3) it is possible to show that

$$G = K_I^2 / 2E^* \qquad (14)$$

Applying the condition $G = W$ results in

$$p'_o = -\sqrt{2E^*W / \pi a} \qquad (15)$$

Furthermore, by combining (9),(10),(12) and (14), the energy release rate is

$$G = (P - P')^2 / 8\pi E^* a^3 \qquad (16)$$

where $P' = 4E^*a^3 / 3R$. Combining (5), (6) and (11) results in

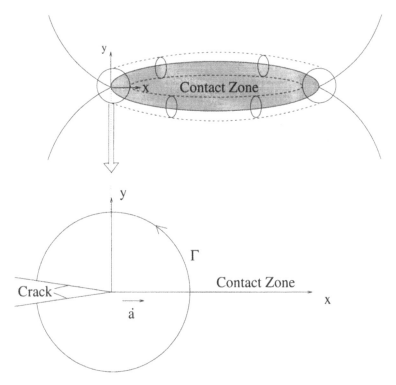

Figure 3. The edge of contact is modeled as the front of an external crack. The energy release rate G is computed by using a path independent line integral J_Γ, where the path Γ is chosen to be a circle enclosing crack tip.

$$P = \frac{4E^*a^3}{3R} - 2a^{3/2}\sqrt{2\pi E^*W} \qquad (17)$$

which can be rewritten in the form (1).

In a force controlled JKR test, the work of adhesion W is obtained by fitting the applied force P versus the contact radius data to (1). In a displacement controlled test, the appropriate expression is obtained using (8), (9) and (15), i.e.,

$$\delta = \frac{a^2}{R} - \sqrt{\frac{2\pi Wa}{E^*}} \qquad (18)$$

The pressure distribution inside the contact zone is found by combining (6), (9) and (15), i.e.,

$$p(r) = \frac{2E^*a}{\pi R}\left[1 - \frac{r^2}{a^2}\right]^{1/2} - \sqrt{\frac{2E^*W}{\pi a}}\left[1 - \frac{r^2}{a^2}\right]^{-1/2} \qquad (19)$$

3.2. Two dimensional JKR theory

Two-dimensional JKR tests have also been carried out using circular cylindrical lenses as shown in Fig.4 [24]. The length of these cylinders, L, is typically much larger than their respective radii R_1 and R_2 and thicknesses d_1 and d_2. For this geometry the contact zone is a strip of width $2a$. Let x denote the in-plane distance from the center of the contact strip (see Fig.5). The pressure distribution inside the contact zone is found to be [24-25]:

$$p(x) = \frac{E^*}{2R}\sqrt{a^2 - x^2} - \frac{\sqrt{2aE^*W/\pi}}{\sqrt{a^2 - x^2}} \tag{20}$$

If F is the total load acting on the cylinders, the force per unit length $P = F/L$ is

$$P = \frac{\pi E^* a^2}{4R} - \sqrt{2a\pi E^* W} \tag{21}$$

An expression analogous to (1) does not exist in the 2D problem. This is because the normal approach cannot be found by consideration of the local contact pressure within the context of the JKR approximation. Specifically, the displacement of a point in an elastic half space loaded two dimensionally (i.e., a line load)

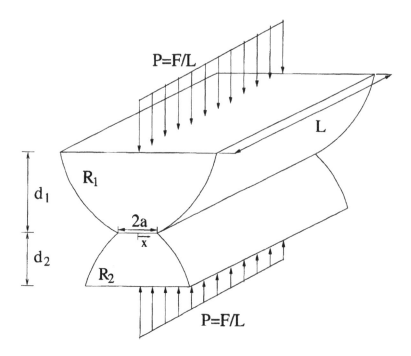

Figure 4. Geometry of a two-dimensional JKR test. The contact region is a rectangular strip of width $2a$.

is proportional to the logarithm of the distance from the line of application of the load. Therefore, the displacement at infinity due to the JKR pressure distribution is undefined, due to the logarithm singularity. In the three-dimensional case this remote displacement was zero, providing a reference point from which to measure surface displacements, but there is no appropriate reference point in the two-dimensional case. In physical terms this means that the approach δ cannot be obtained by the knowledge of the local contact pressure, and the geometry of the specimen must be considered. Due to this difficulty, an expression equivalent to (1) has not been obtained for 2D contact until recently. Hui *et al.* [26] considered the problem of two infinitely long half cylinders in contact as shown in Fig. 5. The flat surfaces of the half cylinders are attached to rigid plates and it is assumed that there is no friction between the contact surfaces. The normal approach δ is, therefore, equal to the change in distance between the plates. If P denotes the load per unit length needed to bring the two surfaces in contact, it is found that

$$\delta = -\frac{a^2}{4R} + \frac{2P}{\pi E^*}\ln 2$$

$$- K_1 P \ln(a/2R_1) - K_2 P \ln(a/2R_2) + 2\sqrt{\frac{2aW}{\pi E^*}}$$

(22)

$$K_1 = 2(1-v_1^2)/\pi E_1, \qquad K_2 = 2(1-v_2^2)/\pi E_2.$$

(23)

In (22), a and P are related to each other by (21).

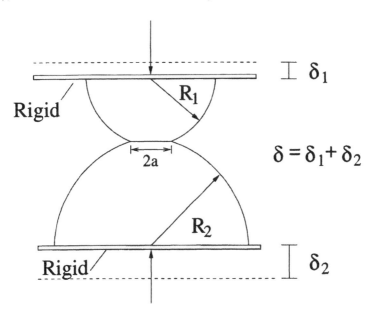

δ_1

$\delta = \delta_1 + \delta_2$

δ_2

Figure 5. In two-dimensional JKR theory, the normal approach can be defined by the change in distance between the rigid plates.

4. VALIDITY OF THE JKR ASSUMPTIONS

4.1. Surface forces

The assumptions of the JKR theory (e.g. D) lead to stresses of the form (19) that are infinite at the edge of the contact zone. In reality, however, cohesive forces will act outside the contact zone, and they will serve to place a limit on the maximum stress that is reached. Because of this, attempts to better model adhesion contact have been made by accounting for surface forces that act outside the contact edge. Derjaguin, Muller and Toporov (DMT) [27] proposed a theory that takes these forces into account, but assumes that they do not change the shape of the deformed surfaces from the Hertzian profile. This approximation leads to adhesion (tensile) stresses that are finite outside the contact zone but are zero inside. It turns out that the DMT theory is valid only for very small contact radii in three-dimensional contact, a situation that is not typically encountered in experiments on elastomeric lenses. A unifying model of elastic contact was proposed by Maugis [28], who used a simple cohesive zone model to represent the surface forces outside the contact zone. Specifically Maugis used the Dugdale-Barenblatt [29, 30] model, illustrated in figure 6, in which the stresses outside the contact edge are

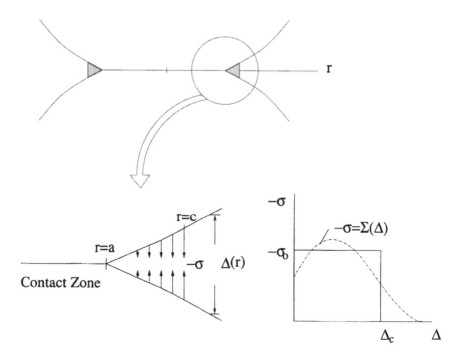

Figure 6. A micromechanical model of the surface forces that act near the contact edge. The inter-surface stresses are given by σ, and the distance between surfaces is Δ. In the Dugdale-Barenblatt model [29-30], σ has a constant value of σ_o until a critical opening displacement is reached. A more realistic model may be similar to that shown by the dashed line.

constant ($= -\sigma_o$) until a critical opening gap Δ_C is reached. This model removes the stress singularity of the JKR theory, and it also removes the stress discontinuity of the DMT theory. Although Maugis only considered the problem of a rigid punch pressing on a linear elastic half space (i.e., $E_1 = \infty$, $R_2 = \infty$), his solution can be easily extended to the general case (i.e., replacing K in Maugis' work by $4E^*/3$ with R defined by (4)).

The JKR theory is based on the assumption that the singular stresses given by (11) provide an excellent approximation to the actual stress distribution near the contact edge, as long as the cohesive zone is small compared with the contact radius (assumption D). This small cohesive zone condition is conceptually equivalent to the small scale yielding (SSY) condition in linear elastic fracture mechanics (LEFM), where the dimension of the plastic zone is small compared with specimen dimensions. Maugis' results can be used to check the validity of this assumption. In the limit where the cohesive zone length is much less than the contact radius, and at distances far enough away from the crack tip (much larger than the cohesive zone length), the pressure distribution of Maugis' cohesive zone model reduces to that given by the JKR approximation.

To demonstrate the concept of SSY, let $c - a = \omega$ denote the size of cohesive zone and a the contact radius (see Fig. 6). The pressure distribution, modified by the cohesive forces, is found to be [28]

$$p(r) = \begin{cases} \dfrac{2E^*}{\pi R}\sqrt{a^2 - r^2} - \dfrac{2\sigma_o}{\pi}\tan^{-1}\sqrt{\dfrac{c^2 - a^2}{a^2 - r^2}} & r \leq a \\ -\sigma_o & c \geq r \geq a \end{cases} \qquad (24)$$

The size of the cohesive zone $c - a$ is determined by the condition that the stresses be bounded everywhere. This results in [28]

$$W = \frac{\sigma_o a^2}{\pi R}\left[\beta + (m^2 - 2)\tan^{-1}\beta\right] + \frac{4\sigma_o^2 a}{\pi E^*}\left[\beta\tan^{-1}\beta - m + 1\right] \qquad (25)$$

$$m = c/a$$

$$\beta = \sqrt{m^2 - 1}$$

where the work of adhesion $W = \sigma_o \Delta_c$. The connection between the JKR theory and Maugis' solution above can be seen by considering the SSY limit of $\omega/a \to 0$ and $a - r \gg \omega$. Equation (25) can be solved in this limit, giving

$$\frac{\omega}{a} = \frac{\pi E^* W}{4\sigma_o^2 a} = \frac{\pi K_I^2}{8\sigma_o^2 a} \ll 1. \qquad (26)$$

In this limit, the second term in the pressure distribution becomes

$$\tan^{-1}\sqrt{\frac{c^2-a^2}{a^2-r^2}} \rightarrow \sqrt{\frac{2a\omega}{a^2-r^2}} \tag{27}$$

Substituting (2) and (27) into (25) results in the pressure distribution of the JKR theory (i.e.,(20)). Thus in the SSY limit, the pressure at material points that are close to the contact zone edge is completely dominated by the asymptotic pressure field (11). *It is in this sense that the stress intensity factor K_I completely controls the deformation field near the contact zone edge, and therefore criteria governing the bonding and debonding may be expressed in terms of the stress intensity factor.* Justification of the SSY concept for the contact of elastic cylinders can be found in Baney and Hui [31].

Maugis' work also shows that for his cohesive zone model the rate of energy flow to the crack tip is still given by (14) and is independent of the details of the local separation or bonding process, as long as the SSY condition (i.e.,(26)) is satisfied. This implies that global experimental quantities such as a, P and δ depend only on the single parameter W. Changing the micromechanical parameters σ and Δ in the cohesive zone model in such a way that the work of adhesion W is kept constant will not affect these global experimental variables. In fact, Rice [23] has shown that this will generally be true for elastic materials where the surface forces depend only on the separation between surfaces Δ and are otherwise independent of the history of contact. Specifically, these surface forces must satisfy a relationship of the form

$$-\sigma = \Sigma(\Delta) \tag{28}$$

of which the Dugdale-Barenblatt model is one simple case:

$$-\sigma = -\sigma_o \qquad \Delta < \Delta_c$$

$$\sigma = 0 \qquad \Delta > \Delta_c \tag{29}$$

This justifies the use of W as the single parameter that describes adhesion properties in many cases. Rice also showed that the work of adhesion for a model of the form (29) was given by the area under the σ versus Δ curve, which for the Dugdale-Barenblatt model is simply $\sigma_o \Delta_c$. More realistic cohesive zone models have been derived using simple representations of intermolecular interactions across interfaces [32, 33], and cohesive elements based on such models have been incorporated into finite element method studies between viscoelastic and elastic bodies [34]. The use of cohesive zone models to represent the separation process along interfaces is common in the fracture mechanics literature, and examples can be found in [35-39].

It must be emphasized that it is only for elastic materials with cohesive zone laws of the form (28) that the cohesive forces can be characterized solely by W without regard to the details of their distribution. We shall see that this is not the case for viscoelastic systems. Additionally, it is not generally true for elastic systems where the cohesive forces depend on the rate of separation.

4.2. Kinematics

The errors in the JKR theory due to kinematic assumptions are (I) large deformation (assumption A), (II) replacing the hemispheres (cylinders) by paraboloids (parabolas), and (III) calculating the deformation by treating the lenses as half spaces loaded by the pressure distribution in the contact circle (strip) (assumption E). The three sources of error could affect the resulting pressure distribution, and could be particularly problematic in the computation of the normal approach. Since the spherical lenses are modeled as linear elastic half spaces in the JKR theory, the displacements of remote points (i.e., at infinity) due to the contact pressure are identically zero. However, for a finite size specimen, δ is not measured with respect to a point at infinity, but relative to the points where the sample is physically attached to the loading machine. Thus the point of zero displacement is not actually at infinity but at some distance roughly equal to the thickness of the sample. The assumption E of the JKR theory, therefore, tends to overestimate the normal approach. The errors resulting from these assumptions (i.e., I, II and III) can be significant since the contact radii can exceed one tenth of the radii of the hemispheres in a typical JKR test.

Maugis [40] investigated the accuracy of the JKR pressure distribution by using the exact initial profile (i.e., a spherical surface instead of a paraboloid) in his analysis. However, his analysis of the local deformation is still based on approximating the hemispheres as elastic half spaces. In this respect his analysis is inconsistent since the errors introduced by these two assumptions are of the same order of magnitude. A detailed analysis of the problem of two cylinders in contact without using (II) and (III) has recently been carried out by Hui et al. [26]. The pressure distribution inside the contact zone as predicted by the JKR theory is shown to be accurate to order ε^2, where ε is the ratio of the contact width to the radius of the smaller cylinder. Therefore, assumptions (II) and (III) are expected to have very small effects on the stress distribution given by (6).

A related problem is that thin lenses are often used in experiments instead of hemispheres. The thicknesses of these lenses can be much smaller than their radii of curvature. In this case, the JKR theory would overestimate the normal approach even if the contact pressure is relatively unaffected by the change in specimen thickness. One way of dealing with this problem experimentally is to insert a ribbon of the same material between the lens and the loading machine [41]. Another option is to modify equation (2) of the JKR theory by the addition of an error term δ_c, i.e., the actual normal displacement δ is

$$\delta = \delta_{JKR} + \delta_c \tag{30}$$

where δ_{JKR} is given by the RHS of (2). Shull et al. [19] proposed that $\delta_c = P\Delta C$, where ΔC represents the loss in compliance associated with the thinner specimen.

Assuming that ΔC is a constant independent of a and P, Shull et al., have been able to use this modified theory to interpret experimental results. A recent analytical work of Hui et al. [26] shows that ΔC is approximately constant over a wide range of contact radii.

The effect of finite strain on the accuracy of the JKR theory has also been examined by Hui et al. [26]. They solved the Hertzian contact ($W = 0$) problem of an elastic hemisphere on a rigid substrate using a finite strain finite element method (FEM). Although large strains were accounted for in this simulation, a linear elastic constitutive law was used to model the elastomer. These FEM results, together with the predictions of the Hertzian solution, are shown in figure 7 for two different values of d/R where d and R are the thickness and radius of the lens, respectively. The theoretical Hertz pressure in these plots is computed using (19) with $W = 0$. The contact radii a in these computations are obtained using (17) and the load from the FEM simulation. In Fig. 7, the applied load is such that the contact radius is about $0.1R$. Even with this contact zone size, there is practically no difference between the theoretical and FEM pressure distributions for both cases. This justifies our earlier statements that the JKR pressure

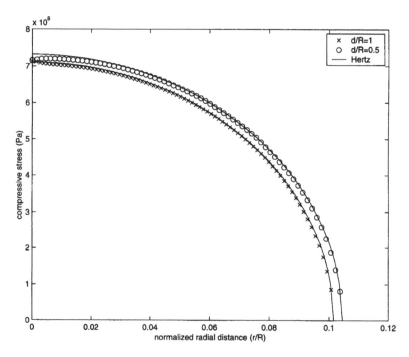

Figure 7. A comparison of theoretical (solid lines) and finite element results (symbols) for the compressive stress field as a function of normalized radial distance (r/R) for lenses of different thicknesses. The results show that the kinematics assumptions of Hertzian theory yield a pressure distribution that is extremely accurate, even when the lens thickness is only one half of its radius and the contact radius is one tenth of the sample radius.

distribution is expected to be much more accurate than the JKR expression for the normal approach. The FEM results for the normal approach are compared with the Hertzian theory in Fig.8 for different values of d/R. These figures show that the normal approach is much more sensitive to the lens thickness than the contact stress. As expected, the Hertz theory overestimates the normal approach.

Numerical results for $W > 0$ are difficult to obtain since the pressure distribution has a square root singularity at the edge of the contact zone. Hui et al. [26] avoided this difficulty by using the fact that the JKR theory was the limiting case of the cohesive zone theory of Maugis [28] where the size of the cohesive zone was small in comparison with the contact radius. Thus, numerical simulations of the JKR theory are carried out by incorporating a small cohesive zone in the FEM program. Details of the numerical implementation can be found in [26]. All simulations are carried out based on the geometry in Fig. 9. The FEM results for the pressure distribution under the contact zone for different values of d/R are plotted in figures 10 and 11, together with the predictions of the JKR theory (19) and Maugis' solution (24). The material constants σ_o and W are chosen to be 0.5 GPa and 0.02 J/m^2. The JKR pressure in these plots is computed using (19) with the contact radius given by (17) and the load from the FEM simulation. Maugis' solution is obtained using (23)

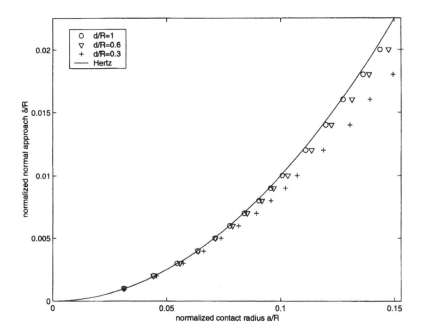

Figure 8. A comparison of theoretical (solid line) and finite element results (symbols) for the normalized normal approach (δ/R) as a function of normalized contact radius (a/R) for lenses of different thicknesses. The results show that the Hertzian assumptions cause the normal approach to be overestimated, and the error increases at larger contact radii. The Hertz theory thus predicts the normal displacement less accurately than the pressure distribution.

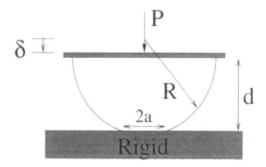

Figure 9. The geometry used for finite element simulation of adhesion contact is illustrated. A lens of radius R and thickness d is pressed into a rigid substrate, and cohesive elements that obey the Dugdale-Barenblatt model are used at the surface to provide the adhesion forces.

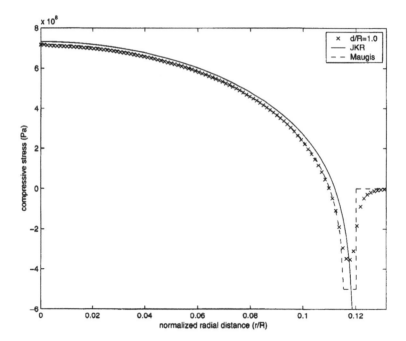

Figure 10. The compressive stress distribution resulting from finite element simulations (symbol) is compared with theoretical model of adhesion contact. Here the sample radius is equal to its thickness. FEM stresses agree well with the Maugis model (dashed line), even for contact radius in excess of one tenth of the sample radius. The JKR theory (solid line) is also shown to be an excellent approximation.

and using the finite element results for the contact radius, In Fig.10, $d/R = 1$ and $\delta/R = 0.01$. The contact radius corresponding to $\delta/R = 0.01$ is about $0.1R$. By (26), the smallest value of ω/a in this simulation is about 0.04, which satisfies the SSY condition $\omega/a \ll 1$. Fig.11 is obtained for $d/R = 0.5$ by using $\delta/R = 0.01$. Since

Maugis' solution takes into account the cohesive forces, it is expected to give better agreement with the FEM results than the JKR theory which are denoted by the solid lines in both figures. Figures 10 and 11 show that this is indeed the case. Note that the FEM results show a normal stress inside the cohesive zone that is slightly smaller than the cohesive strength. This is due to the difficulty of squeezing enough elements inside the small cohesive zone. These figures show that the JKR theory is indeed an excellent approximation and that the kinematic assumptions of Hertz are amazingly accurate, even for contact radii in the range of one tenth of the radii of the hemispheres. Furthermore, the pressure distribution is not significantly affected by the thickness of the lenses.

The FEM results for the normal approach are plotted against the contact radius in Fig. 12 for $d/R = 1$, 0.6 and 0.3. These results are compared with the JKR theory and Maugis' solution. This figure shows that for $d/R \geq 0.6$, the normal approach is well approximated by both the JKR theory and Maugis' theory. The normal approach is overestimated by both of these theories for very thin lenses, as is illustrated by the case of $d/R = 0.3$. The discrepancies are quite small for this case, as less than 10% error is made in using the JKR theory.

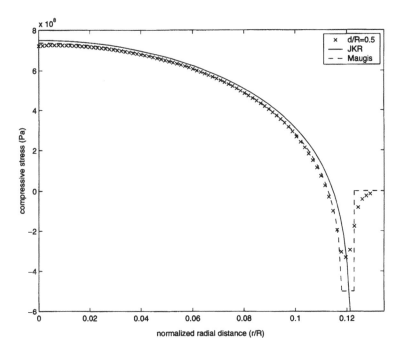

Figure 11. The finite element compressive stress distribution (symbol) is again compared to the Maugis (dashed line) and JKR (solid line) models, this time for a lens of a thickness equal to one half of its radius. The stresses predicted by the theories are again shown to be accurate, indicating that the stress results are not sensitive to lens thickness.

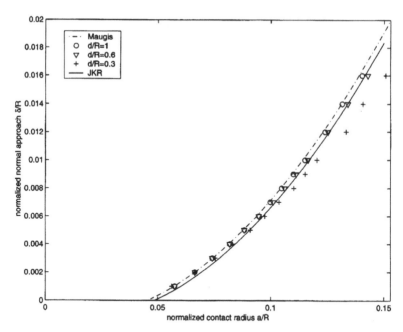

Figure 12. Finite element results for the normal approach as a function of contact radius for lenses of different thicknesses. The predictions of the JKR theory are shown to be quite good for relatively thick lenses, but for the lens whose thickness is 0.3 times its radius, the JKR results overestimate the normal approach. The amount of overestimation increases with increasing contact radii.

4.3. Nonlinear elastic behavior

Assumption B of the JKR theory requires that the material be linearly elastic. Due to the stress concentration near the contact zone edge, the material can exhibit nonlinear behavior, but still be elastic, i.e., the deformation of the material is independent of the loading history. An often overlooked aspect of the JKR theory is that it is valid also for a non-linear elastic material provided that the region of nonlinear behavior is confined to a region small compared with the contact zone (SSY). In this case, the pressure distribution (6) no longer holds as one approaches the contact edge. Indeed, the pressure distribution there may have an entirely different singularity. Nevertheless, because the path independence of the J integral is valid for an elastic material (linear or nonlinear), the amount of energy flow to the crack tip is still given by (14). The argument is as follows: consider a circular path enclosing the crack tip with a radius large compared with the dimension of the nonlinear zone but still small compared with the contact radius. Such a path exists because of the SSY assumption. On this path, the stresses are given by the JKR theory, so that the J integral evaluated on this path is $K_I^2/2E^*$. The path independence of J therefore implies (14) so that (15) holds. Equation (16) is still valid since the actual pressure distribution differs from the JKR pressure given by (8) only in a very small region near the contact zone. We can, therefore, assume that the total load is only slightly affected by this difference in pressure distribution. Thus, the work of adhesion can be determined using the standard JKR technique under these circumstances. Simple as this result is, it has not appeared in the literature.

4.4. Hysteresis

The existence of a work of adhesion implies that the bonding and debonding processes in the JKR theory are completely reversible. While this is true for some model systems [10], most systems exhibit some irreversibility. As a result a lower compressive force is needed to obtain a given contact radius during the debonding phase than is required during bonding. The situation has been termed adhesion hysteresis, and it has been observed for both viscoelastic and elastic systems. Bulk viscoelastic losses provide an obvious reason for irreversible behavior in the former case, and rate dependent surface effects have been offered as an explanation for the latter. Specifically, bond breaking kinetics [42] and viscoelasticity contained in small regions near the contact edges [20, 43] have been named as sources of adhesion hysteresis in materials that are elastic in the bulk. The opinion of the present authors is that it is important to distinguish between surface and bulk sources of adhesion hysteresis, because while the energy approach may be applied to the former case, the stress intensity factor approach is required for the latter.

4.5. Surface effects

Surface effects are different from bulk viscoelastic effects in that equations (5) to (14) hold as long as the surface effects occur in a region of length scale d that is small compared with the contact radius (SSY). On the other hand, equations (15), (17) and (18) no longer hold since the concept of a reversible work of adhesion is not well defined. It is important to note that, based on the stress intensity factor approach, one can only conclude that the descriptions of bonding and debonding (e.g. the rate of bonding or debonding, $\dot{a}$) may be expressed in terms of the history of the stress intensity factor. However, under typical experimental conditions, it is observed that the rate dependent processes at the contact edge are controlled by the *current* value of K_I and do not depend on the entire history of K_I. To determine these conditions, we note that the time scale of the bonding or debonding process τ is

$$\tau \approx d / \dot{a} \qquad (31)$$

Let ΔK_I denote the change of the stress intensity factor when the contact zone extends (recedes) by the amount d, so ΔK_I can be estimated by

$$\Delta K_I \approx \frac{dK_I}{dt} \frac{d}{\dot{a}} \qquad (32)$$

To ensure that criteria governing the bonding and debonding (such as $\dot{a}$) depend only on the current value of stress intensity factor, we must have $\Delta K_I / K_I \ll 1$, i.e.,

$$\frac{1}{K_I}\frac{dK_I}{dt}\frac{d}{\dot{a}} \ll 1 \qquad (33)$$

This condition is defined as the quasi-steady state condition, and is typically satisfied under SSY conditions, provided that the loading is sufficiently smooth. In short, the stress intensity factor changes so little over the time it takes the crack to move one cohesive zone length that we essentially have steady-state crack propagation during that time. This serves as a great simplification when rate dependent processes are present. The quasi-steady and SSY conditions guarantee that the rate of contact growth depends only on the current value of K_I, i.e.,

$$\dot{a} = f(K_I) \qquad (34)$$

where f is some unknown function. The function f is a material property that depends only on whether a is increasing (bonding phase of a JKR test) or decreasing (debonding phase of a JKR test) and is otherwise independent of specimen geometry and loading history. Thus, the same function should be observed for the JKR test and other fracture based experiments such as the peel test.

Since $G = K_I^2 / 2E^*$, (34) implies that, for a given material, there is a universal relation between $\dot{a}$ and G, i.e.,

$$\dot{a} = f(\sqrt{2E^*G}) \equiv F(G) \qquad (35)$$

Under certain circumstances, (34) can be inverted, giving

$$F^{-1}(\dot{a}) \equiv G_c(\dot{a}) = G \qquad (36)$$

This equation says that the amount of energy flow to the edge of the contact zone during the bonding/debonding process is a material property that depends only on the mode of failure (bonding or debonding) and is otherwise independent of specimen geometry (e.g. a peel test) and loading history.

It should be noted that in this work we make a distinction between W, the reversible work of adhesion, G, the energy release upon crack propagation, and G_c, the critical energy release rate necessary for crack propagation. W is regarded as a thermodynamic property of a material that represents the reversible work necessary to create surfaces where none existed before. Specially W refers to the work necessary to overcome the intermolecular attractions arising from van der Waals type forces, and does not include the work necessary to detangle chains that stretch across the interface or to overcome mechanical reinforcements. W is not a property of crack, it is a material property that can be measured through other non-fracture based methods. Because of this, it is nonsensical to say that W depends on the crack propagation rate $\dot{a}$. The energy release rate G on the other hand is not a material property but a mechanics quantity that can be calculated with sufficient knowledge of the stress and strain field in a cracked body. It represents the energy flow to the crack tip per unit area of crack extension. The critical energy release rate G_c is a material property that represents the energy release rate necessary for the crack to actually grow or recede. Since it is a material property, it cannot be calculated from stress and strain fields. Unlike W, G_c depends on crack geometry, so it makes sense to say that G_c depends on the rate of crack propagation. In the rate independent limit of slow crack velocities, $G_c(\dot{a})$ does approach a finite limit

$$G_c(\dot{a} \rightarrow 0) \rightarrow W_o. \tag{37}$$

However, it is not necessarily true that $W_o = W$ since the mechanism of adhesion at the interface can be altered by surface treatment. For example, interfaces may be strengthened by grafting chains on the surface of material 1, and these chains reinforce the interface by diffusing into the second material. For sufficiently short chains, the interface can fail by chain pull-out. Since the frictional resistance to pull-out depends on the pull-out rate, the interface will exhibit rate dependent behavior even if the untreated interface does not. In this case one expects that $W_o > W$ since the frictional resistance to pull-out may not be zero even at zero pull-out rate.

The material function $G_c(\dot{a})$ can be determined experimentally since $\dot{a}$ can be measured and G can be determined using (16). If d can be estimated, the quasi-static condition can be checked using (14), (33) and (37).

Maugis and Barquins [20] first used the JKR technique to characterize surface properties in this manner. In their experiments, performed using glass indentors on polyurethane strips, they obtained G_c as a function of $\dot{a}$ during debonding over four decades of crack speeds. In calculating G_c they explicitly assumed that the materials behaved elastically everywhere, except at the surface where the dissipation was assumed to be given by the product of W and a velocity dependent dimensionless function. Their results seemed to provide justification for this assumption since using several different unloading forces they obtained a master $G_c(\dot{a})$ curve that agreed with similar curves from peeling and flat punch tests. Additionally, they showed that time-temperature superposition [44] could be applied to their observed rate dependence, thus supporting the assumption that viscoelasticit effects were confined to a very small region near the edge of the contact. Since then this technique has been used to characterize other surfaces. Of particular note are the recent results of Deruelle et al. [41] from experiments in which polydimethylsiloxane PDMS lenses were compressed against rigid substrates. By monitoring $\dot{a}$ through the debonding phase of experiments they measured $G_c(\dot{a})$ curves that included measurable rate effects even at such slow crack speeds as 1 nm/sec. Although one might think that such a slow debonding rate would provide a rate-independent measurement of W, this result shows that this is not necessarily the case.

Finally, we point out that for rate-dependent interfaces the energy based approach (i.e., (37)) and the stress intensity factor based approach (i.e., (34)) are entirely equivalent. We think that the stress intensity factor approach is more straightforward, however, as it was used to justify the validity of the energy approach.

4.6. Bulk viscoelastity

The requirement of the JKR theory that the materials be linear elastic severely limits its application since most elastomers are viscoelastic near their glass transition temperatures. For linear viscoelastic materials, the effective creep compliance $C(t)$ corresponding to $1/E^*$ in (2) is defined by

$$C(t) \equiv C_1(t) + C_2(t) \tag{38}$$

where $C_1(t)$ and $C_2(t)$ are the creep compliance functions for materials 1 and 2 in a plane strain test, respectively. The short time limits of $C(t)$ correspond to the instantaneous and long time elastic properties so that

$$C_o \equiv C(0) = C_1(0) + C_2(0) = \frac{1-(v_1^o)^2}{E_1^o} + \frac{1-(v_2^o)^2}{E_2^o} \equiv \frac{1}{E_o^*} \tag{39a}$$

$$C_\infty \equiv C(\infty) = C_1(\infty) + C_2(\infty) = \frac{1-(v_1^\infty)^2}{E_1^\infty} + \frac{1-(v_2^\infty)^2}{E_2^\infty} \equiv \frac{1}{E_\infty^*}, \tag{39b}$$

where $v_1^o, E_1^o, v_1^\infty, E_1^\infty$ denote the short time and long time Poisson's ratios and Young's moduli of material 1. Since most elastomers are nearly incompressible, $C_1(t)$ and $C_2(t)$ are proportional to the extensional creep compliance, with the constant of proportionality being $1-v^2=3/4$.

Attempts to modify the elastic JKR theory by replacing E^* in the elastic theory with some suitable relaxation function $E^*(t)$ have had limited success and are incorrect from a theoretical viewpoint. Such an approach is equivalent to replacing E^* in (2) by $E^*(t)$, which, as will be pointed out below, is problematic. There is thus a need to extend the JKR theory of contact to include linear viscoelastic materials. The two major difficulties associated with such an extension are:

1. For elastic materials the current contact radius is uniquely determined by the current applied load, whereas for a viscoelastic material the size of the contact zone depends on the entire history of the applied load. Thus, the deformation of a material point inside the contact zone during the debonding phase of a JKR test depends also on the bonding phase. This problem has been largely avoided by previous investigators.
2. In the JKR theory, the energy flow to the crack tip is given by (14) and is independent of the details of the bonding and separation processes at the moving edge of contact. Specifically, G is completely determined by the contact radius a

and the applied force F, as well as the elastic properties of the elastomers in contact. However, for viscoelastic spheres in contact, (14) is meaningless in the sense that a unique modulus E^* cannot be defined, since the stress/strain histories of different material points near the contact edge are different. Indeed, the concept of energy flow to the crack tip is difficult to define because of bulk viscoelastic dissipation. These difficulties have been examined in the work of Knauss [45] and Schapery [46] on the viscoelastic fracture of polymers. By interpreting the energy release rate G as the stress work on the surface material points of the cohesive zone per unit crack extension (see figure 13), they showed that G depended not only on material properties (e.g. the creep functions) and the stress intensity factor, but *also on the details of the crack tip separation or bonding process*. Thus the mechanics of viscoelastic deformation is invariably coupled to the separation process, even if the cohesive zone is small compared to the contact radius (i.e., SSY condition is satisfied). We have seen that for elastic materials with rate dependent interfaces, the global deformation decouples from the contact edge micromechanics if the quasi-steady state condition is satisfied. Unfortunately this is not the case for materials with bulk viscoelasticity, and the energy approach is therefore of little use in this case.

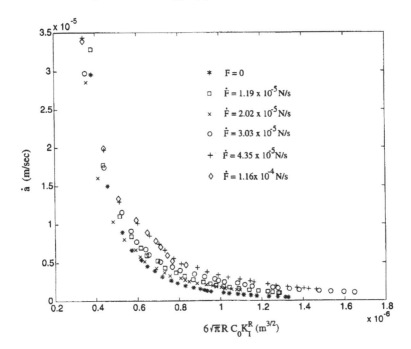

Figure 13. Experimental results showing $\dot{a}$ as a function of K_I^R for a flat layer of plasticized poly(vinylbutyral) in increasing contact with a spherical glass indentor. The results for several rates of loading are shown, and for all rates the relationship between $\dot{a}$ and K_I^R is similar. This provides some evidence for the analysis in [49], which predicts a universal function $\dot{a}$ (K_I^R) for each material that should be independent of loading history.

A simple way to demonstrate that the energy required for crack propagation in a viscoelastic material is strongly coupled to the separation process is to consider the paradox that arises in viscoelastic fracture mechanics [23]. This paradox, in the context of a JKR experiment, can be posed as follows: Consider a JKR experiment on two identical viscoleastic polymeric lenses having work of adhesion W. The lenses are pressed into contact for a sufficiently long period of time so that they are in a relaxed state. The contact radius is then decreased by reducing the applied load on the JKR apparatus. The reduction of load is carried out in such a way so that the contact zone is receding with a constant speed $\dot{a}$ (i.e., the crack is growing inward towards the center of contact). The stress field near the crack tip has the form given by (11) even though the material is viscoelastic [23]. Assuming that the cohesive zone is small compared with the contact zone size, the problem is how to express the energy flow to the crack tip G in terms of K_I. Physically, one would expect that the answer depends on $\dot{a}$ since in the low and high speed limits we essentially have an elastic material, and the long and short time moduli can be used in (14). However, if we assume that G can be computed independent of the separation process, i.e., the crack is perfectly sharp (or the size of the cohesive zone is *exactly* zero), then

$$G = K_I^2 / 2E_o^* \qquad (40)$$

where E_o^* is the effective instantaneous modulus of the materials defined by (39a). Equation (40) is obtained via the following argument: for a continuously growing crack, material elements near the crack tip experience an infinite rate of loading due to the approach of the singular K_I field. Because of this infinite loading rate, the material near the crack tip behaves like an elastic material with the instantaneous modulus E_o^*, and the energy release rate is therefore given by (40). The crack growth criterion $G = K_I^2 / 2E_o^* = W$ would therefore imply a complete lack of crack speed dependence in the propagation of the crack.

The flaw in the above argument is that the loading rate cannot be infinite at the crack tip due to the finite size of the cohesive zone. This means that the material near the crack tip can relax, and the degree of relaxation depends on the cohesive forces and the crack speed, as well as on the characteristic relaxation time τ_r of the material. To gain insight into this relaxation process, let us estimate the time it takes for the crack to move one cohesive zone length. This time is approximately equal to the size of cohesive zone ω divided by the crack speed, i.e.,

$$\tau_c \approx \omega/\dot{a} = \frac{\pi K_I^2}{8\sigma_o^2 \dot{a}} \qquad (41)$$

where Maugis' result (26) is used to estimate ω. If τ_c is large compared with τ_r, which will occur with sufficiently slow crack growth, then relaxation can occur near the crack tip. On the other hand, if τ_c is very small compared with τ_r, then equation (40) is an excellent approximation for the energy flow to the crack tip. It is through this relaxation process that G is coupled to the separation process. Indeed, the energy release rate can be computed for the special case where the separation process is represented by the Dugdale-Barenblatt model. It is [23]

$$G = \left[\int_1^0 C\left(\frac{\pi K_I^2}{8\sigma_o^2 \dot{a}} \lambda \right) f'(\lambda) d\lambda \right] K_I^2 \tag{42}$$

where $C(t)$ is the creep compliance of the material and $f'(\lambda)$ is the derivative of the function

$$f(\lambda) = \sqrt{1-\lambda} - \frac{\lambda}{2} \ln\left[\frac{1+\sqrt{1-\lambda}}{1-\sqrt{1-\lambda}} \right] \tag{43}$$

Note that $\dfrac{\pi K_I^2}{8\sigma_o^2 \dot{a}}$ appears in the argument of the compliance function and that τ_r is implicit in $C(t)$. Since the cohesive strength σ_o is explicitly involved in the expression for G, the energy release rate cannot be determined without consideration of the separation process. It can be shown that, in the limit of very high crack growth rate, $G \rightarrow K_I^2/2E_o^*$, whereas for very slow crack growth rate $G \rightarrow K_I^2/2E_\infty^*$. The ratio of G for slow and high crack growth rates with the same stress intensity factor is E_o^*/E_∞^*, which is very large for many materials.

With regard to the first of the two difficulties mentioned earlier, the analysis of the viscoelastic JKR experiment becomes significantly simpler if we consider only the bonding phase of the experiment. During this phase of the experiment the contact zone is non-decreasing. In this case it has been found [47] that although the stresses near the crack tip can relax and do not necessarily obey the form (11), they can still be described by a single stress intensity factor-like parameter. Schapery called this parameter the reference stress intensity factor [48], K_I^R, and it represents the stress field that would be found in an elastic material (with $1/E^*$ replaced by C_o) of the same geometry under modified loading conditions (with P replaced by $C*P/C_o$ defined in [49]). That the actual stress field in a viscoelastic body can be found by considering a similar hypothetical elastic problem is a result of a mathematical relationship between viscoelastic and elastic boundary value problems [50]. Although this relationship is not at all intuitive and can cause confusion with the introduction of K_I^R, it does present a way to avoid the difficulties associated with the energy balance in the viscoelastic case. K_I^R is denoted by K_I in [49].

The benefit of using the stress intensity factor approach is that K_I^R can be related to the global experimental variables for the bonding phase of a JKR test. For a load controlled experiment, it is found that [49]

$$K_I^R = \frac{4a^3 - 3R(C*P)}{6\sqrt{\pi}RC_o a^{3/2}}$$
(44)

while for displacement control

$$K_I^R = \frac{1}{C_o\sqrt{\pi a(t)}}[\frac{a^2(t)}{R} - \delta(t)]$$
(45)

The operation $C*P$ represents a convolution integral that takes the history of applied load into account, i.e.,

$$C*P \equiv \int_{0^-}^{t} C(t-\tau)\frac{\partial P}{\partial \tau}d\tau$$
(46)

The body is assumed to be undisturbed for $t<0$, and the 0^- in (11) takes the possibility of sudden step loading into account. It is interesting to compare (45) with the corresponding expression for the stress intensity factor in the elastic problem which is (see (14) and (16))

$$K_I = \frac{4E^*a^3 - 3RP}{6\sqrt{\pi}Ra^{3/2}}$$
(47)

Equation (44) shows that, in a load control test, a knowledge of the entire loading history is needed to determine the reference stress intensity factor at any instant, while for displacement control no history dependence comes into the equation. In each case K_I^R can be calculated from the basic experimental data. Furthermore, Schapery [48] showed that for a material with a rate-independent cohesive zone law, the speed of crack healing was determined by the current value of K_I^R alone. That is, even though the material is viscoelastic with an inherent history dependence, all of the history effects come in solely through the calculation of K_I^R via (44), and it is only the current value of K_I^R that controls the crack healing process. Schapery gave the precise form of the predicted relationship between K_I^R and $\dot{a}$ for a power law viscoelastic material that has a cohesive zone model of the form (29). It is possible to adopt Schapery's procedure to determine the relationship between K_I^R and $\dot{a}$ for any linear viscoelastic material with an arbitrary rate independent cohesive zone model, although for realistic models it would be next to impossible to perform the calculation in closed form. The point, however, is that a unique relationship between K_I^R and $\dot{a}$ is predicted. This relationship can be checked experimentally since K_I^R and $\dot{a}$ can be measured in JKR tests. If K_I^R can indeed be uniquely related to the crack healing speed $\dot{a}$, the form of the function $K_I^R(\dot{a})$ can serve as a characterization of the adhesion properties of the system. This is similar to the $G_c(\dot{a})$ characterization used to describe surface effects, but we emphasize that the energy balance becomes impossible to quantify due to viscoelastic dissipation. Since the amount of energy flow to the crack tip cannot be calculated without a priori knowledge of the fracture process, the stress intensity approach is superior to the energy characterization for viscoelastic materials.

There are important assumptions implicit in the analysis of Schapery [48], and these should be stated, since there are restrictions on the cases for which the relationship between K_I^R and $\dot{a}$ is predicted. First, the cohesive zone forces are assumed to be rate-independent, and to be a function of only the separation between the surfaces. This requirement seems reasonable when one recalls that we have only considered the bonding phase of a JKR test, so that chain pull-out or similar failure mechanisms are not present. Second, the quasi-steady state condition is assumed, and again this is typically satisfied throughout most experiments. Third, the compliance functions of the top and bottom materials are assumed to be proportional. Practically, this means either that the materials must be the same or that one material must be rigid relative to the other. It should be noted that even if the assumptions specified here are not met and hence Schapery's analysis is not valid, a relationship between K_I^R and $\dot{a}$ may exist anyway. This can always be checked experimentally using (44) or (45).

Additionally, we have stated that the above analysis and equations (44) - (45) hold only for the bonding phase of an experiment. Similar results have been obtained for the debonding phase, although the equations for the stress intensity factor are much more complicated than (44) and (45). Because of the complexity of the results, the reader is simply referred to [51] for them.

The prediction that the rate of bonding is a function only of the reference stress intensity and is independent of the loading history has been tested experimentally by Baney et al. [52]. Their experiments measured the adhesion between a spherical glass indentor and plasticized poly(vinyl butyral), a commercial material which is used in laminated safety glass. Using several different rates of loading, they measured the bonding rate as a function of K_I^R and results found are plotted in figure 13. As can be seen from this figure, the different loading rates produced $\dot{a}\,(K_I^R)$ curves that were slightly different from one another, but the data generally fell along one master curve. These results support the analysis of [52]. It should be mentioned that the material used in the experiments did not necessarily obey a linear viscoelastic constitutive law over a wide range of stress, and for this reason, the experimental loading conditions were restricted to low values of the average compressive stress.

5. SUMMARY

The kinematic assumptions of the JKR theory are amazingly accurate, even for large contact zone sizes. In particular, the JKR pressure distribution inside the contact zone is essentially unaffected by the contact zone size as well as the lens thickness. For thin lenses, the normal separation deviates somewhat from the theory. This difference can be corrected using the experimental procedure of [19]. An often overlooked aspect of the JKR theory is that (17) is obtained without consideration of the normal separation δ. In other words, within the context of the JKR theory, the measured applied load and the contact zone size completely determine the stress distribution inside the contact zone. This means that a JKR test using load control is likely to have less error because the pressure distribution is relatively unaffected by the specimen geometry.

The problem of large deformation coupled with nonlinear elastic constitutive behavior is still unresolved. It is shown here that if the deformation is assumed to be small and if the nonlinear elastic effects are localized near the contact zone edge, the JKR theory is still valid.

With regard to hysteresis, a special case, which is easiest to treat, is when all the nonlinear effects (including rate effects) are confined to the interface. In this case, a universal relationship of the form $\dot{a}(K_I)$ or $G_c(\dot{a})$ must exist and can be verified by experiments. It is expected that such relationships will be different for the bonding and the debonding phases of a JKR experiment. Further experimental work needs to be done to determine whether this is the case.

Hysteresis due to bulk dissipation is a still an unresolved problem. Here we only considered the simplest case of linear viscoleasticity. To the best of our knowledge, there is no theoretical work on the JKR theory of adhesion based on non-linear viscoelasticity or viscoplastic materials. These problems are extremely difficult due to the fact that there is no consensus on the constitutive behaviour of non-linear viscoelastic materials. Furthermore, analysis of the deformation and stress fields is extremely complicated. For such materials, the stress field near the contact zone may not be characterized by a stress intensity factor, since the stress distribution will no longer have the inverse square root singularity typical of linear problems.

On the other hand, by restricting the experimental conditions to sufficiently low levels of compressive stresses, one should be able to characterize the interfacial adhesion of viscoelastic materials using the stress intensity factor approach. In this regime a relation of the form $\dot{a}(K_I^R)$ is expected to exist during the bonding phase of a JKR experiment.

When performing experiments involving viscoelastic materials, it is necessary to record the entire load and contact zone history during the bonding and the debonding phases of the experiment [52]. If only the bonding phase is considered, then only the load history needs to be recorded in a load-controlled test and K_I^R can be computed using (44). In this case the compliance function of the material must be obtained by separate experiments. It is advantageous to perform a displacement-controlled test for viscoelastic materials since K_I^R is determined by the current contact zone size and normal separation during the bonding phase (see (45)).

REFERENCES

1. K. L. Johnson, K. Kendall and A. D. Roberts, *Proc. R. Soc. London, Ser. A* **324**, 301-313 (1971).
2. M. F. Vallat, P. Ziegler, P. Vondracek and J. Schultz, *J. Adhesion* **35**, 95-104 (1991).
3. A. D. Roberts and A. G. Thomas, *Wear* **33**, 45-64 (1975).
4. D. S. Rimai, L. P. DeMejo, W. Vreeland, R. Brown, S. R. Gabovry and M. W. Urban, *J. Appl. Phys.* **71**, 2253-2258 (1992).
5. M. K. Chaudhury and G. M.Whitesides, *Langmuir* **7**, 1013-1025 (1991).
6. C. Creton, H. R. Brown and K. R. Shull, *Macromolecules* **27**, 3174-3183 (1994).
7. D. Ahn and K. R. Shull, *Macromolecules* **29**, 4381-4390 (1996).
8. R. G. Horn, J. N. Israelachvili and F. J. Pribac, *J. Colloid Interface Sci.* **115**, 480-492 (1987).
9. H. R. Brown, *Macromolecules* **26**, 1666-1670 (1993).
10. S. Perutz, E. J. Kramer, J. Baney and C. Y. Hui, *Macromolecules* **30**, 7964-7969 (1997).
11. C. Creton and L. Leibler, *J. Polym. Sci. Part B. Polym. Phys.* **34**, 545-554 (1996).
12. G. Luengo, J. Pan, M. Heuberger and J. N. Israelachvili, *Langmuir* **14**, 3873-3881 (1998).
13. M. Deruelle, L. Leger and M. Tirrell, *Macromolecules* **28**, 7419-7428 (1995).
14. M. F. Vallat, I. Nebesarova, J. Schultz and A. Pouchelon, *J. Adhesion Sci. Technol* **12**, 433-443 (1998).
15. A. Falsafi, P. Deprez, F. S. Bates and M. Tirrell, *J. Rheol.* **41**, 1349 (1997).
16. H. J. Hertz and J. Reine, *Angew. Math.* **92**, 156 (1882).
17. P. Silberzan, S. Perutz, E. J. Kramer and M. K. Chaudhury, *Langmuir* **10**, 2466 (1994).
18. V. S. Mangipudi and M. Tirrell, *Rubber Chem. Technol.*, **71**, 407-448 (1998).
19. K. R. Shull, D. Ahn and C. L. Mawery, *Langmuir* **13**, 1799-1804 (1997).
20. D. Maugis and M. Barquins, *J. Phys. D: Appl. Phys.*, **11**, 1989 (1978).
21. J. A. Greenwood and K. L. Johnson, *Philos. Mag., A* **43**, 697 (1981).
22. D. Maugis, M. Barquins and R. Courtel, *M'et. Corros. Indu.* No. 605, 1 (1976).
23. J. M. Rice, *Proc. 8th U. S. Natl. Cong. Appl. Mech.*, 191-216 (1979).
24. M. K. Chaudhury, T. Weaver, C. Y. Hui and E. J. Kramer, *J. Appl. Phys.* **80**, 30-37 (1996).
25. M. Barquins, *J. Adhesion* **26**, 1-12 (1988).
26. C. Y. Hui, Y. Y. Lin, J. M. Baney and A. Jagota, *J. Adhesion Sci. Technol.*, accepted for publication.
27. B. V. Derjaguin, V. M. Muller and Yu. P. Toporov, *J. Colloid Interface Sci.* **53**, 314-326 (1975).
28. D. J. Maugis, *J. Colloid Interface Sci.* **150**, 243-269 (1992).
29. D. S. Dugdale, *J. Mech. Phys. Solids* **8**, 100-104 (1960).
30. G. I. Barenblatt, *Adv. Appl. Mech.* **7**, 55-129 (1962).
31. J. M. Baney and C. Y. Hui, *J. Adhesion Sci. Technol* **11**, 393-406 (1997).
32. E. Barthel, *J. Colloid Interface Sci.* **200**, 7-18 (1998).
33. C. Argento, A. Jagota and W. C. Carter, *J. Mech. Phys. Solids,* **45**, 1161-1183 (1997).
34. A. Jagota, C. Argento and S. Mazur, *J. Appl. Phys.* **83**, 250-259 (1998).
35. E. J. Kramer and E. W. Hart, *Polymer* **25**, 1667 (1984).
36. B. Budiansky, J. C. Amazigo and A. G. Evans, *J. Mech. Phys. Solids* **36**, 167 (1988).
37. H. Riedel, *Mater Sci. Eng.* **30**, 187 (1997).
38. C. Y. Hui, D. B. Xu and E.J. Kramer, *J. Appl. Phys.* **72**, 3294 (1992).
39. W. G. Knauss and G. U. Lasi, *J. Appl. Mech.* **60**, 793 (1993).
40. D. Maugis, *Langmuir* **11**, 679-682 (1995).
41. M. Deruelle, H. Hervet, G. Jandeav and L. Leger, *J. Adhesion Sci. Technol.*, **12**, 225-247 (1998).
42. S. Perutz, E. J. Kramer, J. Baney, C. Y. Hui and C. Cohen, *J. Polym. Sci. Part B. Polym. Phys.* **36**, 2129-2193 (1998).
43. K. Kendall, *J. Adhesion*, **7**, 55 (1975).
44. M. L. Williams, R. F. Landel and J. D. Ferry, *J. Amer. Chem. Soc.* **77**, 3701 (1995).

45. W. G. Knauss, in: *Deformation and Fracture of High Polymers*, H. H. Kausch, J. A. Hassell and R. I. Jaffee (Eds.), pp. 501-541, Plenum Press, New York (1973).
46. R. A. Schapery, *Int. J. Fracture*, **11**, 141 (1975).
47. J. M. Baney and C. Y. Hui, *J. Appl. Phys.*, **86**, 4232-4241 (1999).
48. R. A. Schapery, *Int. J. Fracture*, **39**, 163 (1989).
49. C. Y. Hui, J. M. Baney and E. J. Kramer, *Langmuir*, **14**, 6570-6578 (1995).
50. R. A. Schapery, *Int. J. Fracture*, **25**, 195-225 (1984).
51. Y. Y. Lin, C. Y. Hui and J. M. Baney, *J. Phys. D.: Appl. Phys.*, **32**, 2250 (1999).
52. J. M. Baney, C. Y. Hui and C. Cohen, accepted for publication by *Langmuir* (2000).

Adhesion Measurement of Films and Coatings, Vol. 2, pp. 329–340
Ed. K.L. Mittal
© VSP 2001

Revisiting bimaterial curvature measurements for CTE of adhesives

DAVID A. DILLARD* and JANG-HORNG YU**

Department of Engineering Science and Mechanics, M/C 0219, Virginia Polytechnic Institute and State University, Blacksburg, VA 24061

Abstract—Quantifying the coefficient of thermal expansion (CTE), stress-free temperature, and residual stress state is of critical importance for many applications involving bonded systems such as coatings, adhesives, and other laminated systems. This paper revisits Timoshenko's solution for curvature of a bimaterial strip (plane stress) by identifying optimal configurations which improve sensitivity and minimize effects of time dependent changes in moduli. A similar solution is developed from classical lamination theory for plate-like geometries. We find that when the material system is optimized, the curvature calculation is virtually insensitive to the modulus change. For thin layer systems, the bimaterial system is much more sensitive than the bulk material system in the curvature measurement. Such techniques can be easily implemented in modern thermal mechanical analysis equipment which is readily available in most polymer laboratories.

Keywords: Bimaterial system; coefficient of thermal expansion; thermal stress; curvature measurements.

1. INTRODUCTION

The thermal stress induced by the mismatch of coefficients of thermal expansion (CTE) for a multi-layer system is a well-known fact. Beams or plates composed of different material layers may transform into cylindrical or spherical bent shapes under temperature changes due to thermal stress (see, e.g., [1] for references). Timoshenko's pioneering work [2] in the bimaterial system calculated the in-plane stress by measuring the curvature of the material system. Scherer [3] has generalized Timoshenko's analysis and extended it to any type of stress which induces curvature changes. For plate structures, Corcoran [4] laid the groundwork for calculating the residual stress for organic coatings (e.g., paints) on metal substrates. Benabdi and Roche ([5],[6]) have analyzed mechanical properties of

---

*To whom correspondence should be addressed. Phone: (540)231-4714, Fax: (540)231-9187, E-mail: dillard@vt.edu

**Phone: (540)231-7484, Fax: (540)231-9187, E-mail: jayu@vt.edu

thin and thick coatings on various substrates in both residual stress and Young's modulus determination.

The thermal (residual) stress is important in the reliability and durability of material systems when subjected to the condition of thermal cycling. Due to the change in temperature, the thermal stress due to the mismatch of the CTE for a material system is one of the most important factors affecting the performance and reliability of material systems (e.g., micro electrical-mechanical devices [7]). Thermal stresses may also increase due to environmental conditions such as thermal cycling, and it may give rise to mechanical failures in adhesive layers, coatings and laminated systems (see, e.g., [13]). Such effect also has an impact on the dimensional stability and the stress-free temperature (SFT) of the device. Therefore, the importance of obtaining the information on CTE and SFT for adhesives or coatings cannot be overemphasized.

Measuring CTE is often a difficult experimental task, especially when the material cannot be made into a bulk specimen. For example, for coatings that outgas solvents, satisfactory specimens are difficult to make because of porosity. Such difficulty, however, can be overcome by using a bimaterial system.

Curvature measurements on bilayer systems can be traced back to Thomas Jefferson (see [9]) who may have used such technique to design a hygrometer (or a "magic" weather stick) to obtain humidity information by measuring the relative shrinkage and expansion of layers of wood due to the weather change. Ramani and Zhao [10] employed such technique to study the evolution of residual stresses in thermoplastic bonding to metals. This technique is especially attractive in thin bimaterial layers. If we neglect the change in the mechanical properties in the thin coating due to the creation of the interface between the coating and the substrate [11], our theoretical calculations show that the sensitivity of the measurement is of the order $O(1/h)$ higher than that of the bulk specimen where h is the thickness of the specimen.

In Section 2, we seek to determine optimal geometries which improve the sensitivity of the techniques for measuring CTE's for adhesives using beam-like specimens. Such specimens can be easily implemented in commercially available TMA (thermal mechanical analysis) equipment. We find that the important parameters which will affect the resultant curvature include CTE's, α_i, layer thicknesses, t_i, and elastic constants (Young's modulus), E_i. For optimized geometries, the adhesive needs to be several times thicker than the substrate. Under an optimized configuration, the curvature measurement is shown to be insensitive to the modulus change due to viscoelastic behavior of the material.

In Section 3, we use classical lamination theory to investigate CTE measurements for plate-like specimens. We find that, in addition to the above parameters for beams, Poisson's ratios (ν_i) will now affect the results as well in a cylindrical bending case. If both materials are of the same Poisson's ratio, the resulting curvature is $(1 + \nu)$ times larger than that of the beam.

In this paper, we develop a theoretical basis for optimization in measuring CTE, where the edge effect of specimens or time-dependent property of adhesives are not included. Thermal mechanical analysis equipment can quantify the displacements under carefully controlled temperature exposure. We have obtained excellent experimental results using the bimaterial curvature measurements [12],[13]. Applying the bimaterial curvature technique, we are able to detect curvature measurements for adhesives as thin as 10 μm thick.

2. THEORY BASED ON TIMOSHENKO'S BIMATERIAL STRIP

In this section, we first briefly summarize Timoshenko's work on a bimaterial system. We assume that both materials are homogeneous, isotropic, and elastic and their intrinsic stresses are null at the temperature T_0. We also assume a perfect interface where the materials are perfectly bonded and the interface thickness is zero. The radius of curvature of the neutral axis is large compared with the transverse dimension and the transverse section of the beam remains planar before, during and after bending. We further assume the plane stress condition and the material system undergoes circular bending. Denoting p as the magnitude of the normal force acting on both materials and noting that p must be applied in the opposite directions for the two materials, we see that the curvature κ can be calculated via [2]

$$\alpha_1(T - T_0) + \frac{p}{t_1 E_1} + \frac{t_1 \kappa}{2} = \alpha_2(T - T_0) - \frac{p}{t_2 E_2} - \frac{t_2 \kappa}{2}, \tag{1}$$

where at the interface the strain must be identical. Substituting the ratio of thickness $m = t_1/t_2$, the ratio of elastic moduli $n = E_1/E_2$, the total thickness $h = t_1 + t_2$ to (1), the temperature change $\Delta T = T - T_0$, and the CTE mismatch $\Delta\alpha = \alpha_2 - \alpha_1$ will yield the curvature:

$$\kappa_b = \frac{6\Delta T \Delta\alpha(1 + m)^2}{h\left(3(1 + m)^2 + (1 + mn)\left(m^2 + \dfrac{1}{mn}\right)\right)}. \tag{2}$$

Let us define the sensitivity of the measurement $\zeta_b = \bar{\kappa}_b/(\Delta\alpha \Delta T)$, where $\bar{\kappa}_b = \kappa_b \times h$ is the normalized curvature:

$$\bar{\zeta}_b = \frac{6(1 + m)^2}{\left(3(1 + m)^2 + (1 + mn)\left(m^2 + \dfrac{1}{mn}\right)\right)}. \tag{3}$$

Hereinafter, we shall always use the subscript 1 to denote the adhesive layer and 2 for the substrate. In optimizing the specimen geometry, one might want to obtain the greatest sensitivity for measuring the CTE for a given material combination. If one wishes to hold a constant total thickness h, the maximum sensitivity occurs when

$$\frac{d\bar{\zeta}_b}{dm} = 0, \tag{4}$$

which yields the only possible root

$$m = \frac{1}{\sqrt{n}}. \tag{5}$$

This implies that to obtain the optimal sensitivity for a given total thickness h, the thickness ratio should be the inverse of the square root of elastic moduli ratio. Taking an epoxy-aluminum system for example, the ratio of elastic moduli $n \approx 0.05$, from which we obtain $m = 4.47$. This means that the adhesive should be several times thicker than the metal substrate to gain the maximum (normalized) curvature.

A more practical case is when the substrate thickness t_2 is fixed, as when one is working with a given substrate stock. Without loss of generality, we set $t_2 = 1$, and hence $m = t_1$ and $h = 1 + t_1$. Substituting m and h into (3) we find that the normalized curvature $\hat{\kappa}_b = \kappa_b \times t_2$ is equal to

$$\hat{\kappa}_b = \frac{6\Delta T \Delta\alpha (1 + m)}{\left(3(1 + m)^2 + (1 + mn)\left(m^2 + \dfrac{1}{mn}\right)\right)}, \tag{6}$$

where the sensitivity of the measurement $\hat{\zeta}_b = \hat{\kappa}_b/(\Delta\alpha \Delta T)$ is of the form

$$\hat{\zeta}_b = \frac{6(1 + m)}{3(1 + m)^2 + (1 + mn)\left(m^2 + \dfrac{1}{mn}\right)}. \tag{7}$$

Carrying out the procedure as (4), we find that maximum sensitivity occurs when m and n are related by

$$m = -\frac{1}{2} + \frac{n}{2(2n^2 - n^3 + 2\sqrt{(1-n)n^4})^{1/3}} + \frac{(2n^2 - n^3 + 2\sqrt{(1-n)n^4})^{1/3}}{2n}. \tag{8}$$

We will then plot the normalized curvature for both cases.

In what follows, we assume that the elastic modulus of the substrate is higher than that of the adhesive or coating, i.e., $n < 1$. In Figure 1, we plot the normalized curvature $\bar{\kappa}_b = \kappa_b \times h$ vs. the thickness ratio m for different values of n when the total thickness is fixed. Figure 2 exhibits a similar plot for the normalized curvature $\hat{\kappa}_b = \kappa_b \times t_2$ vs. the thickness ratio m for different values of n when the substrate thickness is fixed. Both plots show that when $E_1 \ll E_2$, which resembles adhesives on metal substrates, the resulting curvature is a non-monotonic function of the thickness ratio. If E_1 and E_2 are of comparable values, the increase of thickness ratio does improve the sensitivity for small m's but the gain eventually goes down for large m's. Figure 3 exhibits the plot of normalized curvature $\hat{\kappa}_b$ with respect to

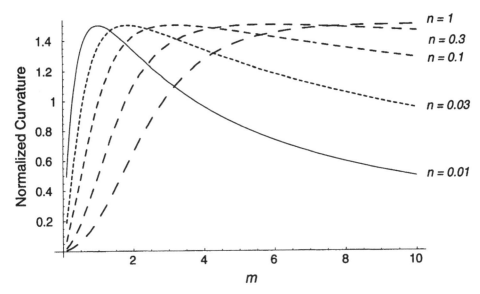

Figure 1. Sensitivity of normalized curvature vs. thickness ratio m for different values of n when the total thickness is fixed.

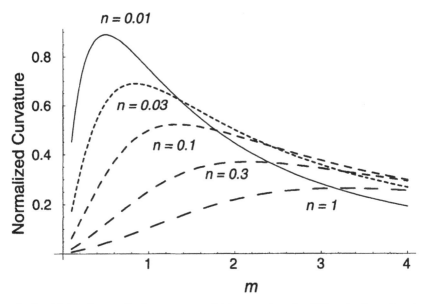

Figure 2. Sensitivity of normalized curvature vs. thickness ratio m for different values of n when the substrate thickness is fixed.

the modulus n. We note that the measured curvature is almost flat for $n > 0.03$ if the adhesive layer is a few times thicker than the substrate. This implies that if the adhesive layer is thick enough, the value $\hat{\kappa}_b$ will be quite insensitive to the viscoelastic effect of the adhesive.

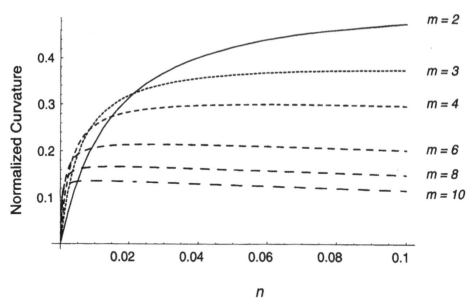

Figure 3. Sensitivity of normalized curvature vs. modulus ratio n for different values of m when the substrate thickness is fixed.

To close this section, we note that when the curvature κ_b is decided, one can easily obtain the information on the residual stress by

$$\sigma_r = \kappa_b \left(\frac{1}{ht_1} \left[\frac{E_1 t_1^3}{6} + \frac{E_2 t_2^3}{6} \right] + \frac{t_1 E_1}{2} \right). \tag{9}$$

The stress-free temperature, on the other hand, can be characterized by TMA where the measured curvature reaches zero, which indicates there is no thermal residual stress.

3. CLASSICAL LAMINATION THEORY

If the specimen cannot be considered as a beam structure, a laminated plate of two layers may be used. Classical lamination theory is well known (see, e.g., [14]), and hence we shall not go into a detailed discussion. Let N_x, N_y, and N_{xy} represent the resultant forces acting on a laminate and M_x, M_y, and M_{xy} the resultant moments, where, for example,

$$N_x = \int_{-t/2}^{t/2} \sigma_x dz \tag{10}$$

and

$$M_x = \int_{-t/2}^{t/2} \sigma_x z dz \tag{11}$$

where σ_x is the normal stress along the x-direction. Following conventional notations, we term A_{ij}, B_{ij}, and D_{ij} the extensional stiffness, bending-extension coupling stiffness, and bending stiffness matrices, respectively. They are related to the laminate stiffness matrix $\bar{Q}_{ij}$ via

$$A_{ij} = \sum_{k=1}^{2}(\bar{Q}_{ij})_k(z_k - z_{k-1}),$$

$$B_{ij} = \frac{1}{2}\sum_{k=1}^{2}(\bar{Q}_{ij})_k(z_k^2 - z_{k-1}^2), \qquad (12)$$

$$D_{ij} = \frac{1}{3}\sum_{k=1}^{2}(\bar{Q}_{ij})_k(z_k^3 - z_{k-1}^3),$$

where z_k denotes the axis coordinate perpendicular to the plate for each layer, and the terms of stiffness matrix $\bar{Q}_{ij}$ for isotropic layers are given by

$$(\bar{Q}_{11})_i = (\bar{Q}_{22})_i = \frac{E_i}{1 - v_i^2}, \qquad (\bar{Q}_{12})_i = \frac{v_i E_i}{1 - v_i^2},$$

$$(\bar{Q}_{66})_i = \frac{E_i}{2(1 + v_i)}, \qquad (\bar{Q}_{16})_i = (\bar{Q}_{26})_i = 0. \qquad (13)$$

The thermal force due to the temperature change may be written as

$$\begin{bmatrix} N_x^T \\ N_y^T \\ N_{xy}^T \end{bmatrix} = \int \begin{bmatrix} \bar{Q}_{11} & \bar{Q}_{12} & \bar{Q}_{16} \\ \bar{Q}_{12} & \bar{Q}_{22} & \bar{Q}_{26} \\ \bar{Q}_{16} & \bar{Q}_{26} & \bar{Q}_{66} \end{bmatrix} \begin{bmatrix} \alpha_x \\ \alpha_y \\ \alpha_{xy} \end{bmatrix} \Delta T dz, \qquad (14)$$

and the thermal moment as

$$\begin{bmatrix} M_x^T \\ M_y^T \\ M_{xy}^T \end{bmatrix} = \int \begin{bmatrix} \bar{Q}_{11} & \bar{Q}_{12} & \bar{Q}_{16} \\ \bar{Q}_{12} & \bar{Q}_{22} & \bar{Q}_{26} \\ \bar{Q}_{16} & \bar{Q}_{26} & \bar{Q}_{66} \end{bmatrix} \begin{bmatrix} \alpha_x \\ \alpha_y \\ \alpha_{xy} \end{bmatrix} \Delta T dz. \qquad (15)$$

Invoking force and moment balance, we have the stress-strain relation for each layer

$$\begin{bmatrix} N_x \\ N_y \\ N_{xy} \end{bmatrix} = \begin{bmatrix} A_{11} & A_{12} & A_{16} \\ A_{12} & A_{22} & A_{26} \\ A_{16} & A_{26} & A_{66} \end{bmatrix} \begin{bmatrix} \epsilon_x^0 \\ \epsilon_y^0 \\ \epsilon_{xy}^0 \end{bmatrix} + \begin{bmatrix} B_{11} & B_{12} & B_{16} \\ B_{12} & B_{22} & B_{26} \\ B_{16} & B_{26} & B_{66} \end{bmatrix} \begin{bmatrix} \kappa_x \\ \kappa_y \\ \kappa_{xy} \end{bmatrix} - \begin{bmatrix} N_x^T \\ N_y^T \\ N_{xy}^T \end{bmatrix}, \qquad (16)$$

and

$$\begin{bmatrix} M_x \\ M_y \\ M_{xy} \end{bmatrix} = \begin{bmatrix} B_{11} & B_{12} & B_{16} \\ B_{12} & B_{22} & B_{26} \\ B_{16} & B_{26} & B_{66} \end{bmatrix} \begin{bmatrix} \epsilon_x^0 \\ \epsilon_y^0 \\ \epsilon_{xy}^0 \end{bmatrix} + \begin{bmatrix} D_{11} & D_{12} & D_{16} \\ D_{12} & D_{22} & D_{26} \\ D_{16} & D_{26} & D_{66} \end{bmatrix} \begin{bmatrix} \kappa_x \\ \kappa_y \\ \kappa_{xy} \end{bmatrix} - \begin{bmatrix} M_x^T \\ M_y^T \\ M_{xy}^T \end{bmatrix}. \qquad (17)$$

Here, following the Kirchhoff hypothesis for plates where the laminate is thin, a line originally straight and perpendicular to the middle surface of the laminate is assumed to remain straight and perpendicular to the middle surface when the laminate is extended or bent. Thus, the middle-surface strains are of the form

$$
\begin{bmatrix} \epsilon_x^0 \\ \epsilon_y^0 \\ \epsilon_{xy}^0 \end{bmatrix} = \begin{bmatrix} \frac{\partial u_0}{\partial x} \\ \frac{\partial v_0}{\partial y} \\ \frac{\partial u_0}{\partial y} + \frac{\partial v_0}{\partial x} \end{bmatrix},
\tag{18}
$$

and the middle-surface curvatures of the form

$$
\begin{bmatrix} \kappa_x \\ \kappa_y \\ \kappa_{xy} \end{bmatrix} = - \begin{bmatrix} \frac{\partial^2 w_0}{\partial x^2} \\ \frac{\partial^2 w_0}{\partial y^2} \\ \frac{2\partial^2 w_0}{\partial x \partial y} \end{bmatrix},
\tag{19}
$$

where u_0, v_0, and w_0 denote the displacement functions along x, y, and z directions, respectively.

For the case of thermal loading where there is no external force or moment, i.e., $N_x = N_y = N_{xy} = M_x = M_y = M_{xy} = 0$ and for the case of cylindrical bending (see, e.g., [1]) with free boundary conditions and assuming the plate is isotropic, we have for the x-dir and y-dir displacement functions $u_0 = u_0(x)$ and $v_0(\cdot) = 0$. The out-of-plane z-dir displacement function is cylindrical where $w_0(x) = \kappa_p x^2$. We assume that both layers are at the same temperature T by neglecting the heat transfer effect. Employing the same notations as in Section 2, we find that the solution for (16) and (17) is of the form

$$
\kappa_p =
$$
$$
\frac{6\Delta T E_1 E_2 t_1 t_2 (t_1 + t_2)(1 - v_1^2)(1 - v_2^2)(\alpha_1(1 + v_1) - \alpha_2(1 + v_2))}{E_1^2 t_1^4 (1 - v_2^2)^2 + E_2^2 t_2^4 (1 - v_1^2)^2 + 2E_1 E_2 t_1 t_2 (1 - v_1^2)(1 - v_2^2)(2t_1^2 + 3t_1 t_2 + 2t_2^2)}.
\tag{20}
$$

Note that by setting $v_1 = v_2 = 0$, the curvature equation (20) will reduce to Timoshenko's solution for the beam curvature κ_b in eq. (2). If $v_1 = v_2 = v$, we have $\kappa_p = (1 + v)\kappa_b$. Again, substituting $m = t_1/t_2$ and $n = E_1/E_2$, and the sensitivity of the measurement $\zeta_p = \bar{\kappa}_p / \Delta T$, $\bar{\kappa}_p = \kappa_p \times (t_1 + t_2)$, we obtain

$$
\zeta_p =
$$
$$
\frac{6(1 - v_1^2)(1 - v_2^2)(\alpha_1(1 + v_1) - \alpha_2(1 + v_2))}{mn(1 + 1/m)^{-2}(1 - v_2^2)^2 + \frac{1}{mn(1+m)^2}(1 - v_1^2)^2 + 2(1 - v_1^2)(1 - v_2^2)\left(2 - \left(\sqrt{m} + \frac{1}{\sqrt{m}}\right)^2\right)}.
\tag{21}
$$

Following a similar procedure as before, we seek the parameters that optimize the CTE measurement. For a fixed total thickness, the maximum sensitivity occurs

(a)

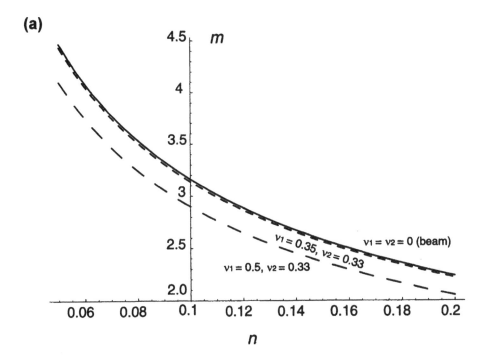

(b)

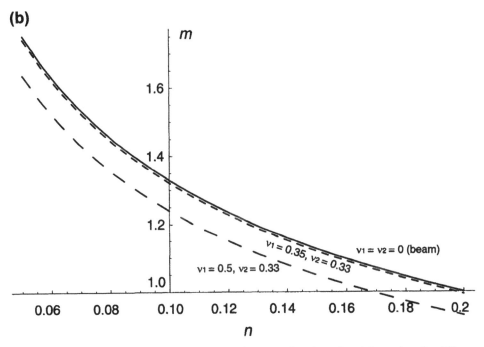

Figure 4. Comparison of optimized thickness ratio *m* as a function of modulus ratio *n* for different Poisson's ratios when the total thickness is fixed (a) and when the substrate thickness is fixed (b).

when

$$\frac{d\zeta_p}{dm} = 0, \tag{22}$$

which yields the only possible root

$$m = \frac{(1 - v_1^2) - \sqrt{n(1 - v_1^2)(1 - v_2^2)}}{-n + nv_2^2 + \sqrt{n(1 - v_1^2)(1 - v_2^2)}}. \tag{23}$$

This indicates that in plate structures, Poisson's ratio does play a role in determining the optimal thickness. If Poisson's ratios are identical for both materials ($v_1 = v_2$), then (23) will reduce to (5). Take an epoxy-aluminum material system, for example, where their CTE's are $\alpha_1 = 70$ ppm/$^\circ C$ and $\alpha_2 = 23.5$ ppm/$^\circ C$, respectively. If the epoxy is in the glassy region, its Poisson's ratio $v_1 = 0.35$ is comparable to that of aluminum $v_2 = 0.33$ or other metal substrates, and we find that there is only a small difference in obtaining optimization results from the assumptions of beams or plates. When the temperature is near T_g for adhesives where $v_1 \rightarrow 0.5$, there will be a larger mismatch of Poisson's ratios and the optimization result will be different.

We have also investigated the case for a fixed substrate thickness. The optimization solution for the relation between parameters m and n, however, is far too complicated to be presented here. Without going into tedious symbolic representations of the solutions, we show in Figures 4a (fixed total thickness) and 4b (fixed substrate thickness) the optimized thickness ratio m as a function of n. We find similar results

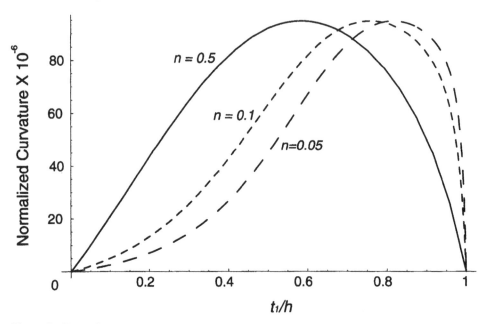

Figure 5. Normalized curvature vs. t_1/h for an epoxy-aluminum system for different values of n when the total thickness is fixed.

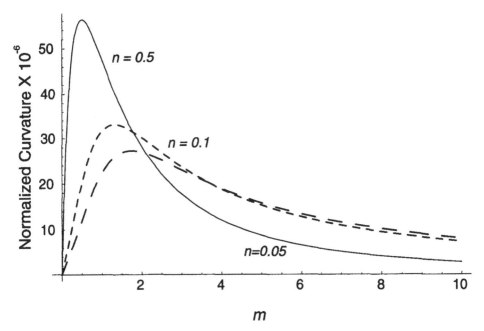

Figure 6. Normalized curvature vs. *m* for an epoxy-aluminum system for different values of *n* when the substrate thickness is fixed.

as for the case of fixed total thickness, i.e., if adhesives are in the glassy region and their Poisson's ratios are comparable with those of metal substrates, there is only a small difference in obtaining optimization results from the assumptions of beams or plates. When the temperature is near T_g and $\nu \to 0.5$ for adhesives, there will be a mismatch of Poisson's ratio and the optimization result will be different.

We summarize the optimization results by showing in Figures 5 and 6 that the normalized curvatures for the epoxy-aluminum system with a fixed total thickness or a fixed substrate thickness, respectively. Both figures show the non-monotonic characteristics of the optimization curves.

Finally, we note that the thermal stresses can be calculated via eqs. (16) and (17). We will not, however, pursue this topic in this paper.

4. CONCLUSIONS

We have provided information as how to optimize the geometric configurations of bimaterial specimens for measuring CTE based on both beam and lamination theories. The optimization results indicate that the adhesive should be several times thicker than the substrate to achieve the maximum sensitivity of the measurement. The optimized geometry also has the benefit of being relatively insensitive to the modulus change due to the time dependent effect of viscoelasticity in the curvature measurement.

Using this technique, i.e., the bimaterial system, we have already tested material systems polyimide-aluminum and epoxy-aluminum. Excellent results have been achieved. The bimaterial technique is sensitive for the polyimide coating down to 10 μm thick. We can also estimate the stress-free temperature for the material system and the physical aging associated with the adhesive.

In closing, the bimaterial system provides an attractive way for the CTE measurement due to its high sensitivity compared to that of bulk specimens. Through this technique, we not only can obtain information on CTE's of adhesives, but also vital information such as SFT. Time dependent effect such as physical aging may also be estimated. Such technique can be easily adapted in material testing systems such as TMA which are readily available in most polymer laboratories.

Acknowledgment

We would like to thank Mr. Chris Smith and Dr. Senthil Vel for their constructive comments. The support of NSF STC funding DMR # 9120004, and Center for Adhesive and Sealant Science, and Engineering Science and Mechanics Department at Virginia Tech is gratefully acknowledged.

REFERENCES

1. M. Dano, and M. Hyer, *Int. J. Solids Structures*, **35**, 2101-2120, (1928).
2. S. Timoshenko, *J. Opt. Soc. Am.*, **11**, 233-255, (1925).
3. G.W. Scherer, *Relaxation in Glass and Composites*, John Wiley, New York, (1986).
4. E.M. Corcoran, *J. Paint Technol.*, **41**, 635-640, (1969).
5. M. Benabdi and A.A. Roche, *J. Adhesion Sci. Technol.*, **11**, 281-299, (1997).
6. M. Benabdi and A.A. Roche, *J. Adhesion Sci. Technol.*, **11**, 373-391, (1997).
7. A.S. Voloshin, and P.H. Tsao, *Proceedings of the 43rd Electronic Component and Technology Conference*, (1993).
8. R. Humfeld, and D.A. Dillard, *J. Adhesion*, **65**, 277-306, (1998).
9. R. Underhill, *The Woodwright's Workbook*, The University of North Carolina Press, Chapel Hill, NC, (1986).
10. K. Ramani, and W. Zhao, *Int. J. Adhesion Adhesives*, **17**, 353-357, (1997).
11. A.A. Roche, and J. Guillemenet, *Thin Solid Films*, **342**, 52-60, (1999).
12. T. Park, C. Buo, S. Guo, J.H. Yu, and D.A. Dillard, *4th Symposium on Experimental/Numerical Mechanics in Electronic Packaging*, Society for Experimental Mechanics (SEM), Orlando, FL, 904-907, (2000).
13. D.A. Dillard, T. Park, H. Zhang, and B. Chen, *Proceedings of the 22nd Annual Meeting of the Adhesion Society*, Panama City Beach, FL, 336–338, (1999).
14. R.M. Jones, *Mechanics of Composite Materials*, Taylor & Francis: Philadelphia, PA, (1999).

Adhesion Measurement of Films and Coatings, Vol. 2, pp. 341–351
Ed. K.L. Mittal
© VSP 2001

Raman spectroscopic determination of residual stresses in diamond films

QI HUA FAN,[1,*] J. GRÁCIO[2] and E. PEREIRA[1]

[1] *Department of Physics, University of Aveiro, 3810 Aveiro, Portugal*
[2] *Department of Mechanical Engineering, University of Aveiro, 3810 Aveiro, Portugal*

Abstract—Diamond films are attractive for many potential applications due to their superior properties, such as a large band gap, wide range of optical transparency, the highest hardness and Young's modulus as well as thermal conductivity of any known material. Most of the film applications require that the films adhere adequately to the substrate. It has been generally recognized that the presence of residual stress is an important factor affecting the film/substrate adhesion. So, a quantitative evaluation and understanding of the film stresses are necessary for optimizing process conditions and designing reliable diamond films. The residual stresses in chemical-vapor-deposited (CVD) diamond films arise from two sources, namely, intrinsic stress due to film lattice mismatch with the substrate, and thermal stress originated from a thermal-expansion mismatch between the film and the substrate on cooling.

In this paper, we report on the determination of residual stresses in CVD diamond films by micro-Raman spectroscopy and by a bi-metal theory and a plate bending theory. It is shown that the residual stresses vary with the film growth. The diamond Raman peak broadening is observed in adherent film, which is actually an effect of biaxial stress on the frequencies of singlet and doublet phonons.

Keywords: Diamond film; stress; Raman spectroscopy.

1. INTRODUCTION

The superior properties of chemical-vapor-deposited (CVD) diamond films make them attractive in many potential applications [1-4]. In order to obtain reliable diamond films, residual stresses must be identified and controlled. Generally such stresses arise from two sources, namely, intrinsic stress and thermal stress. Structural mismatch between the film and substrate and the defects within the film are responsible for intrinsic stress, while thermal stress originates from the difference in the thermal expansion coefficients of the film and substrate.

The residual stresses have been evaluated by a number of techniques, including substrate curvature, x-ray diffraction, and micro-Raman spectroscopy [5-9], among which the micro-Raman spectroscopy is the most commonly used for

---

* To whom correspondence should be addressed. Phone: +351-234-370356,
Fax: +351-234-424965, E-mail: fan@fis.ua.pt

characterizing diamond films. For a perfect diamond crystal without any external stress, its face centered cubic symmetry results in a triply degenerate first order phonon. As the wavevector of the incident radiation is considerably smaller than the Brillouin zone extension, single phonon or first order Raman scattering produces information about phonons near the zone center (k=0). The first order Raman peak for diamond appears at 1332 cm^{-1}. It is a characteristic feature of the diamond structure. The effects of external stresses on the Raman spectra have been discussed by many authors [10-19]. Generally, the width of the 1332 cm^{-1} line is associated with random stress present and is, therefore, found to be sensitively dependent on the conditions of film growth [8]. On the other hand, any directional stress present may give rise to a shift and/or splitting of the line. As-deposited diamond films are usually under compressive or tensile stress due to the difference in thermal expansion coefficients between the film and substrate and due to the growth strain.

It is necessary to note that the residual stress evaluated by the above-mentioned techniques is actually a sum of thermal stress and intrinsic stress. Therefore, the nature of the residual stress is determined by the nature and relative magnitudes of these two stress components. For the same kind of substrate materials, for example Si, the residual stress in CVD diamond films has been reported as both tensile and compressive, depending on the CVD techniques and growth conditions [15-18, 20]. This implies a significant contribution from intrinsic stress.

In this paper, we demonstrate the determination of residual stresses in diamond films employing a plate bending theory and a bi-metal theory and by using micro-Raman spectroscopy. Intrinsic stress induced by the structural mismatch and by the film growth strain, such as graphite inclusion, hydrogen clusters, voids etc., is distinguished from thermal stress. We also discuss the effects of residual stresses on the Raman line broadening.

2. EXPERIMENTAL DETAILS

The substrates used for the diamond deposition were as-received Si (100), mirror polished on one side, and polycrystalline copper (99.95%). The Si substrates were 0.3 mm thick and 5 x 5 mm^2 in size and the copper substrates were 1 mm thick and 10 x 10 mm^2 in size. An ASTeX PDS18 microwave plasma CVD system was used for diamond deposition. The deposition parameters included microwave power, gas pressure, gas flow rate, and CH$_4$ concentration. They were optimized to produce so-called 'white grade' diamond films. Details have been reported in previous works [21, 22]. Diamond films with different thicknesses were obtained by simply varying the deposition time.

A Renishaw 2000 micro-Raman system with a 633 nm He-Ne laser was used to characterize the diamond films. At this laser wavelength, the non-diamond carbon phases scatter more effectively than diamond due to a resonance effect [23]. It is, therefore, useful in characterizing diamond films of 'good' quality. The Renishaw

Raman system works in two modes: main mode and extended mode. In the main mode the gratings are set at a fixed angle and, therefore, a high repeatability of ± 0.1 cm^{-1} in the spectrum can be obtained. Raman spectra obtained in the main mode allow an accurate comparison of the diamond line shifts. All Raman spectra in this work were taken in the main mode and a Type IIa bulk diamond was used to calibrate the peak position. For each sample, five different points along the surface diagonal were measured and the average Raman shift was used to evaluate residual stress.

3. RESULTS AND DISCUSSION

3.1. Distribution of thermal stress and intrinsic stress in diamond films

Diamond films with thicknesses of 1.7 μm to 48 μm were deposited on Si substrates. Typical Raman spectra taken from some of these films and the corresponding diamond line position are shown in Fig. 1 and Table 1. It can be seen that with increase in the film thickness, the diamond phase becomes more pronounced and non-diamond phases decrease. Therefore, a change in the growth strain with the film evolution is expected. In thinner films, the diamond Raman line shifts to wavenumbers higher than 1332 cm^{-1}, corresponding to a compressive stress; while in thicker films it shifts to wavenumbers lower than 1332 cm^{-1}, corresponding to a tensile stress. The change in the nature of the residual stress indicates a possible variation in the nature and relative magnitude of intrinsic stress and thermal stress. It is noted that the samples show slight curvature with the film on the convex side.

It is known that the Raman shift is proportional to biaxial stress σ in diamond films with relationships shown as follows [9]:

$$\sigma = -1.08(v_s - v_0)\,(\text{GPa}) \quad \text{for singlet phonon,} \tag{1}$$

$$\sigma = -0.384(v_d - v_0)\,(\text{GPa}) \quad \text{for doublet phonon,} \tag{2}$$

where $v_0 = 1332$ cm^{-1}, v_s is the observed maximum of the singlet in the spectrum and v_d the maximum of the doublet. In many films, the splitting of the Raman line is not obvious. In this case, the observed peak position v_m is assumed to be located at the center between the singlet v_s and doublet v_d, i.e., $v_m = \frac{1}{2}(v_s + v_d)$ [24]. From Eqs. (1) and (2) we obtain

$$\sigma = -0.567(v_m - v_0)\,(\text{GPa}) \quad \text{for unsplitted peak.} \tag{3}$$

Using Eqs. (1)-(3), the residual stress in the diamond films is obtained, as given in Table 1.

Table 1.
Raman shift and corresponding residual stress in diamond films of different thicknesses deposited on 0.3 mm thick Si substrates. The Raman shift is an average value of five different points in each sample. Curvature radius R is calculated from the bi-metal theory. Bending stresses are calculated from the plate bending theory. Intrinsic stress is derived from the difference between the mean bending stress and the stress evaluated from Raman spectra

Film thickness d_f (µm)		1.7	4	11	23	48
Raman shift (cm^{-1})		1332.43	1332.33	1331.93	1331.73	1331.60
Stress evaluated from Raman spectra σ (GPa)		-0.244	-0.189	0.038	0.151	0.227
Curvature Radius R from bi-metal theory (mm)		6162	2889	1355	908	718
Bending stress components in the film	Stress variation rate a (GPa/mm) $\sigma_f = ax + b$	0.183	0.391	0.833	1.242	1.572
	Stress at interface b (GPa)	-0.417	-0.376	-0.289	-0.210	-0.149
Bending stress at film surface (GPa) $\sigma_{surf} = a*d_f + b$		-0.417	-0.375	-0.280	-0.181	-0.074
Mean bending stress σ_m in film (GPa) $\sigma_m = \frac{1}{2}(\sigma_{surf} + b)$		-0.417	-0.376	-0.285	-0.196	-0.112
Intrinsic stress $\sigma_i = \sigma - \sigma_m$ (GPa)		0.173	0.187	0.322	0.347	0.339

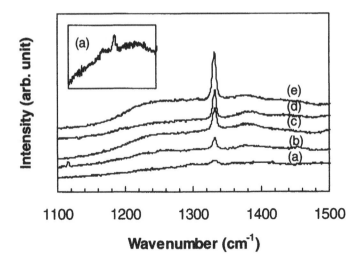

Figure 1. Raman spectra taken from diamond films (a) 1.7 µm, (b) 4.0 µm, (c) 11 µm, (d) 23 µm, and (e) 48 µm thick deposited on 0.3 mm thick Si substrates. Base line shifted in each spectrum for clarity. The Raman spectrum of film (a) is also shown in the insert figure with enlarged scale.

As mentioned before, the residual stress evaluated from Raman spectra is a sum of intrinsic stress and thermal stress. In order to distinguish the two stress components, we consider at first the effect of the thermal mismatch. Assuming the diamond film and the Si substrate behave as pure elastic materials under the residual stresses, the film/substrate can therefore be treated as a bi-metal plate, as shown in Fig. 2. The difference in the thermal expansion coefficients induces stresses in both the film and the substrate and will lead to the film/substrate bending when the sample is cooled from the deposition temperature to room temperature. From the bi-metal theory [25], the radius of curvature, R, is given by:

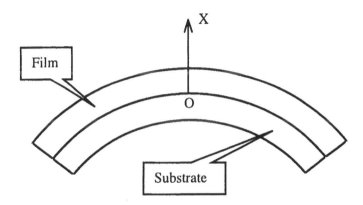

Figure 2. An illustration of the bi-metal bending plate.

$$\frac{1}{R} = \frac{6(\alpha_s - \alpha_f)(T_d - T_r)(1+m)^2}{h\left[3(1+m)^2 + (1+mn)(m^2 + \dfrac{1}{mn})\right]},$$ (4)

$$m = \frac{d_f}{d_s},$$ (5)

$$n = \frac{E_f/(1-v_f)}{E_s/(1-v_s)},$$ (6)

$$h = d_f + d_s,$$ (7)

where α_s and α_f are thermal expansion coefficients for Si and diamond, respectively, $T_r = 25\,°C$ is room temperature and $T_d = 850\,°C$ is the deposition temperature, d_s and d_f are substrate thickness and film thickness, respectively, $E_s = 170\,GPa$ and $E_f = 1050\,GPa$ are Young's moduli of Si and diamond, and $v_s = 0.42$ and $v_f = 0.07$ are their Poisson's ratios, respectively. It should be noted that in the bi-metal theory the thermal expansion coefficients, α_s and α_f, are usually considered as constants. However, it is known that Si and diamond have temperature dependent expansion coefficients in a temperature range of 25 °C (room temperature) to 850 °C (deposition temperature). Therefore, to calculate accurately the curvature radius R, Eq. (4) should be modified as:

$$\frac{1}{R} = \frac{6(1+m)^2 \displaystyle\int_{T_r}^{T_d}(\alpha_s - \alpha_f)dT}{h\left[3(1+m)^2 + (1+mn)(m^2 + \dfrac{1}{mn})\right]}.$$ (8)

From reference [26], we obtain α_s and α_f data for Si and diamond, as shown in Fig. 3. Fitting the data, we obtain the temperature dependent relationships for α_s and α_f as follows:

$$\alpha_s = 2.316 + 7.981 \times 10^{-3}T - 1.413 \times 10^{-5}T^2 + 1.254 \times 10^{-8}T^3 - 4.084 \times 10^{-12}T^4,$$ (9)

$$\alpha_f = 0.981 + 8.144 \times 10^{-3}T - 6.171 \times 10^{-6}T^2 + 2.129 \times 10^{-9}T^3.$$ (10)

Using Eqs. (8)-(10), the curvature radius R is calculated as given in Table 1.

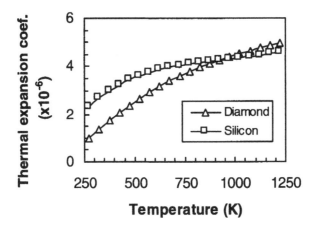

Figure 3. Thermal expansion coefficients α_f and α_s for diamond and Si. Symbols □ and Δ are data from reference [26]. Continuous lines are from the best fittings.

After knowing the curvature radius R, the thermal-mismatch-induced stress σ_f in the film can be obtained from the plate bending theory [25, 27]:

$$\sigma_f = \frac{1}{R}\left[-\frac{Y_f d_f^3 + Y_s d_s^3}{6 d_f (d_f + d_s)} + Y_f \left(x - \frac{d_f}{2} \right) \right], \tag{11}$$

where $Y_f = E_f/(1-v_f)$, $Y_s = E_s/(1-v_s)$. Eq. (11) shows three features:

First, the stress σ_f changes linearly from the film/substrate interface to the film surface, i.e. $\sigma_f = ax + b$, where $a = \dfrac{Y_f}{R}$ is the stress variation rate,

$b = -\dfrac{1}{R}\left[\dfrac{Y_f d_f^3 + Y_s d_s^3}{6 d_f (d_f + d_s)} + Y_f \dfrac{d_f}{2} \right]$ is the stress at the film/substrate interface.

Second, the distribution of the stress σ_f or the two stress components a and b depend on the film thickness, as shown in Table 1. It can be seen that the stress variation rate a increases with increasing film thickness, while the stress component b or the stress at the film/substrate interface varies in an opposite manner.

Third, the stress difference along the film depth is relatively small if we compare the stress at the film/substrate interface, $\sigma_f(x = 0)$, with the stress at the film surface, $\sigma_f(x = d_f)$, as shown in Fig. 4 and Table 1. We, therefore, define a mean bending stress σ_m, which is an average value of the stress σ_f through the film depth, i.e. $\sigma_m = \dfrac{1}{2}\left[\sigma_f(x = 0) + \sigma_f(x = d_f) \right]$. From Eq. (11) we obtain

$$\sigma_m = \frac{1}{R} \left[-\frac{Y_f d_f^3 + Y_s d_s^3}{6 d_f (d_f + d_s)} \right].$$ (12)

The calculated mean bending stress is also shown in Table 1 and Fig. 4. As an approximation, the thermal stress in a film can be considered as constant, as expressed by Eq. (12).

Note that the stress evaluated from Raman spectra is a sum of all residual stresses in the diamond film, while the stress obtained from Eq. (11) or Eq. (12) is only an effect of thermal mismatch. Thus, intrinsic stress σ_i can be derived from the difference in the stresses evaluated from Raman spectra (σ) and from the bimetal theory and the plate bending theory (σ_m), i.e.,

$$\sigma_i = \sigma - \sigma_m.$$ (13)

Obviously, the intrinsic stress σ_i is an average value through the film, as shown in Table 1 and Fig. 4. It can be seen that the derived intrinsic stress is tensile in nature. It increases rapidly when the film thickness varies from 4 µm to 11 µm, while for films thicker than 11 µm the intrinsic stress nearly does not change. We, therefore, deduce that there is a pronounced change in the structure of the diamond film once it reaches a thickness of 4 –11 µm.

It is worth pointing out that the Raman spectra are actually a superposition of contributions from different depths of a film. Since the residual stresses vary along the film growth direction, it is expected that the Raman shift contributed

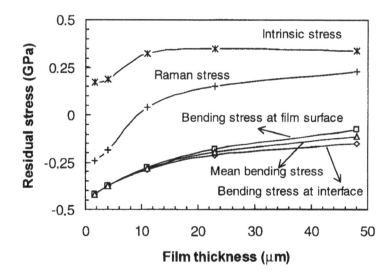

Figure 4. Residual stresses in diamond films grown on Si substrate 0.3 mm thick. × - stress evaluated from Raman spectra, ◊ - bending stress at the film/substrate interface, □ -bending stress at the film surface, △ - mean bending stress, * - intrinsic stress derived from the difference between the mean bending stress and the stress evaluated from Raman spectra.

from layers of different depths is not the same and the signal intensity is also different due to an absorption loss. The details have been discussed in a previous work [21]. However, considering the fact that the diamond films used in this work have a transparent quality, we may assume that the Raman line shift reflects a mean stress in the film. Furthermore, as can be seen in Table 1 and Fig. 4, the relative value of the stress evaluated from Raman spectra is much higher than the bending stress induced by the thermal mismatch. Therefore, the above assumption is expected to yield reasonable accuracy in evaluating residual stresses in the diamond films deposited on the Si substrates.

3.2. Residual stress effects on the Raman peak broadening

In our work, we compared the Raman spectra of the same CVD diamond films in both the adherent state as well as in free-standing state. The free-standing films were obtained by etching off the substrates using nitric acid. An interesting phenomenon was observed: the spectra showed not only a difference in the Raman line shift, but also a difference in the line width. Figure 5 shows an example of Raman spectra measured from an adherent diamond film deposited on Ti/Cu substrate (film A) and from the same film in a free-standing state (film B). The film thickness was ~4 μm. Three features can be observed. First, the characteristic diamond peak of film A appears much wider than film B, as indicated in the figure. Second, film A shows a higher shift in the peak position at ~1336.3 cm^{-1}, while film B shows a peak position at ~1331.7 cm^{-1}. Third, the Raman spectrum of film A exhibits obviously higher background compared with film B. It was noted that the as-deposited sample curved slightly with the film on the convex side.

The difference in the Raman line shifts for film A and film B is easy to understand. In an adherent film, one should expect the presence of both thermal and intrinsic stresses. However, in the free-standing film, only intrinsic stress is expected. Since copper has a larger TEC than diamond in a temperature range of 25 °C (room temperature) to 900 °C (typical deposition temperature), the thermal stress is expected to be compressive in nature. Intrinsic stress can be tensile or compressive, depending on the film growth conditions.

Using Eq. (3), we deduce a tensile stress of 0.17 GPa in the free-standing film and a compressive stress of -2.44 GPa in the adherent film. Therefore, the thermal stress is deduced to be -2.61 GPa in compression. It can be seen that the intrinsic stress in our diamond film is relatively small.

The difference in the Raman peak widths for film A and film B is actually an effect of residual stresses. It has been shown that the frequencies of the Raman phonons of diamond depend linearly on the compressive stress, while the singlet and doublet phonons exhibit different linear relationships, depending on the crystallographic orientation [12]. Using Eqs. (1) and (2), we can deduce that, for the adherent film A, the difference in the Raman line positions between the singlet and the doublet phonons is about 4 cm^{-1}. However, for the free-standing film, the

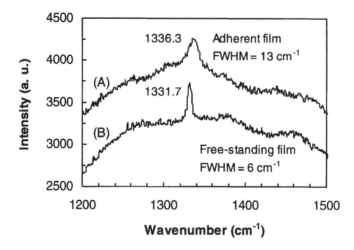

Figure 5. Raman spectra taken from the adherent diamond film and from the free-standing diamond film. The peak position and full width at half maximum (FWHM) are given in the figure.

difference is only about 0.3 cm^{-1}. Therefore, the Raman peak broadening in highly stressed films is expected.

Finally, the higher background in the Raman spectrum of the adherent film is still not clear. It is most probably a contribution from the film/substrate interface as at an early growth stage non-diamond phases and/or a transition layer usually appear before diamond crystals overgrow these phases. Further work to confirm this deduction is under consideration.

4. CONCLUSIONS

Residual stresses in CVD diamond films deposited on Si (100) substrates are evaluated. The shift of the diamond Raman line reflects a combination of intrinsic stress and thermal stress. The plate bending theory and the bi-metal theory show that the biaxial stress decreases linearly along the film growth direction. The stress at the film/substrate interface is compressive in nature and decreases when the film becomes thicker. The stress modeled is quite different from that evaluated by Raman spectroscopy. The difference is, therefore, attributed to intrinsic stress.

Adherent diamond films exhibit wider Raman peaks than the free-standing film. The peak broadening for the adherent film is due to the different frequency dependence of singlet and doublet phonons on biaxial stresses. It is necessary to specify the state of residual stress when using the full width at half maximum to compare the quality of diamond films.

Acknowledgements

The authors would like to thank Mr. A. Fernandes for his help in experiments. NATO research project SFS-PO-OPTOELECT 'Diamond Film Technology' is acknowledged. One of the authors (QHF) would like to thank Fundação para a Ciência e a Tecnologia (Portugal) for financial support.

REFERENCES

1. J. E. Field in J. E. Field (Ed.), *Properties of Natural and Synthetic Diamond*, p.667 (Academic Press, San Diego, CA, 1992).
2. J. C. Angus and C. C. Hayman, Science **241**, 913 (1988).
3. K. E. Spear, J. Am. Ceram. Soc. **72**, 171 (1989).
4. W. A. Yarbrough and R. Messier, Science **247**, 688 (1990).
5. G. G. Stoney, Proc. R. Soc. London Ser. A **82**, 172 (1909).
6. K. Roll, J. Appl. Phys. **47**, 3224 (1976).
7. B. D. Cullity, *Elements of X-ray Diffraction*, p.447 (Addison-Wesley, Reading, MA, 1978).
8. D. S. Knight and W. B. White, J. Mater. Res. **4**, 385 (1989).
9. J. W. Ager and M. D. Drory, Phys. Rev. B **48**, 2601 (1993).
10. S. Ganesan, A. A. Maradudin, and J. Oitmaa, Ann. Phys. **56**, 556 (1970).
11. W. J. Borer, S. S. Mitra, and K. V. Namjoshi, Solid State Commun. **9**, 1377 (1971).
12. M. H. Grimsditch, E. Anastassakis, and M. Cardona, Phys. Rev. B **18**, 901 (1978).
13. R. J. Nemanich, S. A. Solin, and R. M. Martin, Phys. Rev. B **23**, 6348 (1981).
14. S. K. Sharma, H. K. Mao, P. M. Bell, and J. A. Xu, J. Raman Spectrosc. **16**, 350 (1985).
15. H. Windischmann, G. F. Epps, Y. Cong, and R. W. Collins, J. Appl. Phys. **69**, 2231 (1991).
16. J. A. Baglio, B. C. Farnsworth, S. Hankin, G. Hamill, and D. O'Neil, Thin Solid Films **212**, 180 (1992).
17. N. S. Van Damme, D. C. Nagle, and S. R. Winzer, Appl. Phys. Lett. **58**, 2919 (1991).
18. K. H. Chen, Y. L. Lai, J. C. Lin, K. J. Song, L. C. Chen, and C. Y. Huang, Diamond Relat. Mater. **4**, 460 (1995).
19. L. Bergman and R. J. Nemanich, J. Appl. Phys. **78**, 6709 (1995).
20. P. R. Chalker, A. M. Jones, C. Johnston and I. M. Buckley-Golder, Surface Coatings Technol. **47**, 365 (1991).
21. Qi Hua Fan, J. Grácio, and E. Pereira, J. Mater. Sci. **34**, 1353 (1999).
22. Qi Hua Fan, J. Grácio, and E. Pereira, Diamond Relat. Mater. **8**, 645 (1999).
23. J. Wagner, C. Wild, and P. Koidl, Appl. Phys. Lett. **59**, 779 (1991).
24. V. G. Ralchenko, A. A. Smolin, V. G. Pereverzev, E. D. Obraztsova, K. G. Korotoushenko, V. I. Konov, Yu. V. Lakhotkin and E. N. Loubnin, Diamond Relat. Mater. **4**, 754 (1995).
25. S. P. Timoshenko, J. Optic. Soc. Am. **11**, 233 (1925).
26. G. A. Slack and S. F. Bartram, J. Appl. Phys. **46**, 89 (1975).
27. S. P. Timoshenko and J. N. Goodier, *Theory of Elasticity*, p.32 (McGraw-Hill, New York, 1970).